Mycorrhizal Dynamics in Ecc

Mycorrhizae are mutualisms between p .... over 400 million years ago. This symbiotic relationship commenced with land invasion and, as new groups evolved, new organisms developed with varying adaptations to changing conditions. Based on the author's 50 years of knowledge and research, this book characterizes mycorrhizae through the most rapid global environmental changes in human history. It applies that knowledge in many different scenarios, from restoring strip mines in Wyoming and shifting agriculture in the Yucatán, to integrating mutualisms into science policy in California and Washington, D.C. Toggling between ecological theory and natural history of a widespread and long-lived symbiotic relationship, this interdisciplinary volume scales from structure–function and biochemistry to ecosystem dynamics and global change. This remarkable study is of interest to a wide range of students, researchers, and land-use managers.

MICHAEL F. ALLEN is Distinguished Professor Emeritus at the Department of Microbiology and Plant Pathology at the University of California, Riverside. He began his career studying mycorrhizae a half century ago, focusing on the physiology, ecology, evolution, and application of mycorrhizae across the Anthropocene. He was a founding editor of *Mycorrhiza*, President of the International Mycorrhizal Society, and Program Officer at the National Science Foundation. His previous publications include *The Ecology of Mycorrhizae* (Cambridge University Press, 1991) as well as 250 peer-reviewed research papers.

Mycorrhizal Dynamics in Ecological Systems

MICHAEL F. ALLEN

Center for Conservation Biology

CAMBRIDGE
UNIVERSITY PRESS

CAMBRIDGE
UNIVERSITY PRESS

University Printing House, Cambridge CB2 8BS, United Kingdom

One Liberty Plaza, 20th Floor, New York, NY 10006, USA

477 Williamstown Road, Port Melbourne, VIC 3207, Australia

314–321, 3rd Floor, Plot 3, Splendor Forum, Jasola District Centre, New Delhi – 110025, India

103 Penang Road, #05–06/07, Visioncrest Commercial, Singapore 238467

Cambridge University Press is part of the University of Cambridge.

It furthers the University's mission by disseminating knowledge in the pursuit of
education, learning, and research at the highest international levels of excellence.

www.cambridge.org
Information on this title: www.cambridge.org/9780521831499
DOI: 10.1017/9781139020299

First published 2022

Printed in the United Kingdom by TJ Books Limited, Padstow Cornwall

A catalogue record for this publication is available from the British Library.

Library of Congress Cataloging-in-Publication Data
Names: Allen, Michael F., 1952– author.
Title: Mycorrhizal dynamics in ecological systems / Michael F. Allen, Center for Conservation
   Biology.
Description: Cambridge, United Kingdom ; New York, NY : Cambridge University Press, 2022. |
   Includes bibliographical references and index.
Identifiers: LCCN 2021053890 (print) | LCCN 2021053891 (ebook) | ISBN 9780521831499
   (hardback) | ISBN 9780521539104 (paperback) | ISBN 9781139020299 (epub)
Subjects: LCSH: Mycorrhizal fungi–Ecology. | Mutualism (Biology) | Plant-fungus relationships. |
   Plant-soil relationships.
Classification: LCC QK604.2.M92 A45 2022 (print) | LCC QK604.2.M92 (ebook) |
   DDC 579.5/1785–dc23/eng/20211104
LC record available at https://lccn.loc.gov/2021053890
LC ebook record available at https://lccn.loc.gov/2021053891

ISBN 978-0-521-83149-9 Hardback
ISBN 978-0-521-53910-4 Paperback

Contents

Color plates can be found between pages 148 and 149.

Preface

The Périgord Truffle, *Tuber melanosporum* Vittad 1831, as of this writing, could be purchased for US$19.97 per 14 g (0.5 oz), or US$1,426 per kg. These "black diamonds" are one of the ultimate human gastronomic experiences. Yet, despite centuries of study by outstanding scientists, the biology of truffles is so poorly understood that they cannot be commercially produced in consistent, meaningful quantities. Most gastronomic truffles still come from individual truffle hunters, working in orchards or wildlands and selling to expert middlemen, then to the international market often through back doors (see 376). Efforts to collect truffles have been undertaken throughout recorded history, as Romans, Greeks, Babylonians, Sumerians, and Egyptians all wrote about what Aristotle called the fruit of Aphrodite. The complexity of formation, attributed to particular trees, lightning or thunder, or soils, led to extensive research during the nineteenth century, culminating in the funding of the work of Albert Bernhard Frank, a forest pathologist, supported by the king of Prussia. Little did any of the early researchers recognize that the biology of *Tuber* was only a small, yet complicated piece of a story of a diverse type of symbiosis, that plays a major storyline in biological theory, in the application of agriculture and forestry, and holds keys to how carbon was sequestered in the early earth and provides directions to reducing the global CO_2–climate impacts. Truffles, including members of the genus *Tuber*, are mycorrhizal fungi. That is, they are mutualistic fungi, associated with a limited array of host trees, such as oaks, beeches, and hazelnuts. Being a mutualistic symbiont not only means that the ecology of the fungus is complex, but it also means that the ecology of the host is complex. Adding in the complexity of climate and soils that change over time and space, the association falls into the theoretical construct of biology called biocomplexity.

This view contradicts the general approach of research, to find elegance in simplicity; with the simplest explanation being most likely the correct one. This approach led to the view in the 1970s, that the primary function of mycorrhizae was to increase nutrient uptake by exploratory hyphae, especially phosphate and ammonium that are bound by soil particles. Other responses, such as increased water uptake or hormonal shifts, are simply (and only) the result of improved P (or N) concentrations in the plant. This concept is simple and elegant. The problem is that it is wrong, as I will discuss for the remainder of this book.

My first unknowing exposure to mycorrhizae was my ability to find the preschooler's optimal playground – one with grass and dirt! After two decades of digging in the soil, I began working on mycorrhizae in 1974, when initiating my graduate studies in Wyoming, where human impacts were subtle and nature predominant. In one of my first studies, I began working in the short-grass prairie of the Pawnee National Grasslands, Colorado. Mycorrhizal fungi were everywhere, infecting up to 90 percent of the fine roots of the dominant grass, *Bouteloua gracilis*, a C_4 short-statured but nutritious grass. However, the Pawnee Grasslands had moderate to high levels of available P and, being evolutionarily part of the American Great Plains, had high levels of grazing by bison that make N highly available. The major limiting factor to production was water; more water, more production. Could mycorrhizae increase water uptake and therefore throughput, and subsequently carbon fixation? And, could there be multiple factors affecting that uptake, from direct hyphal access to water droplets, to increased root branching and altered stomatal behavior caused by changes in hormonal activity? This formed the basis of my early focus on complexity as a suite of mechanisms regulating mycorrhizae, and therefore plant production.

A second major step occurred when I was writing my dissertation. The first iteration of my literature review focused on the physiological ecology of arbuscular mycorrhizae, especially in *B. gracilis*. But Dr. Martha Christensen, my major professor, said that I needed to write about the entire field of mycorrhizae, including ecto- and endomycorrhizae. I gulped when I realized the magnitude of that literature search but relished the work that followed and the satisfaction of being familiar with the entire field of mycorrhizae. That literature review formed the basis of my first book (36), *The Ecology of Mycorrhizae* (Cambridge University Press, 1991). But even as I wrote the book, the increasing importance of *complexity*, comprised of multiple resources being exchanged and shifting in space and time, and the importance of *connectedness*, multiple plants being directly connected with multiple fungi with each connection living and dying continuously, and all occurring at different rates in different locations, became more and more obvious.

Finally, I wanted to know about the dynamics in field settings, not just in an elegant laboratory experiment. My dissertation was comprised of laboratory studies in pure culture syntheses. While these studies were useful to identify potential mechanisms, they failed to find the rich diversity of alternative expressions of these mechanisms in the natural environment. With training in classical mycology, plant physiology, and ecology, I began observing roots, fungi, and soils wherever I went. Inconsistencies and contradictions continued to accompany every experiment and observation. Subsequently, for nearly a half century in the field, I have looked at mycorrhizae (or at least for them) on every continent, from the Arctic in the Norwegian and Alaskan tundra to the Dry Valleys of Antarctica (where they don't occur), in habitat extremes ranging from the hot deserts of the Sinai, the Gobi, and Death Valley to tropical rainforests of Costa Rica and Taiwan. In witnessing the incredible diversity comprising the natural history of mycorrhizae, what stands out is the necessity to understand change and connections of all partners simultaneously. Then, the attempt to apply these observations to managing natural resources from restoration to global change provided testbeds for accepting or rejecting hypotheses generated from theory and from natural history observations.

I also add one additional note. Some of my conclusions will be wrong and some premature; that is the case with all scientific endeavors. But I have found that if a researcher entering the field can cite my statement, show how she or he will address the concern, and thereby secure funding furthering mycorrhizal research, then my primary goal will be achieved!

To that end, I hope that this book will help the reader enter the field of mycorrhizal studies with eyes wide open, or be of value to experienced researchers to take a second look at how one field of expertise contributes to and is affected by the incredible range of other fields that comprise symbiotic mutualisms, such as a mycorrhiza.

Acknowledgments

It is impossible to adequately acknowledge every person who helped in putting this book together, from reviewing and structuring to thinking and rethinking conceptual frameworks. My former students, postdocs, and technical assistants deserve much of the credit. While I was "mentoring" them, they were changing and improving my own ideas. I also owe thanks to mycorrhizal research colleagues around the world, from molecular biologists to global change scientists. In giving my latest presentation at an international meeting (the European Geological Union) on the Internet (due to the COVID-19 outbreak), I was gratified to discuss my research with colleagues on nearly all continents and continued to gain new perspectives. When I consider the dramatic changes our field has undergone over the half century of research since I started graduate school, we owe much to the increasing breadth of the biological sciences conceptually and technologically. Ideas and the ability to test concepts have changed immensely at all scales of interest.

I also owe specific individuals and institutions for their direct contributions. My wife and colleague, Edith Bach Allen, read and edited multiple versions, and continued to provide loving and intellectually stimulating inputs. Danielle Stevenson reviewed the text, provided many of the illustrations, and also helped assemble the bibliography – not a trivial task. Students in Sydney Glassman's Microbial Ecology Reading Group provided valuable insights for interesting literature, also providing perspectives on the backgrounds and expectations of beginning researchers that I incorporated for this book. Carlos Urcelay, Sydney Glassman, and Ignacio Querejeta reviewed the text and provided valuable insights.

I acknowledge the United States National Science Foundation, which provided financial support from my graduate student days in the mid-1970s, through retirement in 2017. In addition to their financial support I also owe thanks to many program officers, with whom I served, and who managed my grants before and after that service. These outstanding people freely shared their thoughts, critiques, and ideas, all of which contributed to this book. I remember returning to my teaching position at San Diego State University after my stint as an NSF program officer. The first thing I did was to take all of my class notes and deposit them into the round file. My perspectives had changed so much that while the classes were the same, my perceptions had all changed.

I also want to acknowledge the financial and intellectual support of the University of California–Riverside and the University of California Agricultural Experiment Station, and my departmental colleagues in Microbiology and Plant Pathology; Evolution, Ecology and Organismal Biology; Botany and Plant Sciences; and Environmental Sciences. I especially want to acknowledge our Friday Happy Hour group – where more broad-based conceptual ideas were exchanged than at any formal meetings. Finally, I want to acknowledge my colleagues at Rhizosystems LLC, which we formed to develop new technologies for studying mycorrhizae and soil ecology.

Cambridge University Press is especially to be commended for their patience and support. I agreed to write this more than a decade ago, then it went to the back burner while I served as Chair for two departments and Director of the Center for Conservation Biology. It re-emerged upon retirement with support from the outstanding editors at Cambridge University Press, especially Aleksandra Serocka, who oversaw and facilitated my final push.

Glossary of Key Terms

achlorophyllous A plant without chlorophyll. These are either parasitic, or mycoheterotrophic, obtaining both C and nutrients from mycorrhizal fungi connected to photosynthetic plants with chlorophyll.

apoplastic flow Movement of water and solutes moving outside the plasmalemma. Here, water movement is Brownian or in response to a water potential gradient.

arbuscule Complex structure formed within a plant cortical cell wall but with plant and fungal membranes. A site of exchange between plant and fungus.

cheater The process whereby a fungus takes carbon without providing a resource. To be a cheater, this process must occur for long enough for it to reduce the fitness of the plant; it must have evolutionary implications.

clade Organisms that have evolved from a common ancestor.

cluster root Roots that form clusters of tight lateral roots. Generally, from 8 to 250 μm in diameter and 0.6 to 35 mm in length.

coenocytic Not having regular septa. Glomeromycotina and Mucoromycotina form coenocytic, or aseptate hyphae, although adventitious septa may occasionally form.

colonization The process whereby a plant or fungus invades and occupies a new spatial location.

dikaryon Two or more closely associated nuclei within the same cell. Can be a homokaryon or heterokaryon.

heterokaryon A dikaryon with genetically different nuclei.

homokaryon A dikaryon with genetically alike nuclei, or a monokaryon.

infection The process whereby a mycorrhizal fungus invades and occupies a new root segment.

monokaryon A fungal cell with a single nucleus.

mutualism Two organisms involved in a symbiosis that positively affects both partners fitness.

mycoheterotrophic An achlorophyllous plant that obtains both carbon and nutrients from a mycorrhizal fungus also connected to a photosynthetic plant.

mycorrhiza a mutualistic symbiosis *between* a plant and a fungus localized in the roots or rhizoids, in which the fungus obtains carbon from the plant in exchange for resources that the fungus extracts from the surrounding substrate. Pl = mycorrhizae or mycorrhizas. As a mycologist, I use -ae, as in hyphae rather than hyphas, for my writings.

mycorrhizal fungus The fungal partner in a mycorrhiza.

mycorrhizal plant The mycorrhizal status of a plant when the mycorrhizal fungus is present.

mycotrophic A plant genetically capable of forming a viable mycorrhiza.

nonmycorrhizal A plant that does not have a mycorrhiza. This status can result from a mycotrophic plant growing in an environment with no mycorrhizal fungi or can result from a plant being nonmycotrophic.

nonmycotrophic A plant genetically incapable of forming a mycorrhiza. In some cases, nonmycotrophic plants may be invaded by mycorrhizal fungi, but the infection process initiates an immune response and necrotic tissue.

parasexual A process whereby genetic material can be transferred through mitosis, without meiosis. Is controversial.

peloton Fungal hyphal coils growing within plant cell walls. Mostly in orchid mycorrhizae, but also found in many mycorrhizae.

species Originally defined as the group of organisms wherein at least individuals can mate and produce fertile offspring. Since (and including) Darwin, many have concluded that the rank of species was arbitrary. Species concepts have changed. As an ecologist, my view of a species lines up with the view of species as an arbitrary cut-off along the tree of life (65; 507) For purposes of this book, I use the taxonomy of the original author of whatever taxon was studied, however arbitrary.

symbiosis Two organisms living intimately together.

symplastic flow Movement of water within a plasmalemma, through the cytoplasm.

taxon A taxonomic level of any rank.

1 · *Introduction*

Travel to your nearest nature reserve. It could be anywhere, within a human habitation or wildland, on continents or islands around the globe. It could be a forest, shrubland, grassland, forbland, or even desert. Across your view is a diverse array of plants, complete with a complex architecture with several levels from the ground surface to the highest leaf, be it a redwood tree extending hundreds of meters into the sky, or a diminutive, but still complex, structure only 10 cm high of mosses, liverworts, and cryptogams. As you separate the stems and look down, you see roots: tens of meters of roots or rhizoids for every square meter of the soil surface. Take your hand lens and pull up some of those roots. There will be soil hanging from those roots, held together by tens to hundreds of meters of threads of fungal hyphae per cubic centimeter. The vast majority of these hyphae form mycelia from many species of fungi, directly connecting fine roots to the larger soil matrix, serving as a living pathway interconnecting the plant, which is actively fixing carbon, within the micro patches of nutrients and water necessary to fix that carbon. This is the macro-microscopic world of mycorrhizae; the space where many plants interface with many fungi, interweaving a matrix of roots, soil particles, and decomposing organic debris from once living tissue, extending thousands of kilometers across continents. But there are holes in this web. It is dynamic. Individual hyphae may die within hours or live for decades. Mortality may come from a grazing collembolan, an ant colony, an agricultural field, a volcano, or an ice age. What connects all of the organisms comprising this landscape? The first primitive proto-eukaryote established a mutualism with an endosymbiotic energy machine (mitochondrion), a symbiosis that resulted in a creature capable of far more efficient growth and adaptation. Somewhere among these primitive proto-eukaryotes, one cell ingested a photosynthetic cyano-bacterium, forming a group of autotrophic, eukaryotic microorganisms with organelles, thereby changing the globe. Today, symbiosis plays a similar critical global role. The mutualistic arbuscular mycorrhizae form glycoproteins and other complex organic recalcitrant compounds that sequester atmospheric CO_2 and regulate global greenhouse gases. This is the complex, interconnected web of mycorrhizal symbiosis. It is comprised of a mutualism between the vast majority of plants, and fungi from all phyla of the true fungi.

Mycorrhizae are mutualistic symbioses between plants and fungi localized in the roots or rhizoids, in which the fungi obtain carbon from the plant in exchange for resources that the fungus extracts from the surrounding substrate. This definition, which I outlined three decades ago (36), extends the concept of a complex suite of relationships that go well beyond the single plant–fungus symbiosis focus, reasonably well understood, to a *network* connecting many plants and many fungal mycelia and comprising an ecosystem. This view encompasses complexity resulting from two advances in evolutionary and ecological thinking. Before the 1990s, the predominant idea was that mutualisms were abnormal, and generally unimportant in determining the evolutionary history of an organism and ecological functioning. Since then, the study of mutualisms in general and for mycorrhizae specifically have entered mainstream journals and thinking in both ecology and evolution. The second was that the functioning of mycorrhizae could be understood in the context of enhancing the fitness of a plant and a fungus through the supply of external resources. We know that a single plant lives in an environment in which it is colonized by many mycorrhizal fungi, and a single mycorrhizal fungus invades many plants. This connectedness view from new observations complicates and enriches the picture that was the focus for many of us a quarter century ago.

The formal study of mycorrhizae (Latinized plural of mycorrhiza) is well over a century old. The term "mycorhiza," coined by A. B. Frank in 1885 (256), refers to μύκητας-ρίζα, mykitas-riza, a fungus-root. He described the symbiosis as a mutualism in that it is comprised of a distinctive morphological structure in which the fungus encases a root and acts as a "wet-nurse" to the plant, ". . . er funktioniert im Bezug auf diese Ernährung als die Amme des Baumes." This description followed the remarkable observations of F. Kamienski (398) on *Monotropa* in which he classified the fungus–*Monotropa* relationship as a mutualistic type of symbiosis. These were notable advances conceptually. Theodore Hartig (324) had illustrated the Hartig net, which formed in conifer roots. However, he did not recognize this structure as fungal in origin. His son, Robert Hartig, the "father of forest pathology," studied *Agaricus* (=*Armillaria*) *melleus*, the honey fungus, and his observations on the invasion and structure of this disease postulated that it resembled the Hartig net and rhizomorphs of ectomycorrhizae (323), and the Tulasne brothers, in their detailed descriptions of *Elaphomyces granulatus*, noted that the mature sporocarps enclosed the root tips and opined that this fungus parasitized the plant (727). Indeed, research on the importance of fungi as plant parasites had made noteworthy progress during this period, as evidenced by Tillet's (709) experimental demonstration of the disease of wheat kernels (smuts and bunts), Berkeley's (117) discovery of *Botrytis* (=*Phytophthora*) *infestans* as the causal agent of potato blight as opposed to lightning or electricity, and DeBary's detailed studies of rusts and smuts (197).

The evidence for mutualism was initially quite shaky and viewed by many with skepticism. Kamienski (398) described the functioning of the monotropoid mycorrhizae as a symbiotic relationship existing along a parasitism-to-mutualism gradient, concluding that this relationship was probably more mutualistic than parasitic. His hypothesis was based, in part, on the evidence that the fungus invaded a root, but no necrosis of that root was observed and the root remained healthy. In the absence of experimental growth response data, this approach remains a useful observation that at least sets the stage for further study. This evidence formed the basis of Janse's (381) remarkable study of arbuscular mycorrhizae in Indonesia, Lohman's (465) survey of mycorrhizal types in Iowa woodlands, and even more recent work such as our observations of dark-septate fungi found in epiphytic bromeliad roots in the seasonal forests of Mexico (72). These observations contrasted with the views of early researchers such as McDougall and Liebtag (493), who envisioned mycorrhizae as a "mutual reciprocal parasitism," in that both would gain some nutrients (such as organic N) but independently the plant would still grow better without the fungus.

DeBary (198; 199) formally defined a symbiosis as two organisms living intimately together. These observations were based initially on his observations of the cyanobacteria-*Azolla*, and lichens; mutualistic associations as opposed to the known parasitism of many smuts, rusts, and blights. He envisioned symbiotic relationships in a neutral context as comprising interactions ranging from mutualism to parasitism. From his work we can develop a simple but powerful +/0/− suite of relationships between the two symbionts, plant and fungus, that differentiates between a mycorrhiza and a parasite, and other symbiotic relationships. These include parasitism (+/+), amensalism (0/+), antagonism or competition (−/−), neutralism (0/0), commensalism (+/0), and mutualism (+/+) (36).

Frank (256) defined a "mycorhiza" as a functional relationship whereby the underlying carbon needs of the fungus are provided by the tree through its leaves, "*Nahrungsbedurfniss des Pilzes wird sich hauptsächlich auf die assimilirten, kohlenstoffhaltigen Nahrungstoffe beziehen welche der Baum durch seine chlorophyllhaltigen Organs bereitet,*" and a significant quantity of water and nutrients required by the tree from soil can only be supplied via the fungus, "*das ganz für den Baum erforderliche Quantum von Wasser und Nährstoffen aus dem Boden nur durch Vermittelung des Pilzes demselben zugeführt wird.*" This means structural elements that are both internal to the plant for accessing C and extramatrical or external in the soil to access water and nutrients.

Crucially, fungi provide external resources (from the soil, or other growth medium) to the hosts and, in exchange, plants provide carbon to the fungi. This exchange, in which both symbionts are evolutionarily adapted to acquire, provides positive benefits for both (450). Thus, a mycorrhiza has a defined structural aspect. While the nature of a mutualism–parasitism relationship of this symbiosis is sometimes difficult to measure, and fluctuates temporally depending on resource availability, mutualism is a defining character. A mycorrhiza is evolutionarily, a mutualistic symbiosis.

Up to the 1980s, the widespread view of mycorrhizae, and mutualisms in general, was illustrated in Williamson's paradigm (782) that [mutualism] is interesting, but unimportant. He stated that most examples of mutualism are tropical (see discussion in (36)). May (491) further stated that mutualisms are mathematically unstable. Indeed, most mathematical community models suggest that, while parasitism confers stability, mutualisms result in unstable dominance by a limited suite of hosts (e.g., Bever (119)). But experimental evidence (303; 406) continues to show that mycorrhizae increase diversity by supporting lower density species. Molecular evidence now points to both widespread and stable mutualisms going back as far as eukaryotic organisms (480), and mycorrhizal relationships go back to the early invasion of land by plants.

Somewhere between 10 and 85 percent of the net CO_2 fixed in terrestrial ecosystems travels through the mycorrhizal fungal/root interface (239). This simple loss of large amounts of C by a plant requires compensatory procurement of large amounts of soil resources. In many situations, a majority of resources like N and P taken up by the plant may come from the mycorrhizal fungus. But water transport, pathogen protection, and other resources like cations can be just as crucial, under limiting circumstances. Acquisition and exchange of resources creates the complexity that is the core of this book.

A mycorrhiza is a co-evolved mutualistic relationship between plants and fungi in which the fungus extends from the plant into the surrounding substrate, extracting soil resources in exchange for carbon resources fixed by the plant. In this definition, the fungus must extend into the substrate differentiating a mycorrhiza from endophytic fungi, which are important ecologically, but appear to have very different modes of physiological interaction. I also note that the fungus uses carbon directly transferred by the plant. While some mycorrhizal fungi access soil organic C, the majority come from a host plant. This is crucial, as mycorrhizal fungal carbon can be measured isotopically as autotrophic, not saprotrophic carbon, allowing it to be tracked in ecosystem models by age and function (44; 410; 723).

One problem with this definition is that mycorrhizae are also formed by achlorophyllous plants that receive their carbon from the fungus (called mycoheterotrophs). This issue will be discussed in greater detail in Chapters 3, 4, and 6. Especially relevant to this discussion, again, is Kamienski's (398) description of monotropoid mycorrhizae. At this point, we still do not understand what resources the plant provides to the fungus. Is this a case of fungus as host and plant as parasite, or does the achlorophyllous plant provide an unknown resource? Or is the fungus a conduit between the mycoheterotroph and a nearby community of chlorophyllous plants?

Another view is that almost any plant–fungus interaction that appears mutualistic is a mycorrhiza. For example, there were reports early in the twentieth century of *Fusarium* as a mycorrhizal fungus (89). Wilde et al. (780) and Iyer (374) described the "epirhizal" mycorrhizal type, in which almost any fungus can sometimes increase plant growth. Many fungi can be found localized in roots, including numerous facultative parasites, or saprotrophic fungi living on dying roots. Recent versions of the same arguments have emerged in describing members of the Sebacinales, fungi whose DNA are often found in EM tissue, as forming a mycorrhiza. However, the mechanisms of interaction, soil–fungus–plant, may not fit my definition of a mycorrhiza (769). Endophytes are mutualists found in almost every plant part. The fact that a group of fungi, such as the Sebacinales, are often an element of the mycorrhizosphere microbiome, is a very interesting topic for research of a potential fungal symbiotic mutualism (sensu (199)). Moreover, there is a report that these interesting fungi produce pelotons in ericoid plants (651). But at this stage, the mycorrhizal and physiological status needs more work.

Historical Types of Mycorrhizae

Research differentiating the types of mycorrhizal associations goes back well over a century. The origins of the study of mycorrhizae lie in the type called an ectotrophic mycorrhiza and, more recently, with its relative, the arbutoid mycorrhiza. Ectomycorrhizae (or EM) are characterized by an extensive hyphal network growing in the interstitial boundaries between the walls of cortical cells (the Hartig net), and a mantle that covers individual short roots, and whose hyphae extend outward into the soil, often for many meters! This mycorrhiza is the original one described by Frank (256) and Kamienski (398), and occurs between some woody plants and fungi in the Endogonaceae and members of the Ascomycota and Basidiomycota. The ectomycorrhiza evolved independently in many plant and fungal lineages. Arbutoid mycorrhizae were often originally described as ericaceous ectendo-mycorrhizae. Although they penetrate cortical cell walls, they are comprised of fungi that form EM. These arbutoid mycorrhizae may be mostly Basidiomycota, many of which also form ectomycorrhizae with neighboring plants, and with Ericaceae in the Vaccinioideae, Arbutoideae, and Monotropoideae families.

Frank (257) followed his 1887 descriptions of types of EM with the observation that there were also mycorrhizae that penetrated cortical cells, but did not lead to cell necrosis. He called these endomykorhiza. In this work, he described what we now call ericoid and arbutoid mycorrhizae. His student, Schlicht (639), carefully described another type of endomykorhiza associated with herbs now called arbuscular mycorrhizae (or AM), in which arbuscules form a structure capable of resource exchange. Dangeard (189; 190) illustrated both arbuscules and vesicles, also showing the multinucleate status of the fungi forming AM. Janse (381), in studying the endophytic fungi in Java, illustrated AM in many tropical plants, and orchid mycorrhizae from arboreal and ground orchids. Just as interesting, he also reported that achlorophyllous orchids also formed mycorrhizae, just as Kamienski (398) noted for ericaceous achlorophyllous plants. Gallaud (272) clearly described and illustrated the "Arum" type of AM, forming arbuscules independently within cortical cells from intercellular hyphae running between cells, and the "Paris" type of AM, in which the hyphae run from cell to cell, forming intracellular hyphae, arbuscules, and coils, or pelotons.

Only a few of the early studies actually demonstrated a growth promotion of mycorrhizae. Frank (255) showed a growth enhancement in pine growing in organic soil, limited by N. In other studies, a lack of necrosis in the root tissue following fungal penetration was observed (see (381; 639)) and a mutualistic interaction was postulated. A number of studies subsequently demonstrated the growth enhancements by ecto-mycorrhizae (e.g., see (305; 319)). In the 1950s, the groundbreaking studies by Mosse (520) and Gerdemann (282) demonstrated that gloma-lean fungi, classified as zygomycetes and not pythiaceous fungi, trans-ferred nutrients and enhanced growth, characteristics of a mycorrhiza. For an early history, I highly recommend reading the historical section of the Proceedings from the 6th NACOM (509).

As a practical means of organizing mycorrhizal types, two initial foci can be utilized. The first is based on the type of interface between the host and fungus, while the second is phylogenetically-related associations. The first focus is characterized by the location of fungal structures within or outside root cortical cells. In an endotrophic mycorrhiza, or endomy-corrhiza, the fungus penetrates the cell wall, forming an extended fungus membrane–interspace–plant membrane interface creating an enlarged sur-face area between plant and fungus. In the ectotrophic mycorrhiza, or ectomycorrhizae, the fungus forms a network of hyphae between the cortical cells (the Hartig net) and an external mantle (of varying coverage and thickness). The cell membranes of both symbionts remain intact. Because these are morphological characteristics, and often overlap, I prefer to think of these as gradients in types, rather than absolutes.

The second focus is based on phylogenic interactions between plants and fungi. I will explore these in greater detail in Chapters 2 and 3, but I establish some basic relationships here to form consistent terminology. Four primary categories stand out, three of which are endomycorrhizal and one is ectomycorrhizal. The most common is the *endotrophic mycor-rhiza* (or endomycorrhizae), which is comprised of arbuscular mycor-rhizal (AM) or Glomalean mycorrhizae (also called vesicular-arbuscular mycorrhizae (VAM) or, in some older literature, phycomycetoid mycor-rhizae). These occur in some 70 to 80 percent of plants, from basal "primitive" plants through some of the most advanced clades, in symbi-osis with the Glomeromycotina of the Mucoromycota. This association appears to have evolved a single time as primitive plants emerged onto land, and all Glomeromycotina fungi are symbionts.

A second category of AM is called the "fine-endophyte," initially identified as *Glomus tenue* (299), which appears to have evolved about the same time, and is another endomycorrhiza. These fine endophytes appear to be a polyphyletic group of Mucoromycota (614) that can be found in plants ranging from ancient liverworts to modern grasses and saltbush. Little is really known about the group, but they are repeatedly observed in many habitats, especially those under drought stress or under highly acidic or highly eutrophic conditions.

A third category, also of *endomycorrhizae*, is formed by the septate fungi (Ascomycota and Basidiomycota). This is a diverse group, including the orchid mycorrhizae and the ericoid mycorrhizae. Orchid mycorrhizae form associations with fungi in the Cantharellales [note: in the older literature, the fungi were known by their imperfect name, *Rhizoctonia*]. The ericoid mycorrhiza type is found between plants in the Ericaceae, subfamilies Ericoideae, and ascomycetous fungi in the Eurotiales and Onygenales (in the Eurotiomycetidae) and Helotiales (in the Leotiomycetes). There are suggestions that other Ascomycota in the Hypocreales, and potentially even Basidiomycota in the Sebacinales, form orchid or ericoid mycorrhizae, but these are not yet resolved. A type of mycorrhiza, called an ectendotrophic or "E-strain" mycorrhiza, is formed between conifers and Ascomycota, especially in the genus *Wilcoxina*, in the Pyrenomataceae. Finally, there are "dark-septate" mycorrhizae. This is a diffusely defined relationship. An early study by Haselwandter and Read (328) characterized this mycorrhiza between sedges and Ascomycota in alpine ecosystems. But subsequently many "dark-septate mycorrhizae" have been delineated simply through the presence of saprotrophic or weakly parasitic fungal presence within roots. I will discuss this more in Chapter 3.

Many plants form dual mycorrhizal types, in time or space. Many plants form dual mycorrhizae for a variety of ecological and phylogenetic reasons (705).

Often any fungi within roots are described as mycorrhizal (epirhizal, dark-septate, and fungi in the Sebacinales). These relationships will be discussed in more detail as we go through further chapters. Some may be mycorrhizal, some other mutualistic symbioses, some simply present, and there are many that we simply do not understand. Because mycorrhizae have evolved independently several times, and because the interface structure exists as a gradient, communities of mycorrhizae form more of an array of functioning entities than of any single function. For example, one fungus may secrete high levels of lignocellulases releasing and taking up organic N, where a second largely searches out NH_4^+ bound to organic surfaces. A community with multiple mycorrhizal types and functions is an attribute of complexity that is essential to forming a basic understanding of what mycorrhizae are and how they work.

Resource Dynamics and Mycorrhizae

At its most basic, the importance of a mycorrhiza is to increase the interface between the root and the external growth medium. This interface is comprised of a root cortical cell–fungal wall interstitial zone in which materials are exchanged between the host and fungus. Frank (256) observed that the fungus often completely encased the host short roots, requiring that a large fraction of the nutrients and water, especially N, must go through the fungus to be taken up. The fungus extends outward from the root into the soil, forming a second interface between the fungus and growth medium, generally soil. The fungus provides an increasing surface area for uptake and transport of limiting nutrients. Stahl (678) noted greater throughput of water in a mycorrhizal plant. His postulated mechanism was that more water moving brought more nutrients to the interface, increasing plant nutrient uptake. Hatch (334), in an elegant suite of studies, demonstrated that those resources, such as orthophosphate, that are not dissolved and carried by mass flow, are especially reliant on mycorrhizal relationships for their uptake. Hyphal transport from the soil to root cells is the basic structure that forms mycorrhizal functioning. This basic structure also differentiates a mycorrhiza from other mutualisms, such as leaf endophytes, in which the fungus does not infiltrate the surrounding growth matrix.

The extramatrical fungal hypha is the external functional feature of the mycorrhiza. Individual hyphae range in size from 2 to 20 μm in diameter. This is smaller than fine roots (20 μm in fine grasses to 300 μm in tree roots), root hairs (as thin as 7 μm in grasses), or even cluster roots (8 to 250 μm). This size difference allows the fungus to penetrate soil pores that roots cannot. Many fungi also form cords of multiple hyphal strands, allowing them to grow long distances across the soil. There can be up to a kilometer of hyphae per square centimeter of soil. This small diameter and high density allow the fungi to reach nutrients bound to organic or clay particles and penetrate soil micro- and even ultramicropores in search of nutrients and water. These hyphal tips can pick up nutrients and water away from the host (up to several meters) and, wrapping into cords, transport these resources between and along hyphae apoplastically as well as within (symplastically) hyphae to the mycorrhizal interface. This movement can occur through complex "vessel-like" elements (213) in some EM fungi. Even AM fungi can form arterial hyphae that will wrap together, creating an interstitial space within a small network.

Although mycorrhizal fungi do not carry out photosynthesis, by entering the root, they have direct access to C fixed by the plant. As a means to obtain C, the fungus exchanges nutrients for C with the host. This provides a much more direct pathway to plant C than for pathogens or saprotrophs. This basic structure also creates a unique feature, differentiating it from all other symbioses, in that a mycorrhiza is the only mutualist that can directly and simultaneously connect two hosts. A plant may have multiple rhizobia, or leaf endophytes, but only a mycorrhiza can physically connect two plants at the same space and time.

The formation of hyphal connections has led to a controversial hypothesis that mycorrhizal fungi redistribute resources among plants (e.g., (567)). Alternatively, Fitter (246) argued that this merely means that a fungus can maximize its own C gain by taking advantage of many plants in its environment. I will explore these arguments later in the book.

The nutrient sink of the plant and C sink of the fungus represents a mycorrhiza at its most basic (36). But nutrient and carbon allocation do not necessarily coincide (Figure 1.1). In general, remember that if a mycorrhizal fungus is present, it will try to invade any potential host. From the plant perspective, infection depends upon whether the plant rejects the fungus or not. We do not know if the fungus has selective mechanisms, other than C transport to pre-select the plant. At the

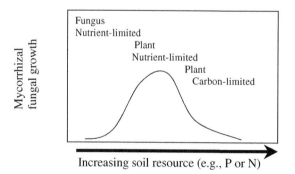

Figure 1.1 The interaction of plant and fungal nutrient limitation on the biomass of mycorrhizal fungi. At high nutrient levels, fungi will receive little carbon from plants and will be C-limited. At lower nutrient levels, plants will be N- or phosphorus P-limited and will allocate C to mycorrhizal fungi. At the same time, if N or P concentrations are sufficient for fungal growth, mycorrhizal fungi will proliferate. At the lowest nutrient levels, both fungi and plants should be nutrient limited, and fungal biomass will be low or lost regardless of C allocation to the fungi by plants. Derived from Treseder and Allen (719).

extreme low end of nutrient availability, mycorrhizae may not facilitate nutrient acquisition, and do not form. I have observed mycotrophic plants with no mycorrhizae growing among recently deposited pumice particles with no organic matter and extremely low measurable nutrients, although presumably there are enough diffusing in the water to support these scattered individuals. This condition is also the case for proteoid cluster roots that grow in low-nutrient sandy soils. As nutrients become limiting to the plant, but available to the mycelial network, the exchange of C and nutrients mediated by the mycorrhiza is maximized. On the opposite extreme, as nutrients become more available the plant can access them readily, reducing the transfer of carbon and, consequently, mycorrhizal fungal growth. Importantly, nutrients and carbon have different transfer proteins at the membrane level. What this means is that the C deficiency of the fungus and P (or other nutrient) deficiency of the plant must cross-talk, but how remains unknown.

A critical point to understanding mycorrhizal complexity is that the Treseder and Allen model is dynamic. For example, at the upper end of the soil resource curve, when soil nutrients are readily available and plant densities are low, mycorrhizal fungal growth is reduced as the plant allocates less C to the fungus. At that point, plant densities increase, driving the curve toward a lower nutrient availability per plant. Mycorrhizae will increase as each plant attempts to maximize its share, and plants will allocate more C for fungi to that end. I will explore this relationship in more detail in Chapter 4.

The mycorrhizal relationship is comprised of organisms with unique, individual physiologies, not just pipes. For instance, the fungi do not function well in saturated, O_2-limited soils, and often the plant rejects the symbiosis under high nutrient conditions. Organisms have growth, metabolic, and reproductive requirements and have many means to acquire these requirements. Water flow, mineralization, transfer of growth-regulating chemicals, and complex enzymatic dynamics all regulate mycorrhizal associations. These all contribute to a diverse array of interactions within a mycorrhizal symbiosis.

Distribution of Mycorrhizae

Mycorrhizae may be the most widespread type of terrestrial mutualistic symbiosis. They can be found across the globe. They can be found in nearly every terrestrial habitat, including many wetlands (see (36)). Mycorrhizae also exist in extremely deficient to extremely rich growth media. The tropical forests are often regarded as optimally conditioned to support mycorrhizae (high rates of production, low soil nutrients), but plants at high altitudes and high latitudes form mycorrhizae as well.

Three conditions in which mycorrhizae are limited stand out. One is in aquatic environments when a lack of O_2 inhibits aerobic respiration. A second is under extremely high fertility in which the plant can readily obtain all nutrient and water conditions necessary for growth. Under these conditions, CO_2 is the limiting factor for growth, and the plant initiates rejection mechanisms to reduce C loss (the extreme right end of the Figure 1.1 curve (24). These conditions today appear to exist largely in glasshouse pot culture and hydroponic conditions, and in sandy, heavily fertilized, wet (or irrigated) agricultural soils. Importantly, these are also highly leaky conditions and can lead to groundwater pollution (e.g., nitrates) and are an undesirable extreme.

The third is severe disturbance. Extreme cases include newly-formed, almost sterile substrates such as volcanic eruption materials and retreating glaciers. This can also include extreme conditions such as the Antarctic Dry Valleys, where there are no plants, and habitats dominated by extremophiles without plant growth, such as caves, hot springs, and highly contaminated, heavy-metal soils. While many of these are natural, human disturbances often reduce or eliminate mycorrhizae through soil loss caused by erosion, strip-mining in which the topsoil is discarded, and high-intensity agriculture. Mycorrhizae become an essential element in the restoration of a self-sustaining, desirable ecosystem.

Hierarchy and Complexity

Just as a mycorrhiza became more complex when multiple fungi and plants became involved, understanding the basic functioning of mycor-rhizae in communities or ecosystems also became more than simply a nutrient uptake mechanism for individual plants. One idea is that groups of organisms, growing and reproducing independently, nevertheless interact to create additive ecosystem processes – that is they exhibit "emergent properties" (495), where the interaction is greater than the sum of the parts. Previously, I proposed a simple hierarchical structure for studying the ecology of mycorrhizae based on the "individualistic" concept (36). This idea was based on the predication that an individual is the entity selected against (471). Rillig and Allen (609) realigned community and ecosystem hierarchies dependent on relationships to changing environments. More recent ideas have complicated this view even further. Organisms can exist in "metapopulations" that are relatively isolated, but occasionally interactive. This concept has been extended to "metacommunities," repeating groups of organisms that interact and support one another (783). It remains an important theoretical topic as to whether the interactions among organisms under shifting conditions represent true "emergent properties," but these have the potential to create complexity. For this reason, an alternative is to envision a more complex hierarchy that cross-talks both up and down within a hierarchy, and across hierarchies at co-existing temporal and spatial scales (Figure 1.2).

Read and Perez-Moreno (599) portrayed another hierarchical perspective, focusing on scaling as a trade-off between precision of measurement and relevance of process. As one measures the individual physiological interaction between a mycorrhizal plant and fungus, accurate determinations of elemental exchange of elements can be made, largely in the lab or glasshouse (occasionally in the field). However, as one moves to the scale of the ecosystem, measurements of exchange rates become less precise, but more accurately reflect conditions in the field or forest. They make the point that most studies have been done at short-time intervals, across small spatial scales. This is especially true for measuring "ecosystem processes" in glasshouse pots or growth chambers. But how representative are these studies to understanding how mycorrhizae affect forest production across a landscape where perturbations are applied?

One result of complexity is that the highly relevant end of the Read and Perez-Moreno curve, may not actually be a reduced precision, but may simply reflect a family of individually-precise outcomes, rather than a single outcome that results from a small-scale study. This represents adding multiple variables, rather than a single-variable experiment. For example, in a pinõn–juniper stand (49), water and N were the limiting factors to production. When NH_4NO_3 was added, the EM pinõn pine shifted C allocation from EM to increased needle production to enhance even more production. The juniper simply increased both root growth (no change in AM) and shoot growth. However, P then became limiting. The AM juniper was quite able to continue taking up P; but without the EM fungi, P declined in the pinõn pine leaves. When seasonal drought hit, the AM juniper was quite able to tolerate. But because the pinõn now had larger leaves (and thus greater evapotranspirational area), and reduced EM to search for water, the pinõn died whereas the juniper increased productivity. These interacting

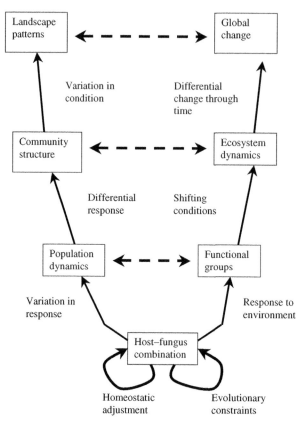

Figure 1.2 Hierarchical interactions among levels of mycorrhizal interaction. For the purposes of this text, I envision two distinct hierarchies that interact between similar levels creating a complex suite of mycorrhizal controls. Mycorrhizae themselves are sensitive to homeostatic adjustment by the environment, and constraints based on the individual lineages. From Rillig and Allen (609).

mycorrhizal and resource factors created a complex web of interactions, the understanding of which required past laboratory–glasshouse studies of nutrients and water, coupled with newer technologies of isotope signatures and continuous monitoring of changing environmental (soil and atmosphere) conditions.

Networking Topology and Connectivity Dynamics

One of the more important recent developments in ecology resides in studying networking topology to study interactions across a community.

Network theory links complexity, because of the multiplicity of entities involved, and connectivity, because of the interconnected exchange of a unit between them. One example is in the application of network theory to evaluating power grids, and directionality and susceptibility to disruption. A power grid exchanges energy between a power generation hub (say a tree in a mycorrhiza), across a power grid cable link (a mycorrhizal fungal hypha), and to a metropolitan area (maybe a mycoheterotrophic plant). Increasing use of network theory to understand vulnerability holds interesting clues to describing mycorrhiza behavior (98; 511; 659). In the case of mycorrhizae, we have traditionally focused on mycorrhizal fungus and host plant; this sets up a conceptual hierarchical relationship, a small parasitic or mutualistic symbiont, and the larger (and more important) host. However, mycorrhizae consist of extremely complex networks of small and large fungi, and small and large plants. Carbon preferentially moves from plant to fungus (except for mycoheterotrophic plants), but other elements, such as N and P, primarily move from fungus to plant. Fungi range from small individuals of *Cenococcum graniforme* to a *Leccinum scabrum* that may extend across a plant stand. The plants can be large, such as a clone of a single aspen (*Populus tremuloides*) that will have connections with thousands of individual stems and of both EM and AM fungi and hundreds of taxa of mycorrhizal fungi, down to a small, herbaceous columbine (*Aquilegia formosa*) located within the clone and tapping the *Scutellospora calospora* that is connected into the aspen. Networking theory provides an interesting perspective looking at both the number and structure of linkages to assess the stability of systems (e.g., (102)).

This approach is just beginning to be applied in mycorrhizal research (674), but provides a unique opportunity to tease apart complex interactions of mycorrhizal communities. For example, Southworth et al. (674) proposed that, for a mycorrhizal network, the node is actually the fungus, whereas linkages are the plants. This perspective reverses the usual thought and research process wherein the plant is studied as the focal organism contributed to by an array of symbiotic fungi.

What we do know, at this beginning point, is that stability comes not from a single mycorrhiza, but from a complex network of hubs, nodes, and linkages across a community. This will become clearer as we examine topics ranging from mycorrhizal communities to mathematical ideas of stability.

Complexity and Stoichiometry

Complexity itself is a complex topic. What comprises complexity, and how does it relate to understanding mycorrhizal dynamics? Complex derives from *complexus*, literally a weaving or twining together. Webster's dictionary defines complex as that which is made up of many elaborately interconnected parts. Complexity is then the condition or quality of being complex.

Researchers working in cybernetics developed a view that derives from Schrödinger's famous cat, where elements of a complex cannot be examined without destroying it (351). Biologists have not been quite so pessimistic, but ecological theory is clearly as diverse, disjointed, and fragmented. Heylighen (351) goes on to organize complexity into two dimensions that may help to study mycorrhizae: distinction, which implies variety, encompassing disorder, chaos, heterogeneity, and entropy; and connection, which implies constraint, encompassing order and negentropy.

How can we utilize this approach to better understand mycorrhizal dynamics? One way to focus our thinking is to visualize complexity by building on the original framework of Mosse (521) of a mycorrhiza as an integrating overlap of soil, fungus, and plant forming a series of interacting resource exchanges. Complexity derives from the intertwining of distinction and connection. For example, each of the three spheres is comprised of many distinct players, each of whom is competing internally and externally. A patch of seasonal tropical forest 6 meters in diameter may contain 18 species of plants, all capable of tapping into the hyphal networks of some or all of the 15 taxa of AM fungi present (21). In another analysis, out of 129 EM root tips extending from a single tree, 42 taxa were differentiated, the same morphotypes found in the surrounding trees (695). Just as importantly, the "soil" is comprised of an almost infinite number of patches of physical, chemical, and biological entities, each of which affect the mycorrhiza.

All of these entities are interconnected, transporting materials and energy. Thus, the connectivity of complexity derives from the chemical reactions that must occur for each component to persist. Ecological complexity can be simplified conceptually to a suite of well-known chemical reactions comprising a living plant and a living fungus. For example, photosynthesis is well described:

$$6CO_2 + 12H_2O \rightarrow C_6H_{12}O_6 + 6\,H_2O + 6O_2$$

In order to take up 6 molecules of CO_2 through the plant's stomata, approximately 300 times 6, or 1,800 molecules of H_2O are transpired.

The fixed C is respired by the leaf, or translocated to the roots and respired by the fungus, again by a well-known suite of pathways:

$$C_6H_{12}O_6 + 6O_2 + 3ADP + P_i \rightarrow \text{glycolysis} \rightarrow$$
$$\text{tricarboxylic acid pathway} \rightarrow 6CO_2 + 12H_2O + 3ATP$$

The complex part is that to fix the CO_2 requires a large amount of N, in the form of enzymes, particularly RuBPCarboxylase, and Fe and Mg forming the core of the chlorophyll molecule, and P to serve as the recipient and carrier of the energy fixed to convert the CO_2 and H_2O into $C_6H_{12}O_6$ (potential energy) and O_2. All of these elements must be present in appropriate amounts and they form a distinct stoichiometry. These processes create the famous "Redfield" ratio, wherein the optimal values of C:N:P of 106:16:1 typify the values necessary for growth in an autotroph algal cell. Plant cells normally have more C because of the need for structural compounds. If we focus only on the seven critical macronutrients, our C:N:K:Ca:P:Mg:Fe ratio is approximately 512:17:11:2.7:2.4:2.2:1 (using table 6.1 of Salisbury and Ross (631)). If one of these elements is in short supply, photosynthetic rates decline to a level based on the lowest common denominator – Leibig's law of the minimum. For example, in nature, N-limited plant leaves can have a C:N ratio as high as 50, whereas fungi may have a C:N ratio of 7–10. If the fungal N uptake is constrained to a concentration below 7–10, then the ability to transport to the plant element declines. There is a feedback then to the plant to reduce C production and allocation, and the overall symbiosis is affected. We know that the mycorrhizal fungus and plant interact intimately to acquire adequate amounts from their respective sources. The fungus acquires N, P, K, Ca, Fe, and Mg from the soil, providing those to the plant. The plant fixes C, critical for energy and building structure, and provides energy-rich C compounds to the fungus in exchange. Complexity, therefore, arises from the incredible number of requirements of each with a different source, sink, and exchange process and rate across spatial and temporal patches.

   A tertiary level of complexity in mycorrhizal relationships comes at the inter-species level. This perspective is based on the co-evolutionary, mutualistic relationship as initially envisioned by DeBary (199). However, instead of an endpoint, take the +/0/− interactions described by DeBary and place these relationships into a web instead of a single linear exchange. We can build the web using two axes. First, add a second plant. This simple action creates a conceptually complex view of mycorrhizae – a plural view of symbioses (Figure 1.3). In this case, the

In separate environments, i.e., glasshouse pots

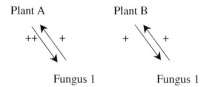

Plant A   Plant B

++   +   +   +

Fungus 1   Fungus 1

In an interactive environment, i.e., same pot or in the field

Plant A ⟷ Plant B

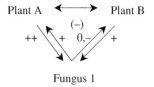

(−)
++   +   0,−   +

Fungus 1

Figure 1.3 Changing relationships between symbionts as a function simply of adding species.

mycorrhizal fungus in separate environments or locations forms a mutualistic relationship with both host plants. However, plant A provides somewhat more C to the fungus in exchange. The fungus still simultaneously infects both hosts to enhance its own C gain. Consequently, when both plants are in the same location, the fungus preferentially sends resources to plant A versus plant B. Over the life of the plant, plant A gains even further competitive capability over plant B because of the mycorrhiza. The relationship between the mycorrhizal fungus and plant B begins to become less mutualistic. As more plants and fungi are added, varying outcomes can occur.

Both the environment and the organisms are dynamic, not static. Temperature may initially be limiting, then N to increase the photosynthetic machinery, then P for energy storage and use, and finally water as a dry season begins. If the form of N available is NO_3^-, which can move by mass flow, then most of the N may bypass the fungus. If the dominant form is NH_4^+, which binds to clay and organic matter particles, mycorrhizae will play a large role. If organic N predominates, EM will serve better than AM, and so on. Growth of roots and the fungal network can be very rapid (38) or very slow (723) depending on the organisms involved and the dynamic nature of the environment. The simple +/0/− becomes an intricate interplay of biotic and abiotic forces that underlie a great deal of the complexity inherent in shaping plant communities and the ecosystems they comprise.

In 1991, when the *Ecology of Mycorrhizae* was published, it was rare that ecologists considered mutualisms important. They were considered fascinating oddities. May (491; 492) calculated that mutualisms are mathematically unstable. For this reason, he postulated that there are few natural examples of importance. Great strides have been made since then, such that most general textbooks acknowledge the importance of both mycorrhizae and mutualisms in general. Mycorrhizal ecologists no longer have to justify the importance of studying our favorite subject! I do not believe that particular issue could have been adequately addressed in 1991. However, I believe that our newer understanding of complexity in mycorrhizal associations specifically addresses this hypothesis. Specifically, what is stable is far more challenging than the mathematics used by May. We will address the specifics in Chapter 3, to show that, evolutionarily, mutualisms are not only stable mathematically, but also stable ecologically and evolutionarily.

Natural History, Theory, and Complexity

In studying mycorrhizae, like all of the sciences, there is a search for universals. Organisms, connectivity, and complexity exist in an intricate dance, both structurally and temporally. But in the end, scientists search for simple, definable universals, preferring those that are elegant, as outlined in Occum's Razor (726).

Alternatively, Natural History, defined broadly, is scientifically those observations that link the multitude of organisms, chemistry, and physics, as organized into communities and ecosystems (see (106)). Natural history often reflects complexity even when initially viewed as simple interactions. Natural History is messy!

A search for universals is an appropriate approach, especially at the cell or molecular level for how exchanges occur and are regulated. However, highly complex behavior emerges at the organismal scale as the dynamics play out. While AM appear to be a monophyletic group of fungi (Glomeromycotina), other mycorrhizal groups such as EM are clearly polyphyletic, and very different morphologically. Different plant and fungal taxa have very different nutrient, water, and light requirements that shift, in both time and space. Mycorrhizal fungi are not microorganisms, despite the necessity to use microscopes to observe structure and behavior. While individual hyphae can be only a few micrometers in diameter, individual organisms can extend many meters in length. I estimated that a single organism, such as a fairy ring, 10 m across with a highly branched mycelium, could weigh 3 g (dry mass). Many can be far larger, extending through many different environments, and sometimes living to a great age. The very size of both organisms results in a high degree of connectivity, well beyond a simple 1:1 relationship. Thus, we are dealing with two highly complex symbionts mutually dependent upon each other. Because a plant has many roots, with many mycorrhizal fungi, and a fungus attaches to all compatible roots encountered, there is a potential for many types of relationships in the field. *Here, I use a natural history approach, with as many independent field observations, compiled over a human lifetime, to describe the mycorrhizal symbioses living around us.* Natural History observations fundamentally result to increase complexity. However, with enough observations, they provide meaning that can form basic relationships leading to theoretical universals, which in turn, reduce complexities.

Summary

- Prior to the 1990s, studies of mycorrhizae focused on the physiology and morphology of the mycorrhizae of individual organisms.
- During the 1990s, the advent of molecular technologies allowed us to identify the associations in the field, and also to realign phylogenies of fungi, recognizing the relationships that characterize different mycorrhizae. We also developed the ability to differentiate different functional groups, which indicate that mycorrhizae represent a range of responses, not just a single function.
- Connectivity is not 1-to-1, it is networks of nodes and links with networks comprised of multiple fungi interacting with multiple plants. With this understanding models can then begin to scale to global level processes.
- Mycorrhizal fungi sequester and respire CO_2, release methyl bromides that affect the Antarctic ozone hole, immobilize and sequester carbon that can make global contributions to the carbon budget, and determine the relative efficiencies of different element immobilization and mineralization. However, the relative contributions depend on the composition of plant and fungus and the rapid shifts between them in an ever-changing environment.
- Connectivity and complexity are the reality, not just an interesting sideline.

2 · Structure–Functioning Relationships

From the beginning of mycorrhizal research, understanding of functioning was based on the morphological structure of the plant–fungus interface. The structure of the fungal hyphae within the root, which regulates the exchange of resources between plant and fungus, and the extramatrical hyphae, the extent and locations of which dictate the ultimate flows of resources between soil, hyphae, and root, characterizes a mycorrhiza, determines the type of mycorrhiza, and, to a large extent, determines the functioning of that mycorrhiza (255). Overall, knowing these two structural components, internal root colonization and extramatrical hyphae, allows us to make a number of predictions about the mechanisms and quantity of resource exchange between the two symbionts. In this chapter, I describe the morphology of the fungus and the modifications to the morphology of the plant caused by the formation of the mycorrhiza and discuss how these determine the basic functioning of the mycorrhiza. I emphasize illustrations of elements as can be seen in the field. I highly recommend other works, particularly from Larry Peterson et al. (563) and Mark Brundrett for outstanding detailed illustrations and photographs of mycorrhizal structures. This extends into the entire life cycle: the extent of the developing hyphal network, spore production, and spore dispersal.

Features in Common

The multiple types of mycorrhizal symbiosis described in Chapter 1 have a number of differing features. However, they all have transport structures within the plant, and they all have an external hyphal network that extends into patches of substrate. One way to understand mycorrhizae is to start with those plants that do not form mycorrhizal associations, that is, nonmycotrophic plants. These consist of about 15 percent of plants worldwide (e.g., 319; 389). Plants that do not form mycorrhizae have been known since Stahl (678). Two common nonmycotrophic groups, certain species occurring in the Proteaceae and the Amaranthaceae, are illustrative and found in distinctly different habitats. One (Proteaceae) thrives at the extremely low nutrient end of the Treseder–Allen model, and the other (annuals in the Amaranthaceae) at the high end (Figure 1.1). The Proteaceae predominate in extremely old, nutrient-deficient, and often sandy-textured soils. These are especially common in the ancient soils of Australia (such as the Kwongan Plains (552)) and South African Fynbos (183). P concentrations are extremely low (available P is often less than 1 mg kg^{-1}), and soil texture is generally sandy, with a predominance of meso- and macropores (greater than 10 to 30 μm). Proteaceae form cluster roots from 8 to 250 μm in diameter and from 0.6 to 35 mm in length (439). Although low in concentration, when soil moisture is available, nutrients that are present can often readily diffuse to the root.

Nonmycotrophic annuals of the Amaranthaceae and Brassicaceae (many of the perennials of these families can form mycorrhizae) tend to predominate in nutrient-rich patches, and again predominate in silty to sandy soils with some micropores (5 to 30 μm), but are accessible to root hairs (generally 10 μm to 30 μm in diameter). These soils are highly disturbed. Disturbances come from grazing in the steppes of Asia or North America, or from tillage and fertilization associated with agriculture. Disturbance can also appear as eutrophic patches from deposition at the high tide line and upwelling streambeds. Nutrients, especially P, tend to be less limiting in these habitats. Both groups of plants also have adaptations, including the production of high concentrations of organic acids that weather inorganic P and increase availability. Here, mycorrhizae provide little benefit. Thus, mycorrhizae largely exist within a sweet spot of plant C:N:P nutrient stoichiometry (see Chapter 1), which covers most of the terrestrial world.

A further clue to the structure–functional relationships of mycorrhizae is that they are rare in aquatic environments. There are two issues that emerge in aquatic environments. The first is that O_2 is limiting to fungal growth. The second issue is that some nutrients readily move by mass flow when water is continuous. Unless bound in clays or organic matter, nutrients, such as K^+, Ca^{2+}, NO_3^-, and organic $-PO_4$, move readily, diffusing from areas of high concentrations to those of low concentrations. Again, an extensive hyphal network provides minimal value in these environments.

Carnivorous plants were long thought to be nonmycotrophic. In part, this is because they occupy wet environments where O_2 is often limiting.

Nutrients readily diffuse, and their prey provide N and P. However, more recently, reports of arbuscular mycorrhizae (AM) carnivorous plants have emerged (592; 593).

Mycorrhizal fungi are comprised of fungi with hyphae that have several characteristics that make them ideal mutualists for plants. First and foremost, mycorrhizal fungi are really not microorganisms; they are macroorganisms of connected microscopic units (hyphae). As such, they can form extremely small structures, hyphae that can be as small as 1 μm in diameter that can penetrate even larger ultramicropores. The fungal mycelium, comprised of a network of hyphae, extend outward from a root surface from a few millimeters to several meters, searching for nutrient–rich patches. In conjunction with their small size, hyphae often produce and excrete enzymes that mineralize organic particles. Plants largely produce acid phosphatases (e.g., 73; 796), but mycorrhizal fungi produce both acid and alkaline phosphatases. This allows the fungal hyphae to access P from a different source – alkaline soil patches. In some cases, mycorrhizal fungi produce many other enzymes that release nutrients from organic matter. Mycorrhizal fungal hypha respire CO_2 and produce organic acids that locally reduce pH, thereby weathering clays and releasing phosphate. Finally, the difference between mycorrhizal fungal hyphae and saprotrophic or pathogenic fungi is that they transport the acquired nutrients directly to the host in exchange for host carbon, instead of depending on enzymes that mineralize complex carbohydrates from their environment.

There is a common structure delineating all types of mycorrhizae (Figure 2.1). This consists of an external (extramatrical) fungal mycelium network (a structure where fungal cells become intimately intertwined with root cells), an interface with a fungal membrane (an exchange space), and a plant membrane (within the root cortex), all outside the plant endodermis and stele. Soil resources, such as N, P, or H_2O, needed by the plant are transported from soil into the root cortex through fungal hyphae. This transport can be symplastic or apoplastic. The root cortex is then coupled to the plant's vascular system via the endodermis, forming a symplastic transport step in resource exchange. Carbon is fixed by the host plant in photosynthesis and transported into the root stele, out through the endodermis and cortex, and into the fungal hyphae found within the cortex. The majority (if not all) of the C for the fungus comes from that host plant. That C is then transported to the extramatrical hyphae for fungal growth and reproduction. Importantly, to function as a mycorrhiza (as opposed to a root endophyte), the fungus must have both

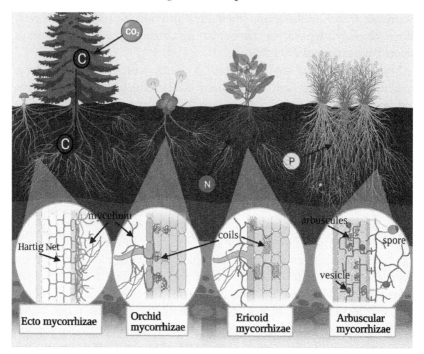

Figure 2.1 Common structure of mycorrhizae, from substrate (usually soil) → fungus → root → leaf (see text for details). C flows from the atmosphere to plant to fungus to soil (losing CO_2 through respiration at every step). Through the mycorrhizal network, nutrients and water flow from the substrate, usually soil, to fungus, across a fungal membrane, across the intercalary space (in EM, including a cell wall), across the plant membrane, and into the vessels. Figure drawn by Danielle Stevenson, with permission.

extramatrical hyphae and (as opposed to epirhizal fungi) hyphae within the root cortex. There is a fungal membrane running parallel to a plant membrane where exchange occurs (272). There may or may not be a plant wall separating the membranes (differentiating an ectotrophic from endotrophic mycorrhiza). The fungus does not penetrate the endodermis of the plant. Thus, the plant does have some ability to regulate the direction and rates of exchange between the cortical cells (recipients of fungal resources) and the resources in the plant vascular cylinder.

Mycorrhizal fungi growing in the soil invade the root in a type-specific pattern and then radiate outward into the soil, forming a mycelial network. There are well-defined architectures for both internal and external mycelial networks, and well-defined structures for each. The spatial extent becomes critical. I have observed roots of large AM tropical forest trees extending for over 100 m from the trunk. As nearly every fine root tip forms an AM, there are millions (or more) of connections with numerous taxa of AM fungi. Simultaneously, each fungus connects across many trees. For example, in an abandoned Spanish mining site, fruiting bodies of the ectomycorrhiza (EM) fungus *Pisolithus tinctorius* were seen for over 10 m from individual saplings of *Pinus halepensis*, and many sporocarps and saplings could be observed all across the site. The ensuing *common mycorrhizal network* (CMN) has both plants and fungi forming a connected network stretching across several hectares.

There are two distinct structural–functional processes that interface in a complex way. First, the plant fixes CO_2 in the leaf tissue and transports sugars via the phloem to the roots and then to root cortical cells. These cells then transport simple organic C molecules to the mycorrhizal fungus. Second, the fungus picks up resources (e.g., N, P, H_2O) from the rhizosphere soil through a complex hyphal network, transporting them in a simple transportable form to the host. While there is an exchange, the molecular enzyme and transporter proteins for C versus each nutrient are different. For example, AM plants appear to convert sugars to lipids using at least two lipid enzymes, FatM and RAM2, whereupon these C sources are transported to the AM fungus using an ABC transporter (138). For P, the AM system uses the Mtpt4 P transporter (322). Somehow, these must co-occur and must be inter-dependent stoichiometrically. Otherwise, the relationship would have to shift toward parasitism or commensalism. EM appear to be far more complex (274), likely because EM derive from many independent evolutionary events in different taxa, both plant and fungus (see Chapter 3).

Structures of Mycorrhizae and Their Phylogenic Distribution

Frank (255; 259) distinguished two morphological types of mycorrhiza based on whether the fungus penetrated the cell walls of the root cortex. He described two broad categories: the first is the ectomycorrhiza (EM), in which the fungus fails to penetrate the individual plant cortical cell walls. These are widespread mycorrhizal associations between many woody plants and fungi in the Basidiomycota, Ascomycota, or Endogonales. The second form is the endotrophic mycorrhiza, or endo-mycorrhiza, which does penetrate the cell walls. Most plants form arbuscular mycorrhizae (AM), a type of endomycorrhiza. Orchids and Ericaceous plants forming orchid and ericoid mycorrhiza were also observed to be mycorrhizal, in which the fungus penetrated the cortical root cells or endomycorrhiza.

Arbuscular mycorrhizae (AM) are the most widespread of the mycorrhizal associations. Schlicht (639) described these endomycorrhizae on a number of herbaceous plants. He described the arbuscular structure, internal to the plant for resource exchange, and also noted a lack of a necrotic response, used to detect pathogen activity. The plants forming AM represent the basal plant lineages and can be found in all plants where the environment is conducive except a few plant taxa in more recently diverged families. A monophyletic group of fungi, the Glomeromycotina, form this type of mycorrhiza. (Previously, Glomeromycotina were classified as Glomeromycota and, before that, as Glomeromycetes.) All members of the Glomeromycotina are symbionts, and all but one group, *Geosiphon* (associating with cyanobacteria colonies), form arbuscular mycorrhizae. The AM symbiosis (Plate 1) is characterized by an extramatrical hyphal network of arterial (or runner) hyphae, with hyphae from the mycelial network or germ tubes from spores that penetrate the root cortex, weave through the outer cortical layers, and then penetrates the cell walls (but not the cell membranes), creating a complex, highly branched exchange structure called an arbuscule. In some cases (such as maples), the fungus does not form arbuscules, but forms coils, also called pelotons.

Internally, the hyphae penetrate the root cortex near the tip just behind the region of elongation. A penetration peg, or appressorium, forms between cortical cells, analogous to those formed by fungal pathogens of plants. Intracellular hyphae then grow toward the inner cortical layers. Within the inner cortical layers, two types of architecture form, which can be dependent upon the taxa of plant in which the mycorrhiza is formed, or the conditions. We really don't know what regulates the formation of these types. These are termed Arum- and Paris-AM (272; 562). The most common is the Arum-type of AM, where the fungus penetrates the cell wall of individual cortical cells and forms an intricate branching "arbuscular" (from tree) structure with an extensive fungal: plant cell membrane interface. These locations have high ATPase activities, P concentrations, and other enzymes and structures supporting massive transport between plant and fungus. In the Paris-type of AM, the fungus passes through cells forming successive pelotons and arbuscules. In both types, the fungus does not penetrate the endodermis.

Importantly, an appressorium penetrates a single point then occupies only a single region of the root cortex, comprising an infection unit. Importantly, these are not mutually exclusive and may form a continuum (207).

The hyphal architecture of AM fungi is unique both internally to the plant and externally in the soil (562), and anyone interested can be readily taught to assess activity. One way to understand AM dynamics is to envision an *infection unit* (e.g., 31; 38; 633) forming from a penetration from the runner network. From the penetration peg, hyphae form an internal network and, simultaneously, an external absorbing mycelial network. Each infection unit then forms an external (extramatrical, or extraradical) mycelial network that expands outward into the soil for capturing soil resources and where reproduction of the fungus occurs. That extramatrical arterial or runner hyphal network of coarse hyphae initiates new AM infections in compatible roots.

From the penetration point, the infection unit forms an equally structured external mycelium. The extramatrical hyphae themselves are aseptate and have a knobby, irregular appearance at a microscopic scale (520; 531). The hyphae radiate out from the infection point in a bifurcating network, decreasing in size with each branch. These hyphae extend beyond a developing depletion zone that forms surrounding a growing root tip to access soil resources. The hyphae branch approximately every millimeter but simultaneously become thinner with each branch (264). More complex variations on this structure have been shown (e.g., 96), but all variants result in a structure with both a coarse hyphal network capable of transporting C to fungal tips and a fine absorbing network capable of penetrating even the larger-sized ultramicropores, down to 2 μm, transporting nutrients and water to the host. During this period, nutrients are taken up by the external hyphae, transported to the arbuscules, and exchanged within cortical cells for C. These radiating hyphae can both bind soil particles and organize soil aggregates (502). If you pull up a root in a sand dune of an AM plant, sand particles stick to the AM fungal hyphae radiating from the roots (690).

With time, root infections are successively formed behind the region of elongation in the expanding roots. The arterial hyphae that infect new root segments are in the 10 μm to 20 μm diameter range, sometimes wrapping around each other to form a very simple rhizomorph-like structure and extending several centimeters from the fine root. New infections in the plant only appear to occur just behind the region of elongation, and in front of suberizing tissue. Infected roots seem to be largely limited to fine roots, but with cortical layers more than three or four layers deep. Cluster roots that are 8 μm to 250 μm in diameter do not appear to support mycorrhizal infections, and exceedingly fine roots, from 20 to 100 μm, appear to support a lower number of internal hyphae (610). As an infection unit ages, the branching external mycelium reaches out to some maximum (in my observations, I estimated 120 cm of hyphae), then dies back from the fine tips. Simultaneously, the arbuscules deteriorate, but vesicles or auxiliary cells are formed, and the arterial network expands to take up and store the C from the host. This process can be rapid. *In situ* AM extramatrical hyphae have been observed preferentially growing in the afternoons (346), likely at the time of maximal allocation of photosynthetic products. During peak growth, hyphal expansion sometimes exceeds 150 μm mm^{-3} soil h^{-1}. Individual arbuscules persist for only a few days (184; 420), depending upon the conditions. The arterial hyphae connect multiple points along and between roots, and even between plants. These hyphae may last an entire growing season (725). The question of whether and how much in the way of resources (C, nutrients) are exchanged as a Common Mycorrhizal Network (CMN) between plants is probably specific to the ecosystem and density of plants and roots. But, hyphal lengths range from as little as 3 m cm^{-3} (62) in an arid shrubland to upward of 60 m cm^{-3} (503) in a tallgrass prairie. The mycelial network connects multiple plants and can even switch hosts. In a mixed grass prairie, the same fungus shifted from a C$_3$ grass, *Agropyron smithii*, to a C$_4$ grass, *Bouteloua gracilis*, as the season shifted from spring into summer (45).

In all Glomeromycotina, internal or external vesicles are formed that store carbon. Members of the Gigasporaceae form auxiliary cells generally distinguished from vesicles, probably also as a carbon storage mechanism. AM fungi also form multinucleate chlamydospores of various sizes. Although these fungi have never been definitively shown to produce sexual reproductive structures, there is considerable new evidence that sexual, and possibly parasexual recombination exists (620; 793). In the past this group of fungi was called phycomycetoid mycorrhizae, but the Class Phycomycetes was discarded. By the 1960s, this type became known as vesicular-arbuscular mycorrhiza (VAM) because the most common order, the Glomales, form vesicles. The Gigasporales do not form vesicles, but still form AM. However, just as maples form mycorrhizae but with coils instead of arbuscules, we still refer to them today as AM. I suspect that the terminology will change again during the next decade. All of these terms, AM, VAM, phycomycetoid, glomeroid, and even endomycorrhizae are synonyms across the historical literature.

Other endomycorrhizae also associate with plants in the families Orchidaceae, and in the Ericaceae, and Epacridaceae (these latter two families are now synonymous). The orchid mycorrhizae are formed between orchids and a group of fungi in the Cantharellales and potentially many other fungi (228). Prior to the development of molecular technologies, the fungi were placed in the genus *Rhizoctonia*, imperfect basidiomycetous fungi. The known orchid mycorrhizal fungi are in the Cantharellales, but from a group, the Tullasnellaceae, many of which are saprotrophic (the jelly fungi) or form a pathogenic relationship with other plants. Internally, the fungi form coils or pelotons for exchange of resources between plant and fungus (Plate 1). The extramatrical hyphae utilize their enzymatic capacity and surface exploration to access nutrients such as N and P (in the case of green orchids such as *Platanthera*). When symbiotic with orchid protonema, both C and P are transported from fungus to plant, until a green leaf emerges (308), while in achlorophyllous orchids, C and N are parasitized from neighboring plants by the mycorrhizal fungi (see, for example, 800).

Ericaceous mycorrhizae were previously called ectendomycorrhizae, because they were observed to form both an external mantle and penetrate cortical cells. Although there remains controversy, today I limit my call of the ectendomycorrhizae to the "E-strain" mycorrhiza, those formed between conifers and Ascomycete fungi such as *Wilcoxina*. We know that there are two distinct types of mycorrhizae in the Ericaceae that are differentiated phylogenetically. The first is a variant of the EM called the arbutoid mycorrhiza, in which many of the same fungi forming mycorrhizae with EM trees will form a thin and rather incomplete mantle and Hartig net. But the hyphae will penetrate the cortical cell walls forming coils for exchanging resources. These mycorrhizae are in the basal lineages of Ericaceae, the Arbutoideae, and Monotropoideae. The fungi forming these mycorrhizae appear to be largely basidiomycetes.

The Ericaceae also form an endomycorrhiza called ericoid mycorrhizae. Ericoid mycorrhizae are formed with a narrow suite of Ascomycota that may have a limited extramatrical hyphal network but an extraordinary enzymatic capacity for breakdown of complex organic materials. These associations are found in the advanced sub-orders Vaccinioideae and Ericoideae. The associated fungi, in the Eurotiales and Onygenales (in the Eurotiomycetidae) could well be monophyletic. A disjunct group of Ascomycetes in the Helotiales might also form ericoid mycorrhizae, but more work on this is needed. These fungi are known for their enzymatic capabilities. The Sebacinales, Basidiomycota, may also form ericoid mycorrhizae (754). Internally, these fungi form pelotons where N and P are exchanged for C. The extramatrical hyphae are generally viewed as constrained, but have the capacity to break down organic matter, extracting nutrients critical to host plant growth. In this scenario, the external network is considered to be essential, but not forming an extensive mycelial network. However, this hypothesis is based on the fact that when a root is extracted, few hyphae can be observed. Early researchers had no way to distinguish ericoid mycorrhizal hyphae from saprotrophs, and the hyphal lengths appeared low, in the bogs and sand plains, where ericoid mycorrhizal plants are abundant. When we were able to directly observe hyphal networks *in situ* in an ericaceous bog, extensive mycelial networks were observed. I suspect that the extramatrical hyphal network is more extensive than previously postulated.

Dark-septate mycorrhizae are a problematic group that might or might not be classified as mycorrhizal. If they transport external soil resources into the roots, and have an evolutionary history of mutualism, then they fit my definition of mycorrhiza. Some appear to fit this definition. Haselwandter and Read (328), in a landmark study, reported an asco-mycetous fungus that penetrated sedges. The fungus decomposed organic matter, transferring N to the plant in exchange for C. Subsequently, there have been many dark-septate "mycorrhizae", so defined in that no sign of necrosis was found in the host, an early criterion for defining a mycorrhiza (e.g., 72; 272; 639). The excellent review by Mandyam and Jumpponen (478) summarizes the challenges and some perspectives in better understanding dark-septate endophytes and their ecological roles.

There are many plant–fungal interactions. Plant pathogens, a parasitic symbiosis as defined by DeBary (197), have been studied since well before Tillet's classic experiments on smut fungi. Epirhizal, "dark-sept-ate" fungi, and Sebacinales fungi have all been described as mycorrhizal fungi at some point. But based on the definition of a mycorrhiza in which resources are exchanged (256), coupled with the perspectives of Harley (320) that a mycorrhiza is a co-evolutionary relationship, these relationships, while they may exhibit mutualisms and should be studied in greater detail, should not necessarily be categorized as a mycorrhiza.

Ectomycorrhizae (EM) were first illustrated by Hartig in 1840 (324). Root tips, and even individual cortical cells are encased in a fungal network. The basic structure of an EM is an external hyphal network, often extensive, a sheath partially or completely covering the tip of a short root (Plate 1) and a fungal matrix embedded within the outer root cortex between individual cells, called the Hartig net. For detailed images and descriptions of each of the relevant structures, see Peterson et al. (563). As implied by its name, this type of mycorrhiza does not penetrate the cell walls, as do AM and ericoid mycorrhizae. Between the cortical cells the net can comprise a large fraction of the EM short root. It appears that generally (but not always), only a single species of EM fungus infects a single short root. But as the long root grows down the soil, multiple fungi appear to occupy the multiple short roots, creating a diverse community along a single long root.

There are a number of variants on the EM. Ericaeae are not limited only to ericoid mycorrhizae. The Vaccinioideae emerged in cooler climates to form an arbutoid mycorrhiza, a type of EM. EM developed in *Gnetum* in the tropics, a primitive conifer, and other EM emerged especially in the Pinales, advanced conifers, and Fagales, advanced angiosperms. More detail is discussed in Chapter 3. Structurally, the monotropoid mycorrhiza is an arbutoid mycorrhiza that connects the mycoheterotrophic plants with EM of surrounding plants for transporting C.

Cell to Mycelial Structure and Connectivity

The cell structure of mycorrhizal fungi has several properties that make fungi uniquely positioned to connect soil resources to plants. Mycorrhizal fungi extend from the surface of the root outward into the soil, and inward within the root. AM Glomeromycotina fungi and *Endogone* EM fungi, both members of the Mucoromycota, are coenocytic or aseptate; that is, there are no regular cross walls or membranes constraining flow. But even Ascomycota- and Basidiomycota-EM fungi have central pores in the septa between cells through which materials can flow. The primary mechanism driving flow is dependent on water potential (ψ) gradients creating flow directionality, as water always flows from points of high ψ to low ψ, unless restricted by membranes. Water and materials also move randomly (Brownian movement) if there is no ψ gradient. Hyphae penetrate deep into the cortical layer. In AM, the majority of the transport structures, the arbuscules and pelotons, predominate in the inner cortical cells, next to the stele. Thus, in AM, the fungi bypass many plant walls and membranes, allowing materials to flow simply along a ψ or concentration gradient. In EM, the intramatrical hyphae intertwine between cortical cells, again bypassing many plant walls and membranes that could act to restrain flows of materials. In AM fungi, as hyphae radiate outward from the root surface, they bifurcate, getting smaller with each branch. This allows the fungus to explore a larger area of soil space but does constrain flow in any individual hyphae. Tinker and colleagues showed, using ^{32}P transport along hyphae, that the flux at the entry point of the root was 38 nmol cm^{-2}s^{-1} (633) in the thickest hyphae compared with 0.3–1.0 nmol cm^{-2}s^{-1} (556) in the soil, near the tips of the absorbing network. The entry point hyphae were greater than 10 μm, where the hyphal diameters at the tips can be as little as 2 μm (264).

The cell wall itself also has an important structure facilitating resource transfer. The hyphal tip is comprised of a layer of chitin, approximately 50 nm thick. As one travels back from the tip, layers of protein and glucans are added, creating a thickness of 125 nm (39; 107). This makes the tips hydrophilic, and capable of taking up and releasing water and nutrients, depending on ψ and element gradients. Symplastic flow, occurring through the protoplasm, occurs within the cells, but apoplastic flow, that occurring between cells, can also occur in the elastic hydrophilic region between the wall and the membrane. These tips occur both in the soil and in the inner cortical layer within the plant. Between these two points the main body of the mycelial network is hydrophobic, where apoplastic transport may well predominate. This will become critical in understanding the physiology of the relationship (discussed in Chapter 4) but here it means that the structural relationship dictates how, when, where, and how much flow can occur.

Not only is the structure important but the chemistry also plays a crucial structure–functioning role. The walls of mycorrhizal fungi are comprised largely of a skeleton of chitin, with chitosans and glucoaminoglycans, and with a matrix of polysaccharides, including glycoproteins. Importantly, while an individual chitin molecule is relatively simple, $(C_8H_{13}O_5N)$, it forms long chains in multiple layers. One notable glycoprotein in that matrix is a family of heat-shock proteins called glomalin. Glomalin is a sticky soap-like substance that binds particles into soil aggregates. Also found in this matrix are melanins, that provide protection from UV radiation and reactive oxidation, among other functions. This structure creates the architecture for mycelial networks to connect multiple plants and multiple patches of soils. The AM fungi, in particular, by exuding glomalin, alter the soil structure by helping to form soil aggregates that in turn alter the availability and distribution of metals and nutrients and determine the flow patterns of water and the structure and friability of the soil. Finally, chitin (because of its layering), glomalin (with its complex chemistry), and melanins (which have a high degree of polymerization) all decompose slowly and have likely contributed to carbon sequestration at different times during the earth's history (see Chapter 3 for more detail).

Connectivity of Mycorrhizae

Mycorrhizae have shown a pattern of convergent evolution since the first land plants. I know of no descriptions of the complex structuring for the external hyphae of the Mucoromycotina AM, but this might be due to the recent phylogenic studies and the difficulty of study for these fine hyphae. Glomeromycotina AM fungi have arterial hyphae (or "runner" hyphae, depending on the source paper) from which new infections develop, either along a growing root already infected or connecting a compatible root of a neighboring plant. We do not know the extent of a single expansion. Mosse et al. (522) estimated expansion of a network at up to 1 cm outward per week during the growing season. Hernandez and Allen (346) found that, within a meadow, the growth rates of individual hyphae could be up to 1 cm per day. However, the difference could be related to directionality of a network versus a single hypha. The rates and pattern may be dictated by the ability of the arterial hyphae to sense CO_2 and other exudates emanating from a newly growing root tip that could provide a region for infection. Koske and Gemma (428) suggested that germinating hyphal tips of AM fungi could sense volatiles from roots up to 15 mm and water-soluble exudates up to 4 mm away. Assuming that AM associations are not host species-specific, then a network could extend long distances. I do not know how far (or if) any mycelial network extends in ericoid, orchid, or dark-septate mycorrhizae. The EM type of mycorrhiza independently emerged multiple times, in very different groups of fungi, including Endogonales (Mucoromycotina), Ascomycota, and Basidiomycota. Probably because of these independent events, the external architectures vary greatly. Some, such as *Cenococcum* spp., appear to produce largely concentrated hyphae around the infected root. Others, like *Amanita* spp., seem to only infect a few roots, but form networks that extend long distances. Ogawa (541) identified three architectures, including fairy rings (e.g., *Tricholoma*), irregular mats (such as *Hysterangium* and *Gautieria*), and dispersed colonies (e.g., *Amanita, Russula*). Agerer (3) defined a vast array of functional features of the EM itself, and Hobbie and Agerer (355) expanded these features to describe the architectures of exploration types. These include mats of short-, medium-, and long-distance, and either smooth or fringe. Each defines different absorbing outcomes.

The extent of a clone of mycorrhizal fungi shows a degree of connectivity beyond the classic one plant–one fungus pot experiments that characterize most functional research. But how connected is connected? What is the extent of a single clone? The answer is we don't know. Studies using small windows, ranging from glass plates in the lab to minirhizotron observations in the field, show many interconnections among neighboring root tips. Stable isotopes including C, N, P, and O move between plants via fungi, and autoradiography can illustrate plant–plant linkages of ^{14}C and ^{32}P (see the extensive discussion in Bledsoe et al. (124)). Molecular sequence analyses provide evidence that fungi of apparent single clones appear as the mycorrhizae of trees scattered across sites of multiple hectares. This pattern has been shown in many vegetation types, ranging from dense forests (e.g., Simard et al. (658)), to dispersed pines (e.g., Gehring et al. (278)) and oaks (e.g., Southworth et al. (674)). From the multiple fungi existing on large, mature trees, the mycelium foraging outward from these trees form new infections on the surrounding seedlings (Figure 2.2), existing across a stand of tens of meters, or even farther. If the plant clones are compatible with a fungus, this network would include not only plants of the same species, but other taxa as well.

Another concern is the longevity of connectivity. Interestingly, the data suggest that the lifespans of EM and AM fungal hyphae are longer than plate-growth patterns might suggest. EM root tips can live three years (545), ranging from a single growing season to six years, and averaging 108 days (with an annual growing season of 116 days) in Alaskan taiga (627). In a *Pinus edulis* stand, δ^{14}C studies of root tips suggest that the carbon of EM root tips was fixed four to five years previously, with some up to six years (723). The rhizomorphs radiating from EM tips were found to persist for 11 months in the *P. edulis* stand (720). However, some individual hyphae die back as fast as they grow (346). Indeed, at different times of the year, or even of a day, overall growth can be either positive or negative. While some hyphae survive for only a day, others persist through the entire growing season. The average hyphal lifespan in old-growth forests ranged from 25 days in a tropical rainforest (AM) (692) to 40 days (EM) in a temperate mixed conifer forest (59). For the arterial hyphae, survival averaged 145 days in a

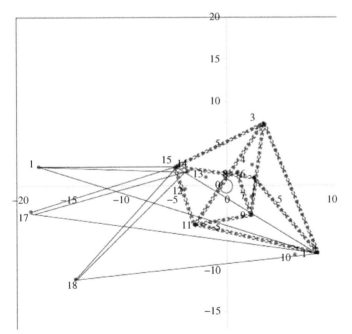

Figure 2.2 The expanse of potential fungal clones connecting a mature oak located at the 0,0 point with seedlings planted at various distances (in meters). Shown are two EM fungal taxa, Thelephora terrestris (dashed lines) and Amanita sp. (solid lines) (40).

California shrub–annual grassland site (725). A modeled AM hyphal absorbing network lives an average of two weeks, with one expanding outward and one dying back simultaneously (75).

While together these observations show connectivity across relatively large areas (greater than several meters squared), mycelia connectivity is dynamic with limited lifespans of many of the individual hyphae but a persistent network. Simultaneously, root production and mortality are also dynamic, with lifespans for fine roots generally of weeks to months. The very dynamic structure of plant and fungus exerts strong constraints on the network.

The next step in creating a structure–function understanding is to understand how these networks form, persist, and breakdown. Common mycorrhizal networks (CMNs) generally allude to a mycelium that connects two individual plants of the same or different species. But, a single plant could also connect two individual fungi of the same or different species. This would also be a CMN. Could network theory help in understanding the CMN structure and functioning? Recently, network theory has been applied to better understanding of CMNs (for example, see the review in Bledsoe et al. (124)). One relevant analog might be the electrical power grid connecting cities. In this network, cities and towns are considered nodes consuming power, and the power lines are the links transferring power. Large cities are hubs, in that there are many links to that node, whereas smaller towns are simple nodes. An assumption is that a single link will always fail, despite all efforts to maintain them. The stability of the entire network functions or fails, depending not only on individual line maintenance, but also on the architecture of the hub–node–link connections.

In the context of connectivity, one outcome is to simply understand what the hub is and then to study the pattern of links. I will explore this topic in greater detail in the community and ecosystem chapters, but two simple examples will show how crucial this theory might be. In walking through a dispersed oak (*Quercus garryana*) stand, visually one might think that the trees are nodes and, the bigger the tree, the greater chance that the large tree was the hub. But, in examining the morphotypes of the EM of the trees, there was no pattern of EM fungal richness: no hub. However, plant linkage to fungal taxa was nonrandom, thus scale-free (674). The fungus appeared to function as the hub, not the plant. Alternatively, in a Douglas-fir forest (*Pseudotsuga menziesii*), the fungal genets were scale-free, suggesting that particular trees served as hubs (112). I would postulate that the structure of the forest determines the networking patterns, which would have important ramifications for nutrient and carbon allocations and survival and establishment of seedlings, especially under stress.

Architecture and Distribution of Mycorrhizae

Mycorrhizae are tied to and governed by the canopy and the rooting architectures of the plants. Being dependent upon carbon allocated from leaf to root, wherever fine roots are located and O_2 is present, mycorrhizae may be present. Mycorrhizae have been observed from the soil surface to between 3 m and 9 m deep in the soil profile, and 100 m or more up in the canopy on roots as epiphytes of redwoods and tropical trees.

Belowground, the architecture is complex and dependent upon the structure of the material. Arbuscular mycorrhizal fungi have been found as much as 9 m deep in the Brazilian tropical forest (528) and EM 4 m deep in a eucalyptus plantation (617) and at least 3 m deep in native oaks (131). They exist in soils above the water table, needing O_2 to respire. Virginia and colleagues (750) found that, in a *Prosopis* shrubland, roots grew as much as 3 m deep into the sandy soil to reach the water table. Within the water table, where O_2 was limiting, N_2-fixing nodules were found. Just above the water table, where O_2 was adequate, AM roots were found. In a complex terrain, where deep *Quercus agrifolia* roots reach the groundwater, moisture supported the EM mycelial network, even extending to nearby seedlings. Where the deep roots could only reach dispersed pockets of winter precipitation, AM roots also persisted into the summer dry periods (590).

In a rock matrix, the pattern is especially interesting. In southern California, dominated by uplifting granites, fractures caused by earthquakes contained both AM and EM roots, extending through fractures at approximately $45°$ angles. AM and EM hyphae penetrated through the decomposing granite (297). In the San Jacinto chaparral, mycorrhizal fungi existed greater than 2 m in the weathered granite (214) and up to 4 m in oak in the coastal range (131). In seasonally arid regions, this process is especially critical. For example, the southern California coastal range has a Mediterranean-type climate with winter precipitation, where water in soil above the bedrock is available only from April until early June. But, here, evergreen oaks and conifers actively photosynthesize all year. Virtually all summer moisture was taken from the deep bedrock (408). Not only are mycorrhizal fungi found deep in the profile, the hydraulic redistribution of the water from the granite matrix sustained the surface mycorrhizae through hydraulic lift, allowing the mycorrhizae to function through the dry summers (589; 590).

The ability of trees to act as structural support affects the distribution of plants such as epiphytes and hemiparasites. This distribution pattern can be both direct and indirect. Directly, the canopy provides a habitat for plants. Many epiphytic orchids, such as a vanilla orchid, grow in a canopy where the orchid mycorrhizae search the bark and tap the water running down the trunk for nutrients. Many Bromeliaceae, such as *Tillandsia* spp., depend upon the mid-canopy for both a support location above the soil surface, but also settle in the canopy for protection from too much solar radiation (294). Other plants, such as many mistletoes, are hemiparasites found high in the canopy. Although hemiparasites tap into their hosts via haustoria and do not have true roots in maturity, they depend upon the water and nutrients extracted by their mycorrhizal hosts.

A basic question then becomes where do mycorrhizae not exist? Mycorrhizae exist in virtually terrestrial habitat, and some fresh-water conditions, and extend to every continent (yes, every continent!). Contrary to earlier reports, AM have recently been found on King George Islands, South Shetland Islands (103). *Deschampsia antarctica* appears to support *Glomus antarcticum* on the Dance Coast of the Antarctica Peninsula (151). However, on the Peninsula of PalmerLand, which extends from the continent of Antarctica itself, no Glomeromycotina AM infections were found (202). Also, dark-septate endophytes were reported (354). In a stand of liverworts, *Cephaloziella exiliflora* in Antarctica, ericoid mycorrhizae of *Hymenoscyphus ericae* and *Rhizoscyphys ericae* were found (159). Although reports are contradictory, there has been a limited sampling of Antarctica habitats. There are locations, particularly in variously controlled zones, especially those with mosses and liverworts, which have not been sampled. Further, the Mucoromycotina-type fine endophyte mycorrhizae have rarely been looked for but could occur in this kind of environment. And, third, AM may be rare because Antarctica currently lacks a ready source of inoculum. Invasion of open habitats with only a sea access is quite limited.

An example is the pattern of re-invasion of newly opening habitats such as glaciers, originating from the Harding Icefield in southern Alaska, which are retreating in response to global warming. At Exit Glacier, retreating from land, all forms of mycorrhizae, AM, EM, ericoid and orchid, were found. I found spores of both EM and AM fungi in the scat of mammals as diverse as grizzly bear (*Ursus arctos*) and moose (*Alces alces*), which roamed the newly opened lands. Where tidewater glaciers are retreating from the sea without a continental connection, only wind-dispersed EM fungi were found, such as species of *Inocybe*, *Lactarius*, and *Cortinarius*. In one location, Dinglestadt Glacier, the peninsula protruded into the ocean and has been exposed at least since the Pleistocene. Scat of bear, lemming, and squirrels were found. A wide diversity of both hypogeous and epigeous EM fungi were found, along with a *Glomus* sporocarp (probably *Gl. macrocarpus*). These fungi are slowly invading up the peninsula, as the glaciers retreat.

These observations denote the importance of propagules, structures that are designed to tolerate stress conditions, disperse, and recombine genetic material. I will discuss the physiology and genetics in later chapters, but crucial structures are those that facilitate dispersal to new habitats and insert new genetic material into existing clones. *Here, I differentiate infection, the process whereby the fungus invades a single root, versus colonization, the process whereby both the plant and fungus invade a new location.* The term colonization is often used for either process, confusing the reader. It is important to remember that, generally, seeds of plants and spores of mycorrhizal fungi are independently dispersed. In a few cases, where plant propagules have rough surfaces where spores persist, there can be co-dispersal. But this is likely rare.

There are two primary mechanisms whereby mycorrhizal fungi disperse to and colonize an open habitat. First, most epigeous-fruiting (above-ground), largely EM fungi in the Ascomycota and Basidiomycota, are wind dispersed. Microscopic spores (1–10 μm) are produced inside gills or pores. They are ejected into wind currents using osmotic pressure such as from a basidium. Only a small fraction of the spores produced actually disperse more than about 40 cm from the mushroom (46). But, given that up to a trillion spores can be produced by a single mushroom (147), a lot of spores can still be dispersed (see also Vašutová et al. (748)). Spores of fungi such as *Thelephora terrestris* can be found many kilometers from potential fruiting zones, such as 5 km to 10 km across the Pumice Plain of Mount St. Helens (33). Dispersal of AM spores requires sufficient wind to entrain spores at the soil surface into the wind stream. Wind speeds of 10 m s^{-1} can generate turbulent eddies with vertical winds of 100 cm s^{-1}, which is greater than the terminal velocity of spores less than 100 μm in diameter. In arid lands, with dry surface soils, smaller Glomeromycotina spores (<100 μm) are readily wind-dispersed (46). Spores larger than about 150 μm are rarely wind-dispersed (46). Importantly, many Glomeromycotina spores have ornamentations (spines or ridges). Just like baseballs with seams, spinning spores with ornamentation have lowered drag compared with smooth spores and may well disperse farther in the winds.

Hypogeous fungi (those fruiting belowground) produce a wide range of spore sizes. Given their belowground fruiting habitat, those with large spores are not wind-dispersed but rather depend on animals for dispersal. Some have asexual chlamydospores that range from a few micrometers up to several hundred micrometers. Nearly all AM fungi form chlamydospores. They can form either singly or in sporocarps. Endogonales, Ascomycota, and Basidiomycota form sexual spores in sporocarps. Many of these spores can be highly stress-tolerant and fungi such as *Rhizopogon olivaceotinctus* form survival strategies to resist fire and recolonize newly growing seedlings of *Pinus muricara* (94; 286). Others form highly nutritious sporocarps (with spores that resist the guts of animals), enticing mammals to ingest. Examples include fungi ranging from *Endogone lactiflora* to *Tuber melanosporum*, the Périgord truffles.

Hypogeous fungi can also be dispersed by animals unintentionally. Neilson et al. (534) postulated that AM colonization of newly-emerging arctic islands was through birds, being attached to their feet (see also Correia et al (180)). *Cervus elaphus* (elk) consumed AM roots by pulling entire shoots from the soil, eating them, then traveling across the Mount St. Helens pyroclastic flow material, depositing corms and bulbs along with Glomeromycotina inoculum in a primary successional site (33). In an interesting twist, animals such as gophers and badgers, and even iguanas (641), will dig in soils, often preferentially into the rooting zones of plants. When they do, spores of hypogeous fungi are also transported to the soil surface in a mound that can catch turbulent winds, where they are dispersed (28). Dispersal has been measured from a turbulent wind several kilometers downwind (56).

Once at the site, inoculum must travel into the soil to find a compatible root tip. Animals facilitate this process by digging into the substrate. Some animals, such as ants, may be carrying inoculum into the soil or moving it across points (28; 263). However, roots are positively geotrophic, whereas fungal hyphae are not (263; 265). The process whereby hyphae on the surface find roots deeper in the soil still needs additional exploration.

Summary

- To summarize, there are a wide array of mycorrhizae that all consist of an external mycelium, a plant–fungal interface, and a plant.
- Most plants form mycorrhizae, but not all. Of those that do not, the fungus can still penetrate and extract carbon from them.

- A mycorrhiza fungal guild has evolved independently many times as mycorrhizal fungi are scattered among many (but not all) lineages. Structurally, if not functionally, they grade into each other, and the separation of mycorrhiza from endophyte, from epirhizal fungus, and even from facultative parasite may be relatively arbitrary and will likely remain a point of disagreement well into the future.
- Any one plant will have many mycorrhizal fungi in the field. The subtle differences between fungi mean that there are a range of multiple benefits to the host plant.
- The fact that mycorrhizal hyphae extend into the surrounding soil means that one fungus can extend between roots and between compatible plants. Together these create what are called a common mycorrhizal network (CMN).
- Networking theory shows a complex array of actors, the stability of which depends on the networking structure. There are many new opportunities to model or measure outcomes. Some simple ones are that there are both fungal and plant hubs that dictate structure–functional relationships on a large scale.
- Interactions exist at the cellular and the ecosystem scale, and different mycorrhizal types involve different structures and different players based on both phylogeny and ecology. Interpolating amongst relationships remains a high priority.
- Finally, as lands open or fill up, mycorrhizal fungi also show multiple dispersal mechanisms, both for altering gene flow locally and for establishing new successional patches at long distances such as to islands or even continents.

3 · *Evolutionary Ecology*

Probably no research topic in mycorrhizae has undergone as much change over the past few decades as the evolution of the symbiosis. The rapid development of techniques and reduction in costs of sequencing, increase in databases and new approaches to sequence database management, data mining, and sequencing analyses has generated a plethora of new phylogenic reorganization, molecular clocks, and theory. Newer sequencing concepts often readily integrate with the fossil record as the field of paleoecology itself rapidly evolves. But for understanding the mechanisms of evolution in a symbiosis, we need to go beyond phylogenetic relationships to understanding both the role of and the shifts in environments that determine how mycorrhizae develop, adapt, and diversify. Here I will summarize the key topic areas relating to mycorrhizal symbiosis, recognizing that there are likely many ideas that will change in the near future. Specifically, I address the hypotheses that: (1) mycorrhizae were crucial to the invasion of land and related to the regulation of atmospheric CO_2, (2) mycorrhizal symbioses are fundamentally stable, and (3) there are both genetic and ecological underpinnings supporting the mycorrhizal symbiosis. Here I explore the four lines of evidence of how evolution has played a key role in the ecology of modern mycorrhizae (36), including (1) paleobiology evidence, (2) extant plant mycorrhizal status, (3) the molecular basis of interaction, and (4) models of mutualism.

It is important here to review the geological time scales. I use the diagram compiled by the Geological Society of America (757), www .geosociety.org/GSA/Education_Careers/Geologic_Time_Scale/GSA/ timescale/home.aspx.

Paleobiology and the Evolution of Mycorrhizae

The formation of a mycorrhiza may well be a requirement to the expansion of complex plant life on land. However, the evolutionary pathway

Geologic Ages

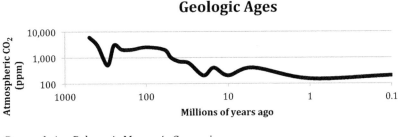

Precambrian Paleozoic Mesozoic Cenozoic

Oligocene Miocene Pleistocene Holocene

Figure 3.1 Estimated atmospheric CO_2 concentrations over the past 500 million years, redrawn from multiple sources. For comparison, the Pleistocene Epoch, CO_2 was as low as 190 ppm. In 1950, atmospheric CO_2 was 320 ppm and, as of this writing, it is currently 410 ppm, and projected to rise to as much as 1,000 ppm, with minimal intervention in fossil fuel burning.

remains unclear. The first theme is that the environment was dramatically different from that of today. Atmospheric CO_2 levels appear to have been as much as 5,000 to 6,000 ppm during the Devonian Period land invasion, which is 15–20 times the current levels (Figure 3.1 (37; 444)). Early substrate (largely mineral, with little clay – hardly soil) would have had little organic matter from which nutrients could have been mineralized and arid environments were expanding (568). The earliest fossil evidence for a mycelial lifeform appears by the Ediacaran (late Precambrian) 600 million years ago (MYa), possibly earlier (212; 273; 466). Although challenging to resolve, molecular clock evidence suggests that the Mucoromycota, with both Mycoromycotina and Glomeromycotina, existed as far back as the Ediacarian, at the end of the Precambrian. The emergence of the other fungal groups may be nearly as old. But this is long before plant life with an organized structure, allowing for the formation of a mycorrhiza, appeared.

Dawn, Long Beginnings, and the Spread of Mycorrhizae: 500 MYa to 145 MYa

Arbuscular mycorrhizae formed the earliest mycorrhizae, likely associated with the first land plants. Two hypotheses for the evolution of the Glomeromycotina exist (Box 3.1). The fossil evidence shows that the group arose during the Ordovician, either 400–500 MYa,

Box 3.1 *Phylogeny of AM Fungi*

Here, I will use the phylogeny published in Spatafora et al. (675). The AM fungi plus *Geosiphon* spp. are all within the Glomeromycotina, a subphylum of the Mucoromycota. The status of this group has been controversial since their discovery and description as members of the Phycomycetes, then Zygomycota (neither of recognized stature in Spatafora et al. (675)). Historically and in the literature, this group has been recognized as *Endogone* (462), *Glomus* (727), *Modicella* (399), *Gigaspora*, and *Acaulospora* (281) in the Zygomycota (e.g., (8)) or even Phycomycota. In some cases, the literature may refer to the group as *Endogone*, a genus (707), Endogonaceae, a family (281), or Glomeromycota, a phylum (646). Within the Glomeromycota, Endogonaceae are an order, Endogonales, of the Mucoromycotina. There is an additional AM-type mutualism, the fine endophyte (FE), or *Glomus tenue* (309). Through a novel approach of sequencing through subtraction and regression analysis, Orchard et al. (544) placed this group in the Mucoromycotina, and subsequent analyses suggest a polyphyletic origin within the Mucoromycotina ((614) and citations therein).

concomitant with the first land plants, confirmed by fossils (601; 602). Based on molecular clocks, the group may have existed beforehand in an unknown trophic status, sometime between 600 and 1,200 MYa (698). Recently, the fine endophyte has been identified as a Mucoromycotina (236) providing another interesting take on land colonization. The emergence of the fungal groups (Endogonales, Ascomycota, Basidiomycota) that led to all other mycorrhizal types likely emerged as either saprotrophs or parasitic symbionts between 450 and 500 MYa (698).

Pirozynski and Malloch (569) hypothesized that AMF first developed between an Oomycete and a chlorophytic alga, but we now know that Oomycetes are, in fact, Heterokontophyta, not fungi. The determination that *Geosiphon* was in fact Glomeromycotina, with its cyanobacterial symbiosis, initially resurrected this idea. But additional molecular work using ribosomal small subunit phylogenies indicates that Geosiphonaceae is derived from an earlier Archaeosporales, within the Glomeromycotina.

At this point, we still do not know the exact origins of the AM structure. A key question thus still remains – when, how, and why did AM arise?

The Glomeromycotina appears to be a monophyletic group. Molecular clock estimates suggest that Glomeromycotina could have emerged as early as 600 MYa, but establishing these estimates is challenging (614; 698), so a 450 MYa emergence based on fossil evidence is more compelling. These data suggest that the fungi that ultimately formed AM appeared and then spread rapidly among the majority of plants first colonizing land. It is likely that the fine endophyte type of AMF (614) spread at approximately the same time.

Liverworts and other early plants began forming primitive lignins, likely for protection from grazing (e.g., (225)). Simultaneously, these compounds provided a structural element essential for the upright growth of the plant and the first development of canopy architecture. These same lignins provide the plant with a belowground structural molecule, facilitating the physical exploration of the soil via rhizoids, and later roots. Lignins allow the plant to constrain the internal growth of the fungal partner. Modern plants control mycorrhizal fungi and pathogens by phenolics and lignins (e.g., (26)). Lignins and other polyphenic compounds have a complex three-dimensional structure that is generated by peroxidase-mediated polymerization of the phenylpropanoid monomers, 4-coumaryl alcohol, coniferyl alcohol, and sinapyl alcohol, catalyzed in the initial biochemical step by the enzyme phenylalanine ammonia lyase (PAL). Other enzymes lead to anthocyanins, flavonoids, and other compounds. PAL was shown to restrict the AM establishment in a nonmycotrophic plant, *Salsola kali* (47). With this step, a mycorrhiza structure as well as function emerges.

The environment would have been quite different from today. The environment was dramatically different in 450 MYa during the Ordovician Period, when AMF first appeared (601). Atmospheric CO_2 was high, and nutrients, especially P, would have been bound inorganically as $CaPO_4$, $FePO_4$ or $AlPO_4$. Liverworts such as *Marchantia* spp., today form both AM with Glomeromycotina, and fine endophyte with Mucoromycotina (Plate 2). P would have been a limiting factor for plant production and C would have been in excess. A mycorrhiza mutualism provides a C sink, while simultaneously respiring CO_2 that in turn lowers solution pH, weathering P (395; 418). In modern elevated CO_2 studies, increases in AM hyphal production and C sequestration with higher atmospheric CO_2 levels have been observed (e.g., (609), see discussion in Chapters 4, 7, and 8).

Weiss (768) first noticed fungal structures that resembled AMF within the roots of a *Stigmaria*, a lycopsid tree from the Halifax Hard Bed, the Carboniferous Period. Kidston and Lang (404) later published a series of photographs showing AM structures in the cortex of *Rhynia* and *Asteroxylon* from the early Devonian Period. Redecker and colleagues (601) published images of AMF in roots of the first land plants, from the Ordovician period, nearly 460 MYa, when bryophytes were the only land plants. Glomalean fungi can be found with many bryophytes today, including hornworts and liverworts. Because these plants do not form true roots, the fungi grow through the rhizoids and colonize even the upper, leafy portions of the plant (Plate 2). Mosses form Glomalean AM (e.g., (798)), but also often have fine endophytes (FE) forming many AM-like structures (594). Arbuscular mycorrhizae can also be found in Lycopodiaceae, Psilotaceae, Equisetaceae, and the ferns. They are even in Selaginellaceae, common in relatively nutrient rich, but often extremely arid microsites (see (759)), implying a role for mycorrhizae in water and nutrient uptake. Others found AM in the early Devonian Period fossils (685), Pennsylvanian coal fossil plants including lycopods, Cordaites, and Psaronius, a tree fern (756) and even arbuscules in *Antarcticycas* from the Triassic Period, 200–250 MYa (686; 687).

Sequencing evidence also supports an ancient origin to AM symbiosis. The symbiosis receptor kinase (SYMRK) gene is a protein kinase gene that appears required for the formation of AM as well as nodule formation for dinitrogen fixation between *Rhizobium* and legumes and between *Frankia* and diverse, actinorhizal plants (e.g., (453; 549)). These special kinase genes appear to have evolved around 400 MYa, similar to the earliest mycorrhizal fossils, and somehow regulate the formation of infections. The later branches of the plant kingdom tree appear to have lost the SYMRK genes for AM formation, leading to EM and nonmycotrophy.

Explosion of Mycorrhizae: 145 MYa to 66 MYa

The late Cretaceous Period was a time of radical change across life forms, including mycorrhizae (Plate 2). Reptiles dominated, birds appeared and began flying overhead, weird crocodilian life forms competed with dinosaurs across warm landscapes, and small mammals ran furtively underfoot. By the Carboniferous, over 300 MYa, so much CO_2 had been sequestered from the atmosphere and deposited it into soils that CO_2 levels declined to less than 1,000 ppm, at least until lignin-decomposer fungi appeared. CO_2 began to increase again during the Mesozoic Era, probably as a function of the volcanism associated with the splitting of Pangaea and Gondwanaland. This period was warmer than today. Under these conditions, mycoheterotrophy emerged as a life form, orchid mycorrhizae appeared, and the first inklings of ectomycorrhizae began.

AM associations began to diversify functionally, sometime in the Cretaceous Period. Mycoheterotrophy may have emerged in the non-photosynthetic gametophyte generation as early as the appearance of liverworts in the Devonian Period (372; 498). Later in the Cretaceous Period, a group of mycoheterotrophic plants in the Corsiaceae, including *Arachnitis uniflora*, split from the other Liliales associated with a limited suite of lineages in the Glomeromycotina (208; 607). During this time period, global CO_2 and global temperatures were both high. We would expect that high atmospheric CO_2 would increase photosynthesis by increasing the partial pressure of CO_2. But, having temperature optima between 20°C and 30°C, C_3 photosynthesis would decline, whereas high temperatures favor respiration, with a temperature maximum between 45°C and 50°C, resulting in lesser soil organic matter accumulation. Mycoheterotrophy would also require a common mycelial network (CMN) emanating from the surrounding vegetation (e.g., bamboo for *Arachnitis uniflora*) which plants could tap. If soil organic matter declined under these conditions, but plant photosynthesis was high, allocation of C to AMF should increase, potentially to a level where if a plant was already tapped into a CMN, it might simply acquire its C like a gametophyte (achlorophyllous) generation, becoming permanently mycoheterotrophic.

Orchids and their mycorrhizae are considerably older than previously thought. They appear to have split from the plant order Asparagales somewhere by the late Cretaceous Period, between 80 and 100 MYa. Orchidales retained their SYMRK genes, promoting mycorrhizal formation (595). But they shed their AM status. Assuming that Ascomycetes such as the Helotiales were already present and either decomposing organic matter or acting as parasites, if the orchids were to tap these fungi as mycoheterotrophs, then a new type of mycorrhiza could emerge. The photosynthetic machinery apparently reappeared in the orchids, allowing these plants to tap both autotrophic and heterotrophic sources of C (800).

Plants in the Ericales emerged around 110 to 140 MYa. The Ericoideae clade retained the SYMRK genes, appearing in tropical environments. But, like the orchids, they shifted their mycorrhizae to an alternate form of mycorrhiza, called the ericoid mycorrhiza. Like orchid mycorrhizae, ericoid mycorrhizae also form structures that penetrate within the cortical cell walls. These mycorrhizae shifted to fungi that are currently capable of breaking down complex organic materials, emerging in several Ascomycota, including the subfamily Eurotiomycetidae, and orders Helotiales, and possibly the Hypocreales. Today, they live in wet, cool habitats, such as northern bogs (e.g., (597)), but also the arid Kwongan Sand Plains and Mediterranean shrublands, habitats where P is especially unavailable and only appears in organically- or inorganic tightly bound forms, below a threshold for AM (Figure 1.1).

The evolution of EM provides intriguing information about the expansion of mycorrhizal symbioses, as the associations evolved independently and multiple times among both plant and fungal taxa (Plate 2). Both plant orders and fungal orders may have evolved 150–200 MYa. Endogonales, within the Mucoromycotina, may have evolved as early as the mid- to late Silurian Period, but the first fossil Endogonaceae appear at the Permian–Triassic boundary (160; 431). There appears to be a hyphal mantle, maybe of *Endogone*, by the middle Triassic Period (431). This appearance coincides with another peak in atmospheric CO_2, when nutrients again would have been limiting to plant production (37). Basidiomycete Agaricomycetes and Ascomycete Pezizales, which form both saprotrophic and EM taxa, appeared 150–200 MYa based on molecular clock estimates (698). Dark septate fungi appear in the roots of plant fossils during the Mesozoic Era, and poorly developed, partial mantles of potential EM appear. Alternatively, the earliest clearly defined EM fossils conclusively appeared in the mid-Eocene Epoch, approximately 50 MYa, and another peak in atmospheric CO_2. These are Endogonales, Ascomycota, and/or Basidiomycota (684).

The appearance of EM plant taxa also coincides with the appearance of some of these incomplete fossils, during the Mesozoic Era, well before the first fossil EM. The Pinales, which form EM, split from the other gymnosperms that form AM such as the redwoods probably in the late Jurassic Period (447). One line of gymnosperms, the Gnetophytes, are especially interesting (see (787)). *Gnetum*, which forms EM, split from *Welwitschia* and *Ephedra*, which form AM, sometime in the Cretaceous Period. *Gnetum* is tropical, with a Gondwanan origin, in tropical Asia, Africa, and central America. *Welwitschia* is found only in the deserts of Namibia. *Ephedra* is found in arid lands in North America and Asia.

A parallel pattern can be found amongst Angiosperms. AM form the basal lineages. As the Ericales migrated to temperate climates, the Vaccinioideae emerged, forming the arbutoid mycorrhizae, that form a partial sheath, an incomplete Hartig net, and sometimes coils that penetrate cortical cell walls. The Proteaceae emerged, apparently without mycorrhizae of any form. Like ericoid plants, they appear to predominate in soils with extremely low nutrient levels, often ancient, leached soils. They developed cluster roots, with the ability to secrete high amounts of organic acids, increasing nutrient availability mimicking AM symbioses. However, Proteaceae can form AM as seedlings, especially following fire under elevated nutrient availability (114; 115), see Figure 1.1). But later, as available nutrient concentrations drop to extremely low levels, the cluster root and nonmycotrophy appear to predominate.

Many of the Rosids, a monophyletic group of plants, retain AM associations, such as *Platanus*. But some began forming EM along with their existing AM, including *Populus*, *Salix*, *Celtis*, *Casuarina*, *Juglans*, *Eucalyptus*, and *Carya*. Many of these became switchers depending on life stage or on soil conditions. But others became predominantly EM, including the Fagaceae and Betulaceae.

But no clear fossil EM has been found, prior to 50 MYa, well into the Cenozoic Era. One hypothesis is that while EM formed, including a Hartig net, the mantle may have been less well developed compared with the clearly defined EM found especially in the Pinaceae and described by Frank (255). In seasonal tropical forests, even though EM fungi are present and isotopic signatures demonstrate the presence of EM, the EM roots themselves are often poorly developed and challenging to observe (330). Due to the warm, moist conditions of the late Cretaceous Period, simultaneous root and mycorrhizal production and turnover were high, resulting in less-than-optimal conditions for EM roots to form fossils.

A Long Decline and Re-emergence? 66 MYa to Today

Sixty-five million years ago, the world changed dramatically. The Chicxulub impact (meteor or other extra-terrestrial rock), just off the northwest corner of the Yucatán Peninsula, hit a globe already in the midst of high volcanic activity as continents continued to split apart. The meteoric fallout deposited a global iridium layer at what is termed the Cretaceous (or Kreide)–Tertiary, or K–T, boundary; at least a partial causal agent in the extinction of dinosaurs. However, plants survived at a distance from the impact and began an evolutionary radiation with their associated mycorrhizae. The globe was already experiencing high atmospheric CO_2, making soil nutrients limiting to plant production, and facilitating the initiation of the many kinds of mycorrhizal interactions that we see today. The radiating blast impacted areas like the Big Bend of Texas and the San Juan Basin of New Mexico. Here, the vegetation was obliterated, and the soil surface sterilized. When the impact force reached Hell Creek in Montana, anything living on the surface was likely destroyed by the time sound waves arrived, some 3 to 4 hours later. Farther from the blast zone, in areas such as the Patagonian plains in South America, depositional zones of ash and dust would have been coupled with a hot, radiating atmosphere, generating spontaneous wildfires. The Chicxulub collision was especially impactful because it hit shallow waters underlain by a substrate (limestone) high in $CaSO_4$ as well as $CaSO_3$. This likely increased CO_2 around an earth already high in CO_2. But for a short period, maybe only a decade, the impact produced H_2SO_4 (sulfuric acid) particulates in the upper atmosphere, and deposited an ash and dust layer that circled the globe, creating a "nuclear winter" where temperatures dropped dramatically, there was little sunlight, and the entire globe acidified. As much as an $8°C$ drop in temperature could have occurred, the combined effects of which would have almost obliterated photosynthesis. Those organisms that survived, fossorial rodents and insects, seed-eating and insect-feeding birds, would have formed detrital-based food chains, largely built on buried plant materials (read Alvarez (78) and Brusatte (144) for a more thorough description of the Chicxulub impact).

While devastating to the existing dominant organisms and that ecosystem structure (see (144)), the scale of disturbance itself facilitated some interesting changes in the biota. The rise of mammals and birds, from the Cretaceous fauna, is well discussed and the subject of extensive research. But plants and mycorrhizal fungi also appear to have undergone significant change. The Mount St. Helens in Washington State, USA eruption makes a small-scale analog. There was a sterile zone across which invasion occurred, the pyroclastic flow on Mount St. Helens, a model for invasion of southern North America and Central America after the Chicxulub impact. However, around the outer edges of the Mount St. Helens blast zone, where the pyroclastic deposition was less than a meter deep, pocket gophers and ants survived, emerging and bringing back seeds and asexual plant reproductive material, along with mycorrhizal propagules. Glomeromycotina spores and infected root pieces were brought to the surface and initiated new mycorrhizae for several years, until those propagules themselves decayed. Larger propagules, including the very large spores of Glomeromycotina, and thick mycorrhizal roots formed by EM tips, were found to survive up to three years on Mount St. Helens, and more than a decade in other stored topsoil analyses (64).

The Chicxulub impact itself would also have created a large, sterile landscape, but not exclusively. There would have been patches within and surrounding large sterile landscapes where soil organisms, including fossorial mammals and insects, and mycorrhizal fungi would have survived. But simultaneous colonization by plants and by mycorrhizal fungi to a point where infection could occur was then, and remains, challenging to plants, in volcanic eruptions and in glacial retreats (53). In observing the time course of invasion of the Pumice Plain on Mount St. Helens, I noted that plants like the biennial *Lupinus lepidus* has interesting adaptations for colonization. *Lupinus lepidus* is sometimes considered nonmycotrophic, based on glasshouse pot studies. These plants are found in the higher elevation meadows of Mount St. Helens above the Pumice Plain and were considered to require cross-pollination. They have low mycorrhizal infection rates in root tips in existing meadows and appear to be facultative mycorrhizal species (63; 712). As soon as the Pumice Plain cooled, in 1980, an individual seed of *L. lepidus* germinated. This individual grew very large, self-pollinated, and produced nearly a dozen offspring. It had no mycorrhizae, but had an extensive, fine root system, some cluster roots, and was highly nodulated with red, active N_2-fixing nodules. By 1983, there were over 1,000 individuals. I inoculated this individual, by experimentally translocating a gopher, a fossorial rodent that was carrying inoculum from a nearby location. The patch that subsequently formed around this individual consisted of other mycotrophic (facultative or obligate) plants such as *Chamaenarion angustifolium*, *Anaphalis margaritaceae*, and even seedlings of the ectomycorrhizal tree *Pseudotsuga menziesii*. With time, new individuals of *L. lepidus* colonized again, were also nodulated and AM, but were smaller and dominated by coarser roots (e.g., (53)).

All of the different types of fungal–plant interactions that constitute mycorrhizae were likely present by this time, although not all structures have been found in the fossil record. Grass phytoliths appear in dinosaur coprolites from the Cretaceous Period but the extensive AM grasslands appear after the K–T boundary. Through the Paleocene, the Eocene, and into the Miocene Epochs, the EM Fagales thrived, and the AM, EM, and NM Caryophyllales began to expand across many of the cooler landscapes. The NM Proteaceae spread into present day South Africa and Australia with the drifting continents and islands.

During the Eocene and into the Miocene Epochs, CO_2 was pulled, once again, out of the atmosphere and buried into great coal beds, such as those of western North America. By the Miocene Epoch, atmospheric CO_2 had dropped to levels as low as 150–250 ppm. Ice ages began stalking the higher latitudes at regular intervals, called the Malinkovich cycles where solar radiation intensity shifted with the eccentricity of the earth's orbit, opening subcontinental-size new, often nutrient rich, nearly sterile lands following glacial retreat. In unglaciated lands with ancient, highly leached soils, Proteaceae defined a life pattern without, or with minimal mycorrhizal formation. Plants in these extreme environments developed cluster roots and produced organic acids. Many of the Caryophyllales appearing between 90 and 50 MYa had diverged by the Paleocene Epoch and include Amaranthaceae and Cactaceae. The earliest Amaranthaceae, which arose between the Cretaceous Period and the Eocene Epoch were perennial and form AM (e.g., (14; 32; 126; 781). During the Miocene Epoch, under a low atmospheric CO_2 environment, these plants developed the C_4 and CAM modes of photosynthesis. Organic acids were also exuded into soils weathering – PO_3 ions that were bound with cations into immobile forms (153). Nearly simultaneously in the Miocene Epoch (401), the tribe Arabidea derived an annual, nonmycotrophic life form out of perennial, AM Brassicaceae (762). *Erysimum* began producing cardiac glycosides and diversified in the Pliocene (2–5 MYa) from other Brassicaceae (367).

Beginning in the Oligocene Epoch and into the Miocene Epoch, and even Pleistocene Epoch, atmospheric CO_2 declined. Now, not only was CO_2 limiting to plant production and mycorrhizae, recurring ice ages began to emerge. This meant that large land areas were repeatedly scraped clean and reinvaded by plants and mycorrhizae. For example, sediment cores at Upper South Branch Pond, in Maine, USA, show that deglaciation occurred some 12,500 to 12,000 years ago. Spores and hyphae of *Glomus fasciculatus* were found in sediments aged between 10,900 and 13,000 years ago, along with pollen from willows, sedges, and grasses, all capable of forming AM in alpine conditions. *Vaccinium*, which forms ericoid mycorrhizae and *Dryas*, capable of both EM and AM, were also found, although the much smaller spores of ericoid-forming and EM fungi were not observed. Alternatively, conifers such as spruce that form EM were found after 10,400 years ago (see (80)). Following deglaciation there would have been extensive land areas without inoculum, opportune conditions for survival of plants capable of surviving and growing without mycorrhizae. Indeed, plant species that are able to form mycorrhizae, facultative mycorrhizal plants, are often found such as *Epilobium, Lupinus, Carex*, and grasses in early seral soils or saturated soils surviving without mycorrhizae (18; 64). Other symbioses, including the dark-septate associations, may also have expanded, especially in glacial forefronts (392).

From the Pleistocene Epoch into the Holocene, CO_2 levels remained low, potentially reducing the importance of mycorrhizae. Some 10–12,000 years ago, horticulture began with the intentional tillage of soil in which to plant selected seeds. By 8,000 BC, animal husbandry emerged, with the fertilization of sown crops. This intentional application of organic fertilizers, particularly N and P, and re-direction of water for irrigation, lessened the need by many plants on mycorrhizae for production. Mycorrhizal responsiveness began the long road to lesser importance that today characterizes plants of many agricultural and forestry practices.

Over the Anthropocene Epoch, mycorrhizae have become of lesser importance with human-driven overfertilization, high rates of nitrogen deposition, soil disturbance with increasing agriculture, commercial forestry, and other development. But, just as human-induced disturbance is increasing nutrient availability, humans are also producing massive amounts of CO_2, increasing atmospheric CO_2 from 310 ppm to 410 ppm within my lifetime! Stoichiometrically, this increase in the partial pressure of CO_2 increases the fixation of C, but makes soil nutrients more limiting, thereby increasing the need for mycorrhizae (37). Thus, over the next century, mycorrhizae will become of even greater importance to human global sustainability.

Mathematical Models, Stability, and Evolutionary Cheaters

The issue of stability becomes crucial in any consideration of complexity. Modelers consider stability as a desirable state; unstable solutions are messy and are prone to extinction. May (490) set the stage for focusing on "unstable" relationships in his discovery of chaotic population growth. He found that if a population of organisms had a rate of increase per individual, or *r*-value greater than 3, then population growth becomes unstable and chaotic. By definition, mycorrhizae increase fitness by creating a positive feedback loop. Positive feedback loops increase the *r*-value, amplifying growth and, presumably, reproduction, thereby ratcheting mycorrhizal plants above stable levels, and resulting in some community extinctions. Negative feedbacks such as pathogens, alternatively, pull successful population growth back, resulting in homeostasis.

Conceptually, this outcome is logical. But like most aspects of mycorrhizal ecology, it is likely too simplistic. Bever (119) illustrated an important example whereby mycorrhizae generated a negative feedback, benefitting a host plant. Using a competitive model between the two plants, this should drive the second locally extinct. But, by altering the relative negative and positive feedbacks, the overall outcome increased community productivity of all players, fungi and plants (see Figure 1.3). Anacker et al. (79) found increasing complexity amongst plant community members with both positive and negative feedbacks. While negative feedbacks to community composition appear to be a better predictor of richness than positive feedbacks, the complex array of multiple interactions combine to make both feedback types players. Bennett et al. (116) reported that AM conspecific trees displayed negative feedbacks whilst EM showed positive feedbacks. Mutualisms as well as parasitism interact to make complex communities. Overall, the mathematical outcomes fail to show that mutualisms in nature promote or reduce stability.

There are cases where mycorrhizal fungi are clear parasites of non-mycotrophic plants. Annual plants, such as *Salsola kali*, are infected by AM fungi, which both triggers defense responses (47) and drains C with no apparent nutrient exchange (44). I would not call this cheating; it is parasitism.

Cheaters have been assumed to be common among mycorrhizae. That is the fungus takes C but does not provide N or P soil resources to a mycotrophic host plant (387; 666). The basis of this hypothesis is from many glasshouse experiments, where plant biomass is lower in mycorrhizal than nonmycorrhizal plants at the end of the experiment. If widespread, this outcome would have dramatic implications to mycorrhizal functioning and evolution questions. Here, I examine this hypothesis.

First, is short-term glasshouse growth an appropriate indicator of cheating or of parasitism? I would argue that there are two factors that question this approach. First is the pot effect; limited soil volume for roots and mycorrhizae to explore artificially creates a lesser response to mycorrhizae, and potentially even a C drain when the pot is fully filled by roots (pot-bound). This is an artifact that is a common practice and needs careful consideration in every experimental design. The second issue is the length of the experiment. Bethlenfalvay and colleagues

Bethlenfalvay Curve

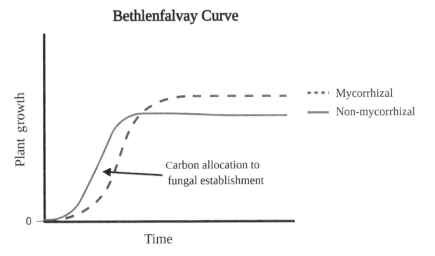

Figure 3.2 The mycorrhizal growth curve (derived from ideas in Bethlenfalvay et al. (118)) compared with the intrinsic growth rate for a plant without mycorrhizae. With mycorrhizae, there is a lag in plant growth due to energy allocation to the fungus. After the mycelial network is in place, the mycorrhiza provides additional resources. If harvest is premature or if the roots are confined (a pot), then there would be an apparent "cheater." Figure drawn by Danielle Stevenson, with permission

(Figure 3.2) showed a clear timeline for mycorrhizal development that, in turn, depending upon the timing of harvest, give a false indication of a "parasitic" relationship. The mycorrhizal plants, if harvested prematurely, would be smaller, simply because the timeline was inappropriate. We have found that, in some cases, the lag phase in plant growth can be years, but, ultimately, there is a positive relationship (75; 606).

Jones et al. (390) outlined a strong conceptual framework for studying cheating in mutualisms, focusing on improving fitness in one partner, and reducing fitness in the other. They note that there are numerous studies citing growth as evidence for cheating. However, there is little evidence that cheating occurs based on *fitness* of the mutualists. This appears to be especially the case for AM, the point of this inquiry.

Another perspective derives from the model of Treseder and Allen (719). This model looks at factors limiting fungal growth (Figure 1.1). Using this approach, we can see that a mutualistic mycorrhiza could, in theory, act parasitically because under high fertilization, low light, or low atmospheric CO_2, the plant becomes C limited, not soil-resource limited. Under these conditions, *if* the fungus remains infective, it only acts as a C drain. But here, the plant reacts by rejecting the fungus. In many instances, the system exists near the interface of the downward part of the curve and fluctuates back and forth across that region because of feedback loops between resources, plant and fungus. The plant is initially nutrient sufficient (using seed resources) and C limited. As it develops photosynthetic machinery and begins fixing C, nutrients become limiting and roots and mycorrhizae increase. These then increase nutrient uptake whereby the plant again becomes C limited, and so on. Thus, the relationship becomes dynamic within the life of a single plant. Depending on the timing of analysis, the relationship could appear anywhere along an apparent mutualistic–parasitic gradient. Only rarely will this become evolutionarily over-riding. This is not really a case of "cheating"; it is a "short-term" condition in which the evolutionary relationship is mutualistic containing ecological disequilibrium conditions.

I can only find one example of "cheating." Seedlings tap into a mycorrhizal fungal network, the CMN, but are not likely to be able to provide an equivalent resource exchange. Seedlings may even gain some of their C from this network. However, as the network is supported by overstory trees, it is likely that this C is not missed by either the true host or by the fungus itself. Further, with time, even the largest tree will die, and likely be replaced by those "cheating" seedlings, themselves becoming the host. To identify when the "cheating" hypothesis as per Johnson and colleagues (387) is explanatory, long-term fitness studies in the field are crucial.

Fungal Genetic Structure

Although the genetics of plants and vertebrates are reasonably understood, fungi have many unique attributes that are different from normally perceived biological dogma. These features are critical to understanding how mycorrhizae initially evolved, and how the symbionts, both plant and fungal, continue to evolve as the environment changes. These features include haploid multinucleate cells, including the formation of heterokaryons, nuclei that are different within a cell, and the lack of visible sexual structures, making attribution of genetic variation to sexual reproduction controversial.

EM are probably better understood, as they appear to follow relatively well-known fungal genetics, at least in the Ascomycota and Basidiomycota taxa. Here, most fungi live as dikaryons (two haploid nuclei per cell). They may be, for the most part, homokaryotic, that is the two nuclei are identical. However, when two compatible hyphae connect, nuclei are exchanged, often leading to the formation of a diploid nucleus (2N), the formation of sexual reproductive structures, recombination, sporulation, and release of 1N spores. *These life cycles can be relatively simple, or highly complex, and are well documented in basic mycology texts.*

The genetics of AMF are both challenging and controversial. Until recently, a basic assumption across the field was that Glomeromycotina do not have sexual reproduction (e.g., (554; 634)). How then do they adapt to differing conditions, and why do we have the different groups that are present? One hypothesis was that sexual structures including zygospores (as evidenced by azygospores in the Gigasporeaceae) do form, just infrequently. Unfortunately, the reports from even excellent scientists have not been reproduced. Nowhere is there confirming evidence for sexual structures (zygospores) in the Glomeromycotina. Again, the advances in molecular biology are writing a new chapter in teasing out genetic diversity and biological functioning of AM fungi. Even the earliest studies included controversies of whether the nuclei within an individual hypha of a Glomeromycotina were homo- or heterokaryotic (e.g., (554)). Today, most evidence tends to support a homokaryotic nature, despite the fact that there can be several hundred nuclei within a membrane-bounded "individual" (see (142; 426)). Evidence showing meiosis genes and genes for mating types analogous to those in Basidiomycota appear to be both present (178) and active in AMF Glomeromycotina (711). Nevertheless, there remains some variation within genomes, indicating a need for additional research (142).

One point to remember is that the species concept of Carolus Linnaeus (463) for categorizing organisms preceded Darwin's concept of evolution (191). A species is a hypothetical construct and variable concept. Indeed, most of the "species" of Glomeromycotina, especially, and likely many fungi, are groups of related individuals, and could be viewed as a clade rather than a species *per se* (507). We will discuss this concept and the implications in greater detail in the population biology section (Chapter 5).

Saprotrophic versus Parasitic versus Mycorrhizal Existence

What are the energetic implications for an organism shifting from either a saprotrophic to biotrophic mutualism or from a parasitic to mutualistic symbiosis? These questions represent exciting concepts. There are energetic costs to shifting in each case. To switch from a saprotrophic to a biotrophic condition, many extracellular enzymes are no longer needed. Especially those enzymes extruded into the environment, such as phenol oxidase and cellobiohydrolase, require energy to produce and disperse. Even if the genes for those enzymes are retained, the energy required to build and disperse the enzymes would be substantially reallocated to growth if not produced. Furthermore, the fungus would need to overcome defense responses by the host plant (see next section). To switch from a parasitic to a mutualistic symbiosis, the plant's defense reactions need to be controlled, to allow the plant to differentiate between a dancing and a consuming partner; not always an easy task.

Even that initial shift has several alternative possibilities. There is still no consensus of the origins for AM symbiosis. The AM fungi have no known relatives that have either a saprotrophic or parasitic lifestyle. The extreme age of the symbiosis combined with the lack of extant plants and environments from these early days in earth's history make understanding of the evolution of symbiosis of this monophyletic group of fungi and the diversity of AM plants challenging at best. EM, on the other hand, evolved from at least 78–82 lineages from saprotrophic ancestors from at least 16 orders of fungi in the Ascomycota and Basidiomycota. Although we know that the Endogonales (Mucoromycotina) form EM, we know little about their evolutionary history. They might be from non-EM symbionts (for a comprehensive review, see (702)). The clades from which orchid mycorrhizae emerged are likely biotrophic, although there are also EM and saprotrophic taxa within these groups.

Molecular Attributes and Fungal Infections

There are several lines of specific genomic regions that regulate mycorrhiza formation and likely have done so since the origin of the symbiosis in early land colonization. These conservative molecular sequences are useful to help to understand the evolution of AM, although not necessarily the more recent mycorrhizal types. I focus on four genetic attributes that might be helpful as examples: the SymRK gene family, the genes for phenylalanine ammonia-lyase (PAL), the involvement of phytohormones in the regulation of the symbiosis, plant structure–function, and the AM fungal protein glomalin.

The receptor-like kinase gene SymRK gene family is found in and plays a role in the formation of nodules for N_2 fixation, and in allowing penetration of AM infection of cell walls (see (283)). There are nod-type homologs in EM plants, but many details remain to be studied. The details of how these genes work is beyond the scope of this book, but the complexity among hormones, sugars, nutrients, and flavanoids makes for an emerging research field in itself (e.g., (274)).

PAL triggers an immune-like response where a browning response indicates phenolic production, killing both the invading AM fungi in a nonmycotrophic host (*Salsola kali* (24; 47)) and the host plant cells. PAL is also triggered, but with a delay, in a mycotrophic AM plant, *Artemisia tridentata* (26). This delay allows the formation of AM in the region of elongation, formation of a functional mycorrhiza, followed by suberization and lignification, and the formation of an efficient transport system for water and nutrients between the root tips (and mycorrhizae) and the leaves.

Plant hormones regulate virtually all aspects of plant growth and development, and probably have since the first land plants. The mycorrhizal symbiosis is dependent upon production, and/or induction of plant hormones for formation, and then for growth and stress tolerance. Cytokinins and auxins are produced by EM fungi, and likely many others. Mycorrhizal plants have higher leaf concentrations of cytokinins, auxins, and gibberellins, and lower levels of abscisic acid, and more complex responses in roots (66; 67). The increase in cytokinins, in particular, stimulates the formation of the root branching pattern of an EM (664) and may stimulate the increased root growth and branching in AM (66). This is important in that AM infection occurs just behind the region of elongation in the root, and the more new fine root branching, the more sites for the formation of new infection units. Increasing molecular evidence shows the involvement of cytokinins and auxins in development of the fungal Hartig net and attenuating host defense responses (481), such as PAL activation.

A key research need is to better understand what the mechanisms are for enabling mutualistic interactions versus recognizing and inhibiting pathogens. Somehow, these had to have evolved with the first land plants, some 400 to 500 MYa. Despite the large and growing information base on molecular-scale defense mechanisms in plants, we still know little about how plants discriminate between the good guys and the bad guys.

Arbuscular mycorrhizal fungi also produce a compound (or compounds) known as glomalin. This compound appears to be a homolog of heat shock protein 60 (271), a complex protein with sugar moieties that facilitate movement through membranes. The heat shock protein 60 appears to be important in the evolution of eukaryotes, regulating membrane interactions of mitochondria and chloroplasts with their cells (304). Possible disruptions of these organelles and their host cells are responsible for a number of disease attributes. It is likely that glomalin has been an important constituent since the initial formation of AM symbiosis, and multiple sources of information indicate that this compound has been important in the sequestration of atmospheric CO_2 at multiple times in earth's history (25; 37; 444).

Summary

- In summary, AM comprise the basal status of the plant kingdom. Traditionally, the fungal group has been seen as the Glomeromycotina (or Glomeromycota, Glomerales, as taxonomic organization shifts through time). However, new information incoming as of this writing suggests that the fine endophyte, or *Glomus tenue*, may well be a parallel lineage within the Mucoromycotina (e.g., (614)), found in the liverworts, mosses, and higher plants across nearly all environments.
- The AM and probably the Mucoromycotina emerged with the first land plants and under a very high atmospheric CO_2, and spread across the globe.
- AM were so effective that they facilitated (or directly bound) a high fraction of drawing down the atmospheric CO_2.
- Sometime during the Mesozoic Era, CO_2 levels again rose, likely associated with the extensive volcanism associated with the splitting of the supercontinent of Pangea. New forms of endomycorrhizae, AM mycoheterotrophic plants, ericoid, and orchid mycorrhizae, emerged during the heyday of mycorrhizal diversification of the Cretaceous Period, and retain forms of the symbiosis SYMRK gene.
- Newly emerging EM (including arbutoid and monotropoid) probably formed during the late Cretaceous Period, but the fossil evidence of actual EM from this period has not yet been found.
- Nonmycotrophy spread, if not emerged, as sterile landscapes emerged from large-scale disturbances ranging from the Chicxulub impact to volcanoes to retreating ice sheets coupled with very low atmospheric CO_2 following the Oligocene Epoch.
- As we look into the future, mycorrhizae are becoming of greater importance, as anthropogenic stimulation of atmospheric CO_2 rapidly exceeds the capacity of the biota to sequester C. A new dawning in the importance of the studies of mycorrhizae is emerging.

4 · *Physiological Ecology*

The definition of symbiosis is two organisms living intimately together, and this chapter examines the physiological basis of the interaction. A mycorrhiza is comprised of two distinctly different organisms, a plant and a fungus, that interface down to the molecular level. Because of this intimate physical closeness, the biochemistry, physiology, and ecology become highly intertwined. At the most basic definitional level, the fungus picks up nutrients and water in the soil, transfers those resources to the host, in exchange for carbon fixed by the plant from the atmosphere. This physical dimension means that resources available to one partner are less available to the other. But both sets of resources are essential to both organisms. Thus, for the purposes of this book, I focus on the unique resource extraction mechanisms of each of the partners, and the exchange currencies that mediate the symbiosis. For the most part, the research reviewed in this section focuses on two symbionts, sometimes living in harmony (see (405; 406)), other times less so (648). As such, most of the research takes two approaches. The first is to define an interaction on first principles of physiology, chemistry, and physics. The second is to compare experimentally, one plant with mycorrhizae and the other without, or nonmycorrhizal (NM). It is important to remember that NM is distinct from nonmycotrophy, in which a plant evolutionarily does not form a mycorrhiza (discussed in Chapter 3).

Creating Demand: Stoichiometry

There are three basic interactive processes that can help to organize an overview of the functional implications deriving from the structures that we examined in Chapter 2. The first is the fixation, transport, use, respiration, and deposition of carbon (C), the energy channel for the mycorrhizal symbiosis, as well as all known life. The host plants fix CO_2 into sugars ($C_6H_{12}O_6$ as a base). They must have water (H_2O), the aqueous environment in which all biochemical reactions occur to convert CO_2 to sugars, and nutrients (especially nitrogen (N), phosphorous (P), iron (Fe), and magnesium (Mg)) that are required for photosynthesis. From the simple glucose, other complex carbons – such as lipids, complex sugars, often with an added P or N, and strung together to form non-water-soluble carbohydrates – are constructed. With the exception of the mycoheterotrophs that do not fix C, the plant provides all, or most, of the C needs of the fungus and supplies those directly to the host. The second is the *passive* movement of H_2O, the matrix in which all life physiological processes take place. Water moves directionally in response to a free energy gradient. However, organisms can control water within cells using membranes made of lipids through which diffusion is restricted. Some nutrients move within mass flow of water from soil to the fungal or plant surface in H_2O, whereas others are bound to charged particles in the soil and must be intercepted physically to be taken up. This third process involves the *active* uptake of nutrients that can require enzymatic breakdown or localized pH shifts to weather or mineralize the nutrient. All nutrients are charged, which requires these molecules to be *actively transported* across membranes. These functions will form the organization for this chapter.

As all cellular processes are active only in the presence of water, we can first think of $C:H_2O$ molar ratios. Most of the water needed by a plant moves from roots to leaves through the vascular system, exiting through the stomata. The first two functions of water are to (1) keep a plant hydrated and (2) keep the stomata open, allowing CO_2 to diffuse into the leaf and O_2 to exit. Only a very small fraction of the H_2O molecules entering the leaf are actually fixed into glucose. Water-use efficiencies vary, but around 300 moles H_2O per mole of CO_2 is a characteristic average for a C_3 plant. All fungi and plants must extract water from soil to form the matrix in which their biochemistry functions. The atmosphere is the ultimate sink as even moist air has a water potential ($\simeq 100$ MPa) lower than that required for any cellular function. This flow gradient is called the soil–plant–atmosphere continuum (SPAC) or the soil–fungus–plant–atmosphere continuum (SFPAC).

The physiological processes leading to photosynthesis and subsequent transport and use of energy rich C compounds themselves can also be better understood in terms of interactions between resources. Photosynthesis requires CO_2 as the C source for fixation; H_2O to maintain turgor, open stomata, and provide electrons, protons, and O for photosynthesis; N for enzymes, including Ribulose 1,5 bisphosphate carboxylase (RuBP carboxylase) which comprises as much as

50 percent of the earth's protein; P for energy-rich ATP and NADPH; Fe as the center of the heme of the cytochromes required for electron transport; and Mg as the center of the porphyrin ring of chlorophyll. Each actively photosynthesizing cell requires a basic elemental composition of C, H, O, and the essential elements of N, P, K, Ca, Mg, S, B, Cl, Fe, Mn, Cu, Na, Sr, Zn Mo, and Ni. These are the building blocks of chemicals comprising cells and cellular processes. Each of these nutrients must pass through membranes and they require active transport to reach leaf cells where photosynthesis operates, or to any cells that are physiologically active. Mycorrhizae have been shown to increase uptake of all of these elements. Three elements are especially critical from a stoichiometric perspective, C, N, and P, which we can use to build our approach. The classical Redfield ratio, of 106C:16N:1P, was developed for ocean phytoplankton that are actively photosynthesizing. Plants, because of their high levels of structural C, tend more towards a ratio of 250C:7N:1P. Fungi have a higher protein content and N-containing chitin cell walls, and thus have a ratio of 136:10:1 (141). Building on the Ågren and Bosatta (4) conceptual framework for building C-nutrient models, the Treseder and Allen model (Figure 1.1) of mycorrhizal activity versus nutrient concentration forms a useful framework in which to construct the interactive relationships (25; 75).

Finally, it is always important to remember that these processes are interactive. CO_2 cannot be fixed except in moist leaves, and using enzymes comprised of N such as ribulose-1,5-bisphosphate carboxylase-oxygenase (RuBisCo), energy in compounds comprised of P such as ATP and NADPH, and all of the other essential elements. Each of these elements shifts continuously within soils, in concentrations and locations at rapid temporal and measurable spatial relationships that vary with soil moisture and plant needs.

Passive Water Mass Flow

Water moves in response to energy gradients, or ψ. It is a passive process and will occur in any direction in which the gradient is established. For our purposes, gradients are created by adhesion to particles or matrix potential (ψ_m), the variable concentration of solutes (osmotic potential ψ_π), and the pressure of water created by solid walls of both plant and fungi (pressure potential ψ_p). We can express the tendency of water to move from point A (for example, in a soil micropore) to point B (in an inner cortical cell wall) as the water potential (ψ) gradient, where

$\Psi_{\text{total}} = \Psi_m + \Psi_\pi + \Psi_P$. Key is the difference in ψ between two locations, ψ_1 and ψ_2. ψ is usually measured in units of pressure, such as megapascals (MPa), where pure water is 0 and humid air is -100. Unimpeded by a membrane, water will always flow from a less negative ψ to a more negative ψ. This flow is termed apoplastic.

The extramatrical structure of mycorrhiza is crucial to the flow of water. Safir et al. (629) found greater water throughput in AM than NM plants. They attributed the extra water flux to improved P nutrition of the host because killing the hyphae with a fungicide (benomyl) reduced P uptake but did not reduce the water flux. However, water flow is passive whereas P uptake is active. In their experiment, the fungicide killed hyphae by inhibiting active transport processes (P transport) but did not break down the structure of the hyphae. Thus, passive transport of water would still occur even though active transport of P was stopped. Others have subsequently shown that dead hyphae increase hydraulic conductivity (L_p) and water movement in soil (2; 591). Some EM form a complex rhizomorph that resembles a plant vessel element. This enables EM to move water from great distances to the root surface (213). AM hyphae can also wrap around each other, forming a very primitive cord-like structure ((39), David Stribley, personal communications) along which films of water can move in response to a ψ gradient (37; 39). Not only can mycorrhizal fungi penetrate ultramicropores in soil, but, after the fungus penetrates the root, it forms an intra-root network that expands even to the inner root cortical layers. This provides an apoplastic flow pathway from the soil several centimeters to a few meters distant from a root, clear to the inner cortical cells of an active root. Membranes dramatically constrain water flows (39) and are termed symplastic flow.

Apoplastic Flux

Movement of water in soil is either by gravity or by diffusion. Following precipitation, gravity pulls water down through soil channels, often created by positively geotrophic root growth (in response to the force of gravity), filling soil pores. Water evaporates into dry air, but most of the water is picked up by roots and root hairs in saturated soils, and transported to leaves, where a small fraction is used for photosynthesis, and the remainder transpired back into the atmosphere. Under saturation, water fills nearly all pores and gaps. As the soil dries, water is used in the pockets with roots. As ψ drops, the water contact within roots declines, and roots shrink back from soil particle contact (765), thereby reducing contact with soil, constraining water flow, as water does not cross air gaps. Larger air gaps form (25) as water disappears from the macropores and then mesopores, still penetrated by fine roots and root hairs. The presence of the hyphae sustains contact between soil particles and roots and provides a linkage across air gaps to maintain flow. Hyphae cross these air gaps, transporting water to the host (25; 37; 588).

With continued drying, water retreats back into micropores, and eventually ultramicropores (see (40) for detailed discussions). As soil dries, $\psi_m = 2\gamma r^{-1}$, where γ is the surface tension and r is the curvature of the pore, starting from approaching 0 in wet soil facilitating contact and uptake. For roots, as water around each root is depleted, air gaps form and contact with the water in each pore is lost. The remaining water contracts, forming a coating (low ψ_m) along the walls of micropores and larger ultramicropores (down to ~1–2 μm). These droplets are still accessible to hyphae, but inaccessible to roots, cluster roots, and root hairs. It is important to remember that fungi can continue to grow at ψ values down to −10 MPa and below by directly intercepting small drops at scales below measurement technologies (300; 301). Mycorrhizal fungal hyphae usually establish in wetter soils, radiating out from the host root. Also important, as water dries, salts and nutrients concentrate in those smaller pores, creating an even greater ψ_π gradient but also concentrating nutrients for uptake. In AM, an absorbing network consisting of bifurcating mycelium starts with large hyphae next to the roots and ends with very fine hyphae near the tips. Because of the branching structure, there are 128 fine tips by the eighth branching order, at 2 μm. Because of their size and number, they can penetrate even the larger ultramicropores and are able to supply the first order branch entering the root, at 10 μm diameter, with water and nutrients (40).

Are the rates of flow adequate to affect plant water status quantitatively? AM increased hydraulic conductivity (L_p), thereby enhancing water throughput (74). Hardie and Leyton (317) also reported greater L_p with AM that they attributed to hyphal uptake. I quantified the water throughput in a mycorrhizal grass (31) and calculated that *if* the difference between an AM and a NM plant was due to hyphal transport alone, water was moving at 100 nl h^{-1}. I compared this value with the rate measured in another fungus, *Phycomyces blakesleeanus*, that can be easily grown and studied in culture. The estimated hyphal water flux was 131 nl h^{-1} in *P. blakesleeanus* (181). Both fungi are aseptate Mucoromycota with hyphae of similar size and under a similar vapor pressure deficit. Hardie (316) then undertook an innovative but tedious task of measuring water uptake in a mycorrhizal versus a NM plant. She clipped off the hyphal connections of half of the mycorrhizal plants. Those plants with the hyphae clipped had the same rates as the NM plant, whereas the mycorrhizal plants with intact hyphae continued transpiring at a higher rate. I further studied the potential rates of hyphal flux (31). Based on a diameter of 10 μm, a transport gradient of 0.5 MPa, an L_p of 10 cm s^{-1} MPa^{-1}, and a wall layer of 100 nm, transport can reach 54 nl h^{-1} (31). Given the extent of the connecting bifurcating hyphal network of 80 cm per entry point, the numbers suggest that direct transport is one mechanism that is altering plant–water relations, here accounting for approximately half of the needed flux. Importantly, one other mechanism is simply an AM structure within the root. A continuous hypha reaches from the appressorium to the inner cortical cells, allowing water to bypass moving from plant cortical cell to plant cortical cell. This internal apoplastic transport plays an unknown role in water transport. Again, these data suggest that a major constraint is the interface between the apoplast and symplastic flows and between the inner cortical cells and the endodermis. More work on L_p of hyphae is needed, but these analyses also suggest that other mechanisms are involved (see Symplastic Transport section).

Bidirectional Flow Patterns and Connectivity: Another Role of Apoplastic Transport

Water always flows based on the ψ gradient. Thus, water taken up during the day, even from deep in the soil profile, always moves from the soil into mycorrhizal fungi and fine roots, into the roots, stems, leaves, and through the plant stomata into the atmosphere. But at night, stomata close, dramatically reducing transpiration, as there is no connection to the atmosphere. When surface soils, neighboring plants, or fungal hyphae have a ψ lower than the plant, water can then move down. Water molecules that would have been lost to transpiration can support the mycelial network in the dry, surface soils. Using dyes, Querejeta et al. (588) showed that deep water taken up by one plant was redirected by apoplastic flow via hyphae extending outward into dry surface soils. This water supported hyphae (408; 589; 590) and mycorrhizal roots of neighboring plants (218). Even dead hyphae, when still connected, could move water (591) by providing a physical structure connecting water films across pores of gas.

The hyphal connection among plants also facilitated nutrient uptake. The main hyphae are hydrophobic, whereas tips are hydrophilic. At night, Lucifer-yellow dyed water and isotopically-labeled water moved from the original host, outward through the mycelial network, to the tips, and was even pulled into nearby soil. The next day, water and dissolved $^{15}NH_4^+$ were taken back up. The large plant thus hydraulically lifted water from deep soils to provide water and dissolved nutrients to the network of fungi and hosts (218).

The apoplastic pathway is likely the dominant flow for increased water uptake into mycorrhizal plant host cells. Symplastic flow must also occur, but it is slow compared to apoplastic flow (see next section, and (39; 40) for detailed discussions).

Symplastic Water Transport

Within organisms, flow is constrained by a semipermeable membrane. One hypothesis is that early in Earth's history, a lipid bubble (like a soap bubble) formed, enclosing a solution of high salts and some reproductive element (likely RNA), within a pool of water with lower salt concentrations. This was the earliest membrane, restricting the exit of critical compounds from the bubble into the pool, and forcing water into the bubble to equalize ψ. In essence, that is a lipid membrane, with embedded proteins, aquaporins, that form micro-channels through cell membranes facilitating water exchange with the cell and reducing the potential for bursting. But if there is a ψ gradient between the soil and the inner cortex next to the stele, then water may flow through and between cell walls. *Apoplastic flow* is the water flow that travels between and through cell walls, and even through plasmodesmata. *Symplastic flow* is water flow across membranes and through the cell contents. The symplastic flow pathway is slow and subject to the ψ and ψ_p status of each cell.

But there is a symplast constraint on water uptake. Once water is within the inner cortex, those molecules must move through the endodermis and into the xylem, from the xylem into the leaf, and from the leaf into the atmosphere. Several studies, particularly with AM, have shown an increase in L_p, which is driven by both an increase in symplastic flow and an increase in aquaporin expression (aquaporins are channel proteins that facilitate the transfer of water between cells) (e.g., (110; 454)). After this interface, water transport has an active transport element, regulated by the leaf stomata. Stomata open allowing water vapor to diffuse into the atmosphere (and CO_2 to diffuse into the leaf) when a proton pump is activated. This pump increases the osmotic pressure (decreasing ψ_p) inside the cell membrane and causing a curvature in the stomatal guard cell. The pump then turns off allowing the osmotic pressure decline (increasing ψ_p) and guard cells to close. Leaf osmotic potential may or may not change with mycorrhizal activity. In *B. gracilis*, little change in either total ψ or ψ_π was found, but stomata were more open, allowing greater water throughput (30). Alternatively, Allen and Boosalis (52) found that one of the AM fungi associated with wheat, *Glomus fasciculatus*, increased leaf osmotic concentration (lower ψ_π) which allowed the stomata to remain open, pulling more water through the SFPAC than in the NM SPAC. By keeping the stomata open, more CO_2 could be fixed, increasing productivity.

In the soil, a number of nutrients that determine osmotic potential are soluble and move depending on diffusion (from high concentrations to low) and mass flow (along with the water movement). In the nineteenth century, researchers (e.g., (256; 678)) believed that an increase in mass flow of nutrients to the EM surface, followed by uptake and transport through the mantle and Hartig net, was the mechanism of nutrient uptake. Indeed, many nutrients move in soil through this mechanism, including NO_3^-, Ca^{2+}, K^+, some forms of organic P, and other resources. Mycorrhizae have been shown to increase the uptake of all of these resources. The increase in salts could be especially critical to plant water relations. E. Allen and Cunningham (19) found that in salt-stressed *Distichlis spicata* (saltgrass), AM plants took up more K^+ to balance the higher Na^+, as well as more P to increase the energy for regulating salt stress. Augé et al. (92) undertook a meta-analysis of 460 studies, showing in general a 24 percent increase in stomatal conductance with AM (especially when heavily infected) versus NM, especially with drought. This pattern could be in response to both increased active and passive nutrient ion uptake; these processes are not mutually exclusive but synergistic.

Mycorrhizae are also known to alter phytohormone balance in ways that increase water throughput. In the roots, cytokinins (which promote cell division and differentiation) are increased, resulting in increased root branching and mycorrhizal infections per plant. In leaves, cytokinins, gibberellins, and auxins are increased and abscisic acid levels are reduced in afternoon peak transpiration. One outcome of this directionality is more open stomata, which increases water throughout the plant and also CO_2 uptake (30; 31; 66; 67). Allen et al. (74) also found that stomata were more responsive in mycorrhizal plants, opening and closing more rapidly with shading in AM than NM plants.

Plant Water Status and Mycorrhizae in the Field

A key question is, does mycorrhiza-induced water transplant matter to the plant? Controversies abound, largely because of two issues: the unrealistic constraints of pot culture studies, and dogma that hyphae are so small and plants so big. Rather than focus on the numerous pot culture studies, I concentrate here on the few comparative mycorrhizal versus NM field studies. There are really two approaches. First, are observations of quantitative differences in water throughput between mycorrhizal and NM plants, which depend upon initial conditions whereby one replicated treatment has mycorrhizae and the other does not. Second, are observational isotopic ratios in mycorrhizal and NM plants, that verify or contradict quantitative or mechanism hypotheses. Field studies are challenging due to environmental stochasticity. Climatic conditions can change by the minute. A cloud drifts by, the light drops, and stomata close, then reopen. Mycorrhizal plants respond more rapidly than non-mycorrhizal plants (74). As Apollo travels across the sky, shadows from trees result in sun flecks that rapidly change leaf and soil temperatures. When cumulus clouds build in the afternoon, solar radiation can bounce between them like mirrors, rapidly increasing solar radiation in short bursts. Roots and hyphae reach perched water tables, or groundwater, tapping water sources not observed by the surface-bound researcher. The extramatrical hyphae themselves improve the hydraulic properties by enhancing water holding capacity, by adding organic matter and increasing aggregation, and improving soil hydraulic conductivity by traversing soil gaps (40; 91; 584). All of the factors simultaneously cause variation from moment-to-moment and point-to-point.

Establishing mycorrhizal and NM plants for field studies is especially challenging because most natural soils already have some degree of inoculum that will invade newly planted seedlings, and inoculum is dispersed by wind and animals. One approach is to add inoculum to plants in a location without mycorrhizae. Severe disturbances make excellent experimental systems. We have used both artificial disturbances such as mine spoils and natural disturbances such as a volcanic eruption for such studies. The alternative approach is to eliminate mycorrhizal fungi from a patch using a fungicide application, such as benomyl, which does not harm the plant.

Using the first approach, we found that the addition of AM significantly improved the water relations of host grasses, *Agropyron smithii* and *A. dasystachum* (12; 13). During the dry early summer of the wet El Niño year of 1983, AM plants had lowered stomatal resistance (r_s) to transpiration (r_s = inverse of higher stomatal conductivity of both H_2O and CO_2 or g_s), and slightly less leaf water stress (less negative leaf water potential, ψ_{leaf}). AM plants also extracted more water from soil showing higher throughput. Importantly, the lowered water stress in the AM plants allowed for a delay in the flowering phenology of up to two weeks compared with the NM plants. This means that AM plants had a higher uptake of CO_2 plus two additional weeks of active growth, significantly increasing productivity of the AM plants. During the next year (1984), a more normal dry season extending through the summer and into the fall, AM plants extracted more water from soil during the growing season and maintained a more favorable water status.

Another experimental system we used was in the wake of a volcanic eruption. On the pumice material (tephra) of Mount St. Helens, where no soil microorganisms survived, the first invading plants were lupines. The biennial forb *Lupinus latifolius* reinvaded several sites. Plants established both on the tephra, and on soil containing inoculum, spread onto the surface by pocket gophers. The plants establishing in the gopher soil were AM, whereas those on the adjacent tephra were NM. g_s was higher in the AM than NM plants, along with leaf N, suggesting greater photosynthesis and greater N_2-fixation. P and leaf water status were unchanged between NM and AM plants (53).

In the alternative approach, using fungicides to eliminate mycorrhizae followed by physiological studies, the results are more difficult to interpret. In part, this is because the fungicide is rather short lived, and fungal recolonization can be rapid. Treatments tend to reduce, but rarely eliminate mycorrhizae. Nevertheless, the beneficial effects of AM to plants during drought stress emerged in an alpine grassland experiment (242), whereas it was inconsistent during wetter periods.

We added benomyl to eliminate the AM fungi, then inoculated seedlings and outplanted them into the experimental plot, in a fully replicated design (see Weinbaum et al. (767) for experimental design). Water relations and photosynthesis were measured for two treatments, plants inoculated with either AM fungi *Scutelospora calospora* or *Acaulospora elegans*. Importantly, both transpiration and photosynthesis increased in the *S. calospora* inoculated plants compared to those inoculated with *A. elegans*, and both were greater than NM plants (352; 606; 767).

The other approach is to analyze the isotopic composition of plant tissue, or in the case of water, the isotopic composition of the transpiration stream, grown under M versus NM experimental conditions. Natural abundance isotope ratios of C ($^{13}C{:}^{12}C$), N ($^{15}N{:}^{14}N$),

O (^{18}O:^{16}O), and H (D (or ^{2}H):^{1}H) are especially useful in assessing the role of mycorrhizae in photosynthetic, nutrient, and water dynamics of the host and of the fungus. Here, ^{12}C (lighter and smaller) is preferentially transformed or translocated compared with ^{13}C (heavier and larger) and ^{13}C remains and preferentially accumulates. This shift can be expressed as the δ^{13}C (or δ^{15}N, δ^{18}O, or δD) ratio, where:

$$\delta^{13}C(\text{‰}) = \left[\left(R_{sample}/R_{standard}\right) - 1\right] \times 10^3,$$

where R = ^{13}C/^{12}C. All natural abundance values are expressed in the same way.

δ^{18}O, δD, and δ^{13}C provide especially useful information to plant–water relations. Ratios of the heavier, but less abundant isotopes are expressed. Querejeta et al. (585; 586) found that AM plants inoculated with a mix of native AM fungi had a lower δ^{18}O signature than NM plants in the field, indicating a greater stomatal conductance (g_s) and a greater throughput of water in several shrub species planted into disturbed soil without inoculum. More recently, Poca et al. (573) demonstrated that water isotopic fractionation by AM occurs, hypothesizing increased aquaporin activity. However, δ^{13}C results were inconclusive, as in other glasshouse studies (e.g., (393)). Nevertheless, greater g_s, as measured by lower δ^{18}O, suggests a greater throughput of water which is typically accompanied by simultaneous higher carbon flux, increasing photosynthesis, but not necessarily a greater efficiency of C gained to H$_2$O lost (water-use efficiency).

Active Transport and Nutrient Uptake

Plant leaf tissue must contain high nutrient concentrations to sustain photosynthesis, particularly N and P. These must be taken up against concentration gradients between available soil N and P, and leaf N and P. Ammonium (NH$_4^+$), a positively charged ion that is the immediate form of available N released by decomposition, attaches/adsorbs to negatively charged clays and humic particles. Phosphate (H$_2$PO$_3^-$ or HPO$_4^{-2}$) becomes unavailable by binding to cations (Ca, Fe, Al). The small amount of bio-available P, often <10 mg kg^{-1}, is largely adsorbed to soil particles.

In a landmark study, Hatch (334) demonstrated that EM hyphae explore soil for those nutrients that do not move readily by mass flow. His work was expanded to AM beginning in the 1960s, with research on

AM by James Gerdemann, Barbara Mosse, and their many colleagues (668). The mycelial architecture (see Chapter 2) outlines the pattern of soil exploration for different mycorrhizae. These include a bifurcated hyphal branching absorption network (e.g., (264; 335)) and the range of exploration models of Ogawa (541) and Hobbie and Agerer (355).

Because nutrients have a charge, they do not diffuse through differentially permeable cell membranes. Furthermore, nutrient concentrations also matter, constituting the free energy gradient that must be overcome; in most cases, there are high concentrations of limiting nutrients in leaf cells, compared with the low concentrations in soil. That means that cells must actively repackage nutrients to reduce the free energy gradients and expend energy to transport nutrients across membranes. This process is complex but is well documented in the nutrient uptake and mycorrhiza literature. Useful detailed overviews of the critical steps in mycorrhizae and plant nutrition can be found in Smith and Read (668), Smith and Smith (669), and Johnson and Jansa (389). These references provide a portal into the extensive literature base on this area. Here, I briefly outline the processes as they become crucial to our later discussions of community and ecosystem processes.

P Uptake

Mycorrhizae are especially well known for improving P uptake (e.g., (570)). Mycorrhizal systems access, repackage, and transport nutrients in interesting fashions to reduce the free energy gradients. We will use P transport in AM as an example. First, AM fungal hyphae respire CO_2 as part of their growth. CO_2 dissolves in H_2O, forming $HCO_3^- + H^+$, weathering $CaPO_4$, and releasing bio-available $H_2PO_4^-$ (395). These hyphae may also produce oxalates, that bind the released cations (Ca, Fe, Al), keeping PO_4^- from being recombined to the soil cations (395). Second, AM fungal hyphae release both acid and alkaline phosphatases and phytases into the immediate surroundings, breaking down organic P into $H_2PO_4^-$ and HPO_4^{2-}, thus making it available (73). Thus, the fungus creates a high concentration of available P in the immediate neighborhood of the hyphal tip.

AM fungi have transporter proteins to take the $H_2PO_4^-$ across the membrane and into the aseptate hyphal tip. The transporter genes and regulation are well-documented and described in detail elsewhere (e.g., (559)). The relevant information for mycorrhizal physiological ecology is that the processes are tightly regulated and tuned to optimize energy efficiency. Data suggest that available P is taken up by the hyphal tip, packaged into polyphosphate granules, then moved by diffusion, mass flow, and Brownian movement from the tip into the arbuscule. Sanders and Tinker (633) and Pearson and Tinker (556) quantified the fluxes of P through hyphae. In the fine hyphal tips, the flow was 0.3–1.0 nmol $cm^{-2} s^{-1}$ in the thin hyphae near the tips (as little as 2–4 μm in diameter) and increased to 38 nmol $cm^{-2} s^{-1}$ in the coarse hyphae entering the roots. These rates demonstrate that protoplasmic streaming through the coenocytic hyphae is occurring at rates reasonable to influence plant P status. Once in arbuscules, the granules break down, releasing a high concentration of available P, where other transporter proteins move the P along a free energy concentration gradient first into the host cells, then the neighboring cells, then into the vascular system for uptake to the leaves (643). Within the leaves, photosynthesis rapidly converts P from inorganic to organic forms, again reducing the free energy gradient for transport (73). In AM plants, most of the internal leaf P is organic, whereas, in NM plants, the internal leaf P is in an inorganic form. This is indicative of greater photosynthesis in AM plants, but also, importantly, reduces the energy costs of P uptake and utilization.

It is important to remember that EM undertake the same processes but may do it more efficiently. EM fungi not only respire but also produce high concentrations of oxalates (55; 250; 298). These acids preferentially bind Ca^{2+} (or other cations such as Fe or Al), keeping $H_2PO_4^-$ in solution for uptake. EM fungi are especially active producers of phosphatases and are even more active in breaking down a range of organic molecules (668). The implications across the broader spectrum will be discussed in Chapter 7 (Ecosystem Dynamics).

N Uptake

The uptake of N by mycorrhizae is even more complicated because of the many different forms by which N can be transported, transformed, and used. Our best understanding of N dynamics comes from EM studies (summarized in Figure 4.1). Nitrate (NO_3^-) moves in solution by mass flow. In general, NO_3^- uptake is not improved directly by mycorrhizae – only secondarily as more water throughput occurs. But, once it is taken up, NO_3^- must be converted within the plant to NH_4^+ and NH_3. Next, they combine with glutamate and use energy (ATP → ADP + P_i) to produce glutamine and build amino acids and proteins, including RuBP

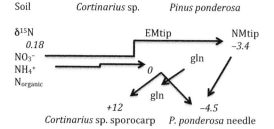

Figure 4.1 The complexity of N uptake and EM. This figure shows the forms of N taken up and transformed and an example of the $\delta^{15}N$ ratios of the different components of a single EM from a California conifer forest (see text for details).

carboxylase for photosynthesis. This pathway has been the focus of most N fertilization research for nearly two centuries, peaking with the use of the Haber and Ostwald processes in the early twentieth century to produce synthetic fertilizers for agriculture and forestry. Later, mycorrhizal researchers found that most forests short-circuited the overall N cycle by taking up NH_4^+ directly from soil particles. EM fungal hyphae search out NH_4^+ bound to organic matter particles, transporting and converting it to glutamine, which can then be transferred directly to the host using C based energy derived from the host (75; 668). Moreover, many EMF are able to break down organic N (N_O) existing as complex proteins into simple amino acids, such as glycine, which can be broken down back into NH_4^+, resynthesized into glutamine, and transferred to the host. Altogether, these processes give EM a complex array of opportunities for improving the N nutrition of the fungus itself, and as currency exchange with the host. Different fungi practice a dizzying array of strategies, symbiotic partnerships, and pathways that improve host plant N acquisition.

In the study of forest ecosystems, Gebauer and Dietrich (276) noted that N isotope ratios differed amongst the different ecosystem compartments, leaves, roots, and fungi of different plants. They pointed out that these different sources determined the resulting $\delta^{15}N$ values, and that these values could help in determining source/sink relations in EM plants. This approach helped show that EM plants short-circuited the N cycle (where plants were presumed to take up NO_3) delinking the long N cycle from tree growth (599). Gebauer and colleagues have subsequently effectively used $\delta^{15}N$ ratios to assess sources and sinks of different plant groups and fungi (e.g., (800)). Hobbie and colleagues (357) reported anomalies in the $\delta^{15}N$ signatures of *Picea sitchensis* foliage. They postulated that fractionation of the N isotopes with each transformation and exchange preferentially favored the higher rates of ^{14}N, leaving ^{15}N. Thus, the differential accumulation of ^{14}N in needles, ^{15}N in the EM fungal fruiting bodies, in comparison with the intermediate values in soil (and in saprotrophic fungi) can be used to determine N uptake rates in plants (49; 358). These represent two competing, but not necessarily mutually exclusive hypotheses that are useful and need additional work both at the individual physiological level and at the ecosystem level (discussed further in Chapter 7).

AM fungi are often presumed to have a simpler uptake pattern, but this may not be the case. Interestingly, AM associations may not fractionate N, at least not as much as EM (185; 356; 358). There is some enhanced NO_3^- uptake with increasing mass flow, but AM hyphae also search out and transport molecules of NH_4^+ (e.g., (794)). Whiteside and colleagues (773–775), using quantum dots, demonstrated the uptake of glycine, amino acids, and even complex organic N by AM fungi. AM also will partner with the mycorrhiza microbiome to enhance the breakdown and uptake of organic N (359) through a complex interactive process of priming and bacterial enzymatic activity. These processes, while simple, become exceedingly complex in the context of rapidly changing wetting and drying, shifting microbiomes, and diurnal growth and mortality of the extramatrical hyphae (346).

Ericoid mycorrhizae can include Ascomycotina species with the ability to break down complex proteins and humic substances to exchange with their host plant partners (598). These processes become especially critical in habitats like bogs, in which NH_4^+, the product of organic matter decomposition, is rapidly nitrified, then lost to the atmosphere through denitrification, or in extremely nutrient-deficient arid soils such as the Sand Plains of Australia, where every molecule is virtually essential. Orchid mycorrhizae are also comprised of Basidiomycotina partners that may have evolved from parasites and are capable of producing complex enzymes to extract amino acids from organic matter.

All plants forming mutualisms with N_2-fixing bacteria form dual symbioses with mycorrhizal fungi, with the possible exception of some lupines growing in extremely nutrient-deficient soils (discussed in more detail later in this chapter). Legumes and actinorhizal plants (forming a symbiosis with *Frankia* spp. Actinobacteria) are highly responsive to mycorrhizae and especially dependent upon mycorrhizae for P (361).

Fe Uptake

The uptake and utilization of Fe is especially complicated and unique. Fe is needed by all organisms and is especially important as the heme forming the center of cytochromes that function as the electron transfer agents in photosynthesis and respiration. Fe is a common mineral in soil, but the vast majority of Fe exists in the Fe^{+3} form and is insoluble at neutral pH. To access Fe, bacteria and fungi, including EM, AM, dark septate, and ericoid mycorrhizal fungi produce siderophores (Figure 4.2). Siderophores are interesting small organic compounds that chelate Fe, allowing these Fe molecules to be transported across cell membranes. Szaniszlo et al. (694) first reported the production of siderophores in EM fungi. Haselwandter and colleagues (e.g., (327)) found hydroxamate siderophores in many types and species of mycorrhizal fungi. These complex compounds form around insoluble Fe^{3+}. The complex is soluble and can be transported into the cell, where the Fe^{3+} is reduced to Fe^{2+}, the available form, so it can be incorporated into needed compounds.

Siderophores of AMF have been more challenging to study. Porter et al. (575) found that in a disease complex of soybeans, symptoms resembled those of Fe deficiency, and also were patches where AM fungi were lost due to high water and low O_2 tension. They speculated that the AMF may produce siderophores facilitating Fe uptake. The ability of AMF to produce siderophores had not been demonstrated until 2017, when Winklemann isolated glomuferrin, a Fe-complexing hydroxamate siderophore isolated from Glomeromycotina (785). This work opens up a new research topic area that has major implications for plant nutrition, especially for agriculture and plant production.

Stoichiometry, Carbon, and Dynamic Exchanges

Photosynthesis is the key process providing energy, in the form of complex C molecules for plants and mycorrhizal fungi (as well as most food chains). Mycorrhizae directly regulate plant photosynthesis through the supply of water and nutrients, and indirectly by producing or affecting regulatory compounds such as phytohormones or other compounds such as aquaporins. These regulatory mechanisms deserve more study, but, for our purposes, we will simply acknowledge that they exist.

If the plant is hydrated, CO_2 is taken up through open stomata, where the size of the opening is determined by the water status of the plant.

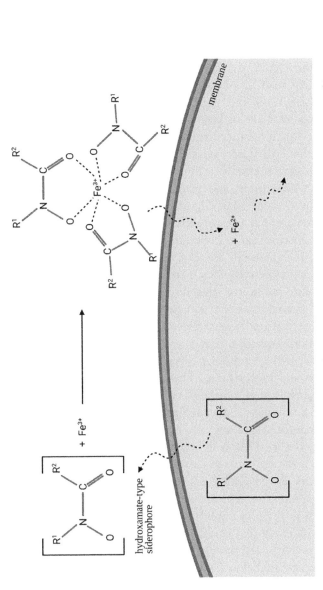

Figure 4.2 Hydroxamate siderophore uptake of insoluble Fe^{3+} and transport across membrane into cells, releasing Fe^{2+}, the useable form of iron (108). Figure drawn by Danielle Stevenson, with permission

Water use efficiency (WUE) is the rate of CO_2 uptake occurring simultaneously with H_2O loss. On average, there are around 300 to 350 molecules of H_2O lost per CO_2 gained. But the efficiency is highly dependent upon a number of factors. First, the pathway for photosynthesis matters. The C_4 pathway, evolved when global CO_2 was especially low, and the CAM pathway, evolved under drought stress, are both more efficient than C_3 photosynthesis. However, both direct measurements of CO_2:H_2O exchange and isotopic analysis using $\delta^{13}C$ values tend to show that mycorrhizae rarely make a large difference in WUE (393; 585; 586). What mycorrhizae generally do is to increase the uptake and throughput of water, resulting in greater photosynthesis, but not necessarily a greater efficiency in water use. This ratio may become more important with current and projected increases in atmospheric CO_2 (discussed in more detail in Chapter 9).

Light level also has a major effect on mycorrhizal activity and formation, generally through its effect on photosynthetic rates and on photorespiration. If light levels are low, the plant simply cannot fix enough CO_2 to support the symbiosis. Hatch (334), Björkman (122; 123), and Hayman (336) all noted that low irradiation, short daylengths, or shading, all factors that reduced light, reduced mycorrhizal activity. But that response was not necessarily linear. At high temperatures and low sunlight, photorespiration may also increase, and even equal photosynthesis. Under these conditions, there would be a reduced flow of sugars to the roots and mycorrhizae. This compensation point, when photorespiration equals photosynthesis, can occur at 300 ppm CO_2 and light levels of 10 μmol m^{-2} s^{-1}, a value that occurs for understory plants within a tropical rainforest canopy. For example, Whitbeck (771) found that AM activity declined below light saturation, <40 percent full sunlight, or about 800 μmol m^{-2} s^{-1}, to 20 percent full sunlight simulating a full canopy. Clark and St Clair (167) found that light limited EM formation in late-successional aspen. In a growth chamber, I found that light <400 μmol m^{-2} s^{-1} can inhibit mycorrhizal formation in a C_4 plant (29; 68). It is likely that there is a biphasic response, with little impact down to somewhat less than half sunlight, with an increasing impact on mycorrhizal activity below plant canopies. However, understory-adapted species may well compensate by a number of strategies, such as greater chlorophyll concentrations to sustain plant growth. More work is needed to sharpen our understanding of light limitations to mycorrhizal activity, especially in understanding understory environments. For example, in a tropical rainforest with a leaf area index of 7, how does an EM *Gnetum* seedling support its *Scleroderma* fungal partner, before the mature plant reaches sufficient light in the canopy? (see Plate 2). Although one possibility is that seedlings are supported by CMNs (361; 706), there are no other EM plants near the *Gnetum-Scleroderma* patches I studied. We also know that stomatal response in mycorrhizal plants can be more sensitive and rapid to changing light levels than in nonmycorrhizal plants (74). How does this dynamic relate to energetic processes such as sun flecks and cloud gaps that reflect additional light creating enhanced photosynthetically-active radiation (PAR)? (370).

C$_4$ plants tend to be more responsive to mycorrhizae than C$_3$ plants (50; 348). Photorespiration and the compensation points may also explain why C$_4$ plants tend to be more responsive to mycorrhizae than C$_3$ plants. During the late Miocene into the Oligocene, and even into the Pleistocene Epoch, atmospheric CO$_2$ dropped to as low as 190 ppm. Under these conditions, C$_3$ plants could have had a higher compensation point, and therefore less C to allocate to their mycorrhizal partners. C$_4$ plants, which do not photorespire because of the spatial separation of photosynthetic steps, would have maintained a positive C fixation and maintained its mycorrhizal relationship. The Miocene Epoch was a time when C$_4$ grasslands expanded. The ratios of carbon to light and H$_2$O matter!

C:Nutrient Relationships

The importance of mycorrhizae was originally ascribed to the ability of mycorrhizal fungi to regulate nutrient uptake. Plants require nutrients at a critical concentration in order to fix CO$_2$, thereby providing the energy that drives terrestrial ecosystems. Most research to date has failed to find a major change in the physiological efficiency of nutrient use with mycorrhizae, but more likely, increased plant biomass responds to increased uptake. The exception could be increased leaf P$_o$:P$_i$ ratios in a C$_4$ grass with AM (73). More work is needed to test this hypothesis, but here we will focus on increased uptake for analyzing C:nutrient ratios. Based on the Redfield ratio, as we increase P or N, C fixation can increase, and thus production. We can build on the models of Bosatta and Ågren (4; 132) and Treseder and Allen (719) (Figure 1.1) to develop an interactive framework for studying growth rates and mycorrhizae. Importantly, this model is built upon internal nutrient concentrations; it is the P within the plant that regulates growth (e.g., (632; 668)). N uptake is linear with C up to saturation (Figure 4.3). However, the

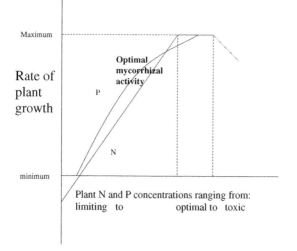

Plant N and P concentrations ranging from:
limiting   to   optimal to toxic

Figure 4.3 Structure of Ågren model for relative plant growth (R_w) against Nitrogen (N) or phosphate (P) at the minimum for growth (N or $P_{,min}$), optimum (N or $P_{,opt}$), and toxic (N or $P_{,tox}$) for N, a linear response, or P, a curvilinear response, and the range for optimal mycorrhiza formation, just below optimal where the plant needs the mycorrhiza to provide nutrient resources (25).

P increment is curvilinear, not linear, as P concentration is generally logarithmically related to C fixation. As mycorrhizae increase nutrient uptake, the plant growth rate can increase relatively, minus the C traded to the fungus for the extra nutrients. Hayman (e.g., (337)) and Mosse and colleagues (e.g., (522)) emphasized the importance of N:P ratios, not just concentrations, as critical to mycorrhizal dynamics (386).

Because the mycorrhiza is a symbiosis (plant + fungus), we can build interactive models that describe how a mycorrhiza allocates resources gained (75). The equations are simple, for example:

in a nonmycorrhizal plant:

$$C_{plant} = 35^* (N_{plant}), \text{ and as plants take up N as } NO_3^-, \text{ then:}$$
$$C_{plant} = 35^* (NO_3^-);$$

in a pasture where plants are AM:

$$C_{plant} = 35^* [(NO_3^-) + (NH_4^+ \text{ acquired from AM})];$$

And in a forest with AM and EM trees:

$$C_{plant} = 35^* [(NO_3^-) + (NH_4^+ \text{ from AM} + EM) + (N_{Organic} \text{ from EM})].$$

In the case of N, C increases linearly to the leaf concentrations of N.

Also important, the N traded by the fungus to the plant is in excess of what it needs for itself in order to utilize the C gained. So that

$$C_{fungus} = 13^* (N_{fungus}) + C_{(from\ plant)}, \text{where}$$

$$C_{from\ plant} = [C_{for\ plant}] - [C_{traded\ for\ (NH4+from\ AM\ or\ EM)}].$$

Interesting physiological response models can be constructed from these relationships for the growth of individual plants (see (75) and citations therein) and will be discussed in greater detail in Chapters 6 (Community) and 7 (Ecosystem).

Just as increased N and P uptake can drive greater CO_2 fixation, so can increasing atmospheric CO_2 drive nutrient deficiencies, and greater mycorrhizal activity. Driving down N or P can drive down R_w (Figure 4.3), but it can also stimulate mycorrhizal response (Figure 1.1). These drivers are, and have occurred within, evolutionary time frames. Under conditions of low atmospheric CO_2, simulating pre-industrial conditions, N and P were less limiting than CO_2 to plant growth, and mycorrhizal activity declined (37; 61). However, with increasing atmospheric CO_2, as N and P became limiting, mycorrhizal activity increased, N uptake by EM fungi increased, and N_2 fixation increased (61). Models projecting impacts of elevated CO_2 imply that mycorrhizae will play a major role in mediating nutrient limitations and in sequestering C (see (579), Chapter 9, Global Change).

C Uptake and Allocation

Under conditions of increased water throughput, increased stomatal conductance, and increased N and P from mycorrhizal associations, rates of photosynthesis and plant productivity increase. Both EM (604) and AM (74; 449) fungi increase photosynthesis in mycorrhizal plants (403; 791). In the case of photosynthesis in a C_4 grass, mycorrhizae increase uptake of CO_2 up to 68 percent, with 51 percent due to gas-phase resistance from increased stomatal conductance, and 77 percent due to a reduction in liquid phase resistance. The reduction in the liquid phase resistance is complex, but an increase in chlorophyll a/b ratio suggests an increased number of photosynthetic units, which could be due to increased P and N, or even the efficiency of P use (higher $P_o:P_i$ ratios (73)). The extent of the influence of mycorrhizae on photosynthetic rates is highly variable, and in need of multi-factorial modeling to separate the particular conditions of when and where mycorrhizae enhance photosynthesis and the variability across both short and long time periods (e.g., (121)).

While mycorrhizae increase photosynthesis in almost every study published, the extent of increased plant biomass is highly variable. There are several C sinks in a mycorrhizal symbiosis that must be integrated to assess mycorrhizal responsiveness. First, improved plant shoot growth has often been observed. In other cases, plant root biomass increases in response to increases in photosynthesis (73), although root biomass is not commonly measured, and assessing root productivity is challenging because of the soil itself. The mycorrhizal fungal hyphae themselves create a C sink, both in hyphal growth and in mycorrhizo-sphere respiration. The mycorrhizosphere doubled in the atmosphere CO_2 of the soil compared with the rhizosphere of a NM grass (418). When the partial pressure of CO_2 increases, CO_2 dissolves into H_2O, creating HCO_3^- plus H^+, thereby acidifying the soil and increasing the solubility of P. Individual AM hyphae can grow up to a rate of 100 μm mm^{-3} soil h^{-1} (60), and with up to tens of meters of hyphae cm^{-3} soil (62; 503), the mycorrhizal C sink could be substantial.

Somewhere between 5 and 40 percent of the C fixed in a mycorrhizal plant is allocated to mycorrhizal fungi, with a great deal of variation based on experimental conditions, taxa of the partners, and type of mycorrhiza (e.g., (239)). We will explore the quantification of C allocation in Chapter 7, but the physiological composition of biosynthesis and transfer is important to understanding mycorrhizal physiology.

For EM fungi, carbon allocation interactions can be extremely com-plicated. Martin and colleagues (482; 483) reported complex biochemical transformations in the sugars transferred from host to plant, going from glucose to mannitol to trehalose. This pattern is important. If the sugars were to stay as glucose, the plant could reabsorb them. But, as trehalose, they must originate from the fungus, as this sugar is not produced by plants (451). Research has long coupled the reciprocal relationship between transport of P from fungus to host, and the transport of C from host to fungus (as an example, see (632)), especially in AM.

For AM fungi, between 5 and 10 percent of the fixed C appear to be exchanged from host to fungus (e.g., (435; 553)). But how that C is transported is controversial. Allen and Allen (44) and Nakano et al. (525) found that the $\delta^{13}C$ values were depleted (more negative) relative to the plant tissue. This means that the C of the AM fungal spores equaled that of root lipids, rather than of leaf photosynthetic C sugars. Lipids are known to be the primary constituent of spores (111). One explanation is that this signature might be due to fractionation during the conversion of hexoses to lipids within the plant followed by transport to the fungus. Subsequent work by Bago (97) indicated that lipids were the primary form of C moved through the hyphae and into spores, and Pfeffer et al. (564) found uptake of glucose and fructose (but not mannitol or succinate) into hyphae for growth, but little or no lipid synthesis by the fungus. Helber et al. (343) reported that the hexose transporter was different from the P transporter, followed by the work of Bravo et al. (138) showing that lipids appear to be synthesized in the plant and transported to the AM fungi. Two key points here are that AM fungi need lipids from the host as a primary C source, which makes them obligately dependent upon the plant, and that AM fungi transport P to the plant along one pathway, while the host plant transports C in the form lipids, all using different transport proteins.

Alternatively, Klink et al. (410) found ^{13}C enrichment of AM fungal hyphae (less negative $\delta^{13}C$). This pattern suggests that simple sugars are likely transported from the plant to construct the hyphae and carry out most physiological processes. Fixed CO_2 simple sugars have an enriched $\delta^{13}C$ signal compared with the whole plant tissue as subsequent biochemical reactions preferentially convert ^{12}C (146). What these data tell us is that we really only marginally understand the basic physiology of mycorrhizal fungi and many open development by physiology questions remain.

Of concern here is that many open questions remain on how the two symbionts maintain a mutualism. How is the mutualistic exchange regulated? Are there different temporal dynamics? For example, growth of the fungus primarily occurred in the afternoon, as photosynthesis cranked up (346). Or are P and C hooked together energetically in some unknown fashion? Or do both individual transporter mechanisms simply occur based on immediate demands, and the overall exchanges either work out, or one partner shuts down activity in response to homeostatic demands? Does the exchange involve different C molecules (lipids versus sugars) for different life stages?

Responses in Extremes

One way to evaluate the physiological functioning of mycorrhizae is to examine the environmental range in which the symbiosis forms or how plants adapt to nutrient acquisition under extreme conditions. NM plants have evolved from AM plants under conditions that promoted or facilitated plant fitness. Fungi have a higher N concentration (lower C: N ratio) than plants. At an extremely low soil nutrient concentration, when the fungus is in an environment of low nutrients, the fungus would itself be N or P limited and would therefore be unable to share with a host. Infection would be limited (based on the Treseder and Allen model, Figure 1.1). Examples of this condition include extreme nutrient-deficient ancient soils, such as found in Australia and South Africa, wetlands, and newly formed volcanic material, such as pyroclastic materials. Many, if not most of these soils, tend to be silty to sandy, with larger soil pores. Here, some plants form proteoid cluster roots that probe soils for N and P (see (438)). Cluster roots also produce high levels of organic acids that mine soils, dissociating P from cations (Fe, Al, Ca) and extracting phosphate (see (133)). Plants in places like the Kwongan Sand Plains in Australia also form ericoid mycorrhizae with fungi that break down organic matter (438). Other moist environments such as the heathlands in England and northern Europe have extensive stands of *Calluna vulgaris* forming ericoid mycorrhizae. In other cases, a plant will acquire N from neighbors, forming mutualisms of AM or EM with neighbors that are symbiotic with N_2-fixing bacteria, *Rhizobium* and *Frankia* (512; 513). There are examples in nature from very different ecosystems that also demonstrate adaptations at this end of the spectrum. At Mount St. Helens, the newly formed pyroclastic flow material consists of Silicon but little else. The initial colonizing plants were lupines that form nodules with N-fixing bacteria to acquire N_2 from the atmosphere, and extensive roots to scavenge for P and other soil resources.

Finally, a plant can become a parasite. There are two types of parasites. The first is a stem epiphyte, penetrating the xylem to obtain water and nutrients flowing up the xylem stream. The second is a phloem parasite, generally splicing into both the xylem and phloem of the roots of neighboring plants to steal C, water, and nutrients. A plant can even become a predator, forming carnivorous glands to capture protein-rich animals such as insects or nematodes!

At the high end of soil resources, the plant can simply acquire those resources without the fungus, and actively rejects the fungus or simply outgrows it. Just as in the low end of the spectrum, infection of the root system by the mycorrhizal fungus is minimal. Examples of this response are common in fertile geologically young soils such as California or Chile, where it is difficult to find an area with a P limitation. Another example is artificially-fertilized agricultural systems and areas of N deposition from atmospheric pollution. Fertilization has long been studied as a mechanism reducing mycorrhizal activity (e.g., (1)). Indeed, there is increasing evidence that plant breeding under P fertilization has led to less mycorrhizal-responsive varieties and crop management systems (e.g., (572)). In wildland ecosystems, recent studies of N deposition have shown a relative reduction in mycorrhizal infections (see Chapter 7 on Ecosystems).

Some plants even evolved the ability to live as nonmycotrophs in disturbed arid land soils, where inorganic nutrients are not leached due to low precipitation, leaving nutrient rich soils for a short period following disturbance. Examples of this are annual plants in the Chenopodiaceae and Brassicaceae that evolved in arid and semi-arid regions and frequent disturbance. These plants play a large role in understanding basic plant physiology. *Arabidopsis thaliana* has become the most important "model" plant for understanding plant genomics. However, it is a nonmycotrophic species, without the mycorrhizal associations common to the vast majority of the world's plants. In studying basic plant physiology, the unique highly evolved trait of nonmycotrophy must always be considered when extrapolating from "model" systems.

Connectivity and Complexity Implications

In this chapter, I have emphasized how one plant interacts physiologically with one fungus. However, both plants and fungi can often be challenging to study in isolation, especially in the field, and a more realistic situation is that multiple fungi interact with each host plant. Often, we cannot even define what an individual is (see Chapter 5). For the purpose of understanding physiological responses between multiple fungal and plant partners, it is crucial that we begin to determine what the extent of those interactions can be.

Because each plant has multiple roots growing in different directions, in the field each plant will interface with multiple mycorrhizal fungi. Interestingly, there is little research in this area. By connecting with different fungi, a plant presumably should be able to differentially tap multiple sources of nutrients such as N and P (75), as different taxa have different exploratory and enzymatic capabilities. Despite this level of understanding, few data actually exist to test this important concept. There are now numerous datasets showing that different fungi confer different benefits on a host, but are these additive or inhibitory? Van der

Heijden et al. (341) and Klironomos (412) showed a curvilinear relationship between plant growth and fungal species richness. All of these studies exist at the community, but not at the physiological level. The only dataset of which I am aware comes from one of our studies of seedlings planted out from a host *Quercus agrifolia* in the field. Here, the N concentration and δ^{15}N of the leaf tissue were measured against the numbers of EM tips and the richness of EM fungi. Leaf tissue N trended

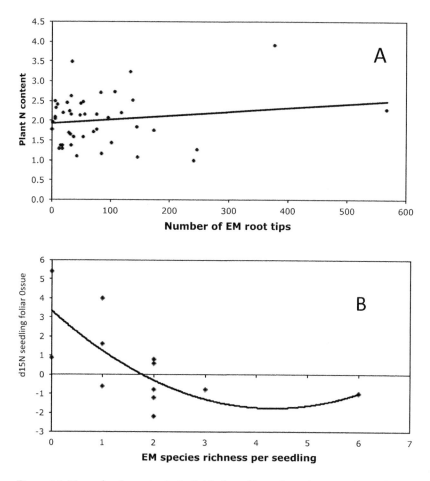

Figure 4.4 N uptake dynamics in individual seedlings planted surrounding a large oak (*Q. agrifolia*). As the number of EM root tips per seedling increased (A), the total N uptake increased ($p = 0.12$), but with the increasing richness per seedling (B), the forms of N transferred significantly changed ($p = 0.047$). See text for additional mechanistic detail. (data from Allen, unpublished)

towards increasing with the total number of EM tips per plant. But when we examined the $\delta^{15}N$ values, those seedlings with no EM resembled the $\delta^{15}N$ of the overall soil. However, the more $\delta^{15}N$ was fractionated the greater richness of EM fungi per seedling. This suggests that a greater proportion of N was taken up by EM fungi, different sources of soil N were being accessed by the different fungi, and/or that greater fractionation of N occurred showing a preferential transport of lighter NH_4^+ to the host and retention of organic ^{15}N by the fungus (Figure 4.4). At the scale of an individual plant, EMF richness matters.

What about the alternative? Does the growth or fitness of an individual fungus improve when connected with multiple plants (445)? Results to date are equivocal (e.g., (445)). Even the well-known cover photo of the 1996 edition of the Sally Smith and David Read (667) book emphasizes that a mycorrhizal fungus is really not always a microorganism. It can be a macroorganism that narrows down into microscopic extensions. By doing so, a fungal individual interacts with multiple plants potentially extending across large land surfaces (see Figure 2.2). There are datasets that imply that individual fungi gain from connecting with multiple plants. I found the same species of fungi living with C_3 grasses early in the spring, and then living with C_4 grasses late into the summer and autumn (50). These fungi were taking advantage of a very long growing season by tapping photosynthesis from plants with both pathways. Isotopic analyses of AMF spores also showed a mixed isotopic ratio (44), showing that spores from the rhizospheres of a mixed stand of C_3 and C_4 plants were using C from plant hosts with both photosynthetic pathways.

Summary

- A mycorrhizal plant is physiologically not simply a NM plant plus elevated levels of N or P. Once a mycorrhizal infection occurs, the plant is physiologically changed in a myriad of complex ways.
- In a simple example, in response to higher nutrient levels, more CO_2 is needed, hence more water throughput is needed to keep stomata open. But then the control must be tighter, altering hormonal balance. To fix more CO_2, a corresponding increase in N must be taken up and incorporated into enzymatic machinery.
- Mycorrhizae increase N uptake by short-circuiting the N cycle, taking up NH_4^+ and organic N, depending upon the environment.

- Mycorrhizae increase N_2 fixation by bacterial symbionts by supplying the plant with additional P, Fe, and Mg, to increase photosynthesis.
- To fix more CO_2, even more Fe and Mg are needed, requiring siderophore production. To get more P, exploratory mechanisms are needed, but the hyphal tip must also extrude enzymes to break down organic P and must increase local respiration and organic acid production to make P available for uptake.
- Because one fungus does not satisfy all the needs of a host in a complicated community, the plant takes on multiple fungal partners, sometimes multiple mycorrhizal types simultaneously. Similarly, a fungus that taps the C resources of multiple plants insulates itself from the vagaries of death of a single individual. From this multiplicity of connections, we generate a complex and interactive community that is very different from the individual physiologies that we have measured in glasshouse pot experiments.
- To understand these complexities, *it is time to move our experiments the field.*

5 · *Population Ecology*

The study of population ecology in plants is as old as the field of ecology but is more complex for fungi. Due to their microscopic morphology, identifying individuals for measuring and modeling is challenging. We have examined the general morphology of fungi and plants comprising the mycorrhizal symbiosis, and we have looked at the larger-scale evolutionary patterns that resulted in the mycorrhizae that we observe and study today. However, selection acts on the individual organism (472). An organism survives to reproduce offspring that in turn reproduce, or it does not: a binary outcome. And, an organism is comprised of a complete genetic code that allows it to survive to reproduction (or not). The structure that the genetic code produces, or the body, has to grow and reproduce based on the genetic expressions of that code. Charles Darwin, Alfred Wallace, and colleagues (193) first described the importance of dynamic change in individual organisms coupled with the environmental impacts of natural selection as the agent resulting in the differential survival in 1858. In the study of mycorrhizae, we need to understand not only the differential changes in gene frequency of taxa of organisms with time but also the differential changes in both organisms comprising the symbiosis or the co-evolution of both taxa within the constraints dictated by the environment. To understand the ecology of a mycorrhizal mutualism, we must focus on the interaction of a plant with a mycorrhizal fungus. Then, because a plant typically forms a mycorrhiza with a second mycorrhizal fungus, which is mycorrhizal with a second plant, and so forth, we immediately have a community. We already know much about the population to community biology of plants. We know far less about fungi. For that reason, I will focus initially upon understanding the dynamics and characteristics of fungal population biology. Then we can scale up to population dynamics and to community interactions, then to landscape patterns and the symbiosis (Figure 1.2).

What Is an Individual versus a Population?

To describe population dynamics of mycorrhizal symbioses, we start with defining an individual, then a population within an environment. Our first issue: what is the selection unit? Our second: what is the selecting force? Within that issue, our first effort is to define what is an individual. For most of the best-studied animals and plants, this appears initially as relatively simple. An organism is an autonomous entity that has the properties of life and largely a 2N (diploid) state. You, the reader, are an individual, as am I. But is an "individual" aspen tree an individual stem, or is it the clone? A single *Pinus longaeva* (bristlecone pine), such as the Methuselah tree in the White Mountains of California, which is nearly 5,000 years old, is considered as a single organism, even though most of the structural material consists of dead tissue, and most of the live cells are likely not more than a few years to decades old. Clones of plants such as *Populus tremuloides* (aspen) can consist of thousands of stems (apparent trees), and individual clones of *Larrea tridentata* (creosote bush, an AM plant) or *Quercus palmeri* (Palmer's oak, an EM plant) have C that has been estimated to have lived as far back as the Pleistocene Epoch, between 11,000 and 13,000 years old (489; 747).

How old are their mycorrhizal symbionts? And, what is an individual symbiont, versus a clone, versus a population, versus a species (see (596) for further discussion)? EM fungi form sporocarps wherein compatible isolates anastomose (fuse) as hyphae, exchange nuclei, undergo meiosis, and, within the sporocarps, produce 1N spores. Spores disperse, either by wind or animals (occasionally by water), and germinate, forming an individual. A population is comprised of compatible units that grow independently. Even EM fungi in the order Endogonales, Mucoromycotina, although aseptate, form zygospores. New individuals arise when two- 1N spores with compatible nuclei germinate, anastomose and grow into a larger organism.

Alternatively, Glomeromycotina, like other Mucoromycota, are aseptate. That is, they only form adventitious septa, or cross-walls, in response to damage or some other need to seal off a hyphal fragment. When an individual hypha is severed, it dies from the damaged point to the tip. The fungus forms an adventitious hypha and either grows from that point or even reconnects with its original old hypha (280). Presumably, these germ tubes and the resulting hyphae largely exist in the haploid (1N) state. In the case of the Glomeromycotina, nuclei are scattered throughout a single spore, and migrate down the germ tube (e.g., (173)) and throughout the hypha (Plate 3). There is considerable variation in the size of nuclei, leading to the hypothesis that there are both haploid and diploid nuclei. Is the population of 1N nuclei homokaryotic (the nuclei are identical) or heterokaryotic (the nuclei are different)? What is the reason for multiple sizes in nuclei? Some researchers argue that since no reports of sexual reproduction have been verified, these are likely ancient, nearly asexual organisms (see (621; 634)). The assumption is that genetic variation within individual hyphae and spores comes from heterokaryosis, caused by mitotic recombination (parasexuality), or even the anastomosis (hyphal fusion) of compatible, but genetically different nuclei. Importantly, Glomeromycotina do have genes for meiosis (442) and a diversity of sex-related genes (620), implying that some form of sexual reproduction has the potential to occur. Certainly, the azygospores in the Gigasporaceae bear a morphological resemblance to zygospores of the Mucoromycotina. Thus, a key first research argument in determining an individual versus a population centers around whether an individual spore, with hundreds of nuclei, is homokaryotic or heterokaryotic. Pawlowska and Taylor (554) provided experimental evidence that four fungal isolates of *Glomus (=Claroideoglomus) etunicatum* from a single field site were homokaryotic. They suggested that the variation in nuclear size could be accounted for by polyploidy, but likely not heterokaryosis. Alternatively, Kokkoris et al. (426) acknowledged that both parasexual and sexual reproduction is likely active.

The Pawlowska–Taylor (see also (699)) view represents a significant change in the perspective of genetics and population biology of fungi broadly. Decades of research has been published on the role of heterokaryons in regulating the ability of parasitic fungi to infect host plants (8), EM fungi to form EM (227), or, alternatively, the effectiveness of the individual EM formed (432). Alternatively, homokaryotic basidiospores of the EM fungus, *Rhizopogon roseolus*, formed effective EM (635).

This debate remains ongoing and is a core arena of research on mycorrhizal fungi, and fungi in general (142; 486). Two questions emerge relevant to our discussion here. First, in what state do mycorrhizal fungi normally exist, as homokaryons or heterokaryons? Historically mycologists presumed that the majority of fungi were largely found in nature as heterokaryons. This idea has been challenged (699). This is crucial, because a homokaryon likely has one genetic response to an environmental cue, with two (nearly) identical 1N nuclei per cell, while heterokaryons could have two different genetic responses, from two different (but compatible) 1N nuclei per cell.

Recently, Kokkoris et al. (424) found that heterokaryotic dikaryons, while infrequent, have a higher number of nuclei and shift with host plant composition. At this point, I will only comment that I find defining the terms of identification of an "individual" and replication of populations and taxa, and the constrained regions of the genome to remain among the most important issues to be resolved for understanding mycorrhizal population biology. Extensive additional genetic work on fungi in general, and mycorrhizal fungi in particular, continues to be needed. Additional work is on-going by excellent scientists, and I expect many exciting new analyses by the time this is published. Second, how are differential gene frequencies of mycorrhizal fungi affected by changing environmental conditions? Theory is well described, but there is a need for experimental evidence.

Genetic Structure

Genetic structure in fungi is not straightforward. While there is a focus on nuclear and mitochondrial genomes to better understand systematics and population biology, extra-nuclear elements must have important impacts on the biology of the symbiosis. All higher organisms carry bins of DNA and RNA from viruses, prokaryotes, and previous other taxa (e.g., I have 278 Neanderthal variants in my genome). Both viruses (e.g., Turina et al. (728) in AM and Vianio et al. (732) in EM) and endosymbionts, presumably prokaryotes (129; 130; 469), must have rather dramatic effects on the biology of individuals infected. In other relationships, for example the chestnut blight fungus, *Cryphonectria parasitica*, a double stranded RNA virus can render the fungus hypovirulent and can transfer horizontally between both compatible and incompatible isolates (266; 314). Viral and bacterial endosymbionts of plants are really only beginning to be studied, and for fungi are almost completely unknown. The endosymbionts illustrated by MacDonald et al. (469) range from 0.2 to 0.5 μm, a size range that may require them to be obligate symbionts, without adequate space for a full genetic compliment necessary for independent growth and survival (421; 730). It is likely that both viruses and smaller endosymbionts are obligate inhabitants of all eukaryotes, including fungi. Moreover, they must have effects on the host populations of both mycorrhizal fungal and mycorrhizal plant-mutualists. In an innovative study, Di Fossalunga and colleagues (203) showed that an endosymbiotic *Candidatus* within the AM fungus *Gigaspora margarita* modulated adaptation to oxidative stress.

Mycorrhizal formation does alter the formation of disease in host plants. This phenomenon is well known, but mechanistically not well understood. Newsham et al. (530) proposed that one major function of AM was to protect plants from disease. I will follow up on this topic more in describing the importance of mycorrhizae in community-scale processes (Chapter 6). However, mycorrhizal infection can, in some cases, trigger PAL synthesis and phenol and lignin formation (26), and PAL protects tissue from many other pathogens. Thus, by triggering phenolic activity, AM protect plants indirectly from a wide range of pathogens.

Disease in fungi must be as prevalent as in animals and plants and an equivalent regulator of populations, but almost nothing about it is known. Species of *Hypomyces*, a genus of parasitic Ascomycete, are commonly observed amongst EM fungal sporocarps. An example is the lobster mushroom, *Hypomyces lactifluorum*, that overgrows a number of EM sporocarpic taxa, including species of *Russula* and *Lactarius*, turning the mushroom orange-red (and often more tasty!). But I have not seen quantification of the impacts of this or any other disease in mycorrhizal fungi.

What Is an Appropriate Taxonomic Unit for Population and Community Analyses?

So, what is an individual or a population that we can use for population or community modeling? The advent of amplification and sequencing technologies revolutionized the ability to define taxa (772). Can we identify directional selection, differentiate mutation from parasexual reproduction from sexual selection, heritability, and genetic drift in fungi if we can't identify individuals? In most cases, these are fertile fields for further research. However, we can begin to address the issues of population differentiation, r and K selection, demography, and predation – characteristics of populations that determine community and ecosystem dynamics. But how should we approach this?

A traditional view builds on the study of a species. But what is a species? In the study of the great apes, we (*Homo sapiens*) are between 96 and 99 percent genetically similar to the chimpanzee (*Pan troglodytes*), which we consider to be a separate and very different species. In contrast, many comparative molecular analyses of fungal communities use a cut off of 96 percent sequence similarity to characterize a taxon between sites, even between two sporocarps that appear

morphologically identical. One key issue is that the hierarchical structure upon which we base taxonomy, the Linnean classification system,
predates the concept of evolution. A species has largely been defined as
the largest group of organisms in which any two individuals can
produce fertile offspring. With most fungi and many plants, we cannot
differentiate an individual. So how can we define a species, especially
without knowing if they can produce fertile offspring? Nevertheless, the
taxonomic system is useful for describing a variable range of similar,
identifiable characters (if morphological) or distinct (to a defined level)
sequences (dependent upon the region utilized). With experience
enough to know better than to extend myself more in both areas, here
I leave these issues to those specializing in taxonomy and evolution to
wrangle through (Box 5.1).

Fungi expand across space and are clonal. Individuals, upon which
selection could act, might occupy merely a single root segment. Other
individual clones, such as those forming fairy rings, may extend many
meters, and might persist for centuries. However, like individual stems of
clonal plants, individual hyphae survive for only short time periods.
Individual hyphae may live for only hours, months, or years. Beyond
individual connected clones, spores ranging from haploid spores with
single nuclei, to chlamydospores with hundreds of nuclei, can be dispersed at least several kilometers by both animals and by wind. The key
question for population biology is at what scale does reproduction
and evolution act? When two germ tubes emerge from a single spore
(Plate 3) and infect two different plants, maybe even two different species
of plant, can they diverge, causing differential responses? When spores are
dispersed long distances, presumably there are different selection forces.
In these cases, how much evolutionary change occurs, and is it measurable? For population genetics, the definition of evolution is simply a
change in gene frequency. But such studies have not been undertaken for
mycorrhizal fungi, and they remain crucial questions to understanding
their evolution.

We can even take this question beyond a single, connected clone.
Many spores of fungi are asexual. That is, they are derived mitotically,
with one to many nuclei that are genetically "identical." I put "identical"
in quotes because mistakes occur in mitosis and mutations inevitably
arise. Is a dispersed chlamydospore of a *Glomus* or a sclerotium of a
Cenococcum or even a conidiospore of an *Oidiodendron*, part of the original
clone, or a new individual? For population modeling, this is an important
question as all models depend upon defining an individual.

Box 5.1 *Systematic Constraints within the Glomeromycotina*

We need some structure to understand how mycorrhizal plants and fungi relate to each other and interact with their ever-changing environment. For the rest of this book, I use a cladistics approach. A clade provides the reader with the name of a group I am referencing, and also implies the limits to the knowledge of phylogeny of the organism. In some cases, especially for EM fungi, a species is reasonably defined and can often be morphologically and molecularly distinguished. Many Mucoromycota and most Ascomycota and Basidiomycota have sporocarps that have been defined by species (see any good fungal identification guide). Others have been differentiated using molecular characteristics ranging from restriction fragment length polymorphisms (RFLP) to newer sequencing approaches using identified regions, such as ribosomal DNA internal transcriber (ITS) regions and ribosomal small RNA or mitochondrial large subunit RNA (rDNA) genes. For example, Kårén et al. (400) differentiated both inter- and intra-specific populations of 44 species (7 genera) of EM fungi across the Fennoscandia region using the number of base pairs (length) of a defined region of the genome after cutting with specific endonucleases. For others, improving libraries of sequences are available. In other fungi, especially for AM fungi, there are few good morphological characters. The Glomeromycotina are difficult to work with for sequence differentiation, and there are even intra-individual issues, as discussed above. Stockinger et al. (683) found that the intraspecific variation makes a species challenging to barcode. Even a morphologically distinct species, for example, many taxa, including *Glomus intraradices*, *Gl. clarum*, *Acaulospora laevis*, or *Scutellospora calospora*, commonly identified across communities, are taxonomically probably species clusters based on their phylogenetic trees. Future research may further define individual populations taxonomically. Using the term *Scutellospora* does not tell the reader all the characteristics of the fungus, but it does convey a level of useful information. It is more specific than saying Gigasporaceae, not as much as referring to *calospora*. Each level is a useful taxonomic grouping and provides useful information. In many cases, the genus into which many individual species best fit remains unknown. Importantly, the creation of these new genera has minimal impact on the descriptions of the species and the presumed critical biologically descriptive unit of taxonomy (e.g., (434)). However, it is important to remember that Linnean classification preceded Darwin and Wallace by a century. As pointed out by Mishler (507), a species is not a "uniquely real biological entity" based on our understanding of evolution. It is a degree of similarity, genetic or morphological, that the author deems appropriate. A clade approach, using clades as the descriptive level, allows the reader to understand the limits of knowledge about the organism that the author is describing (65). As Jim Trappe once noted, erection of a new taxon is a hypothesis undergoing testing. It is important for the reader to understand that significant and continuing shifts in the phylogenic relationships and taxonomy are occurring even as I write this page, as we better utilize molecular systematics approaches and survey the world's diversity. The future of organismal identification rests with these new approaches. But it is equally important to recognize that older taxonomic schema represent the extent of knowledge at the time the research was undertaken. Where possible, I will add new taxonomic information, but a rank-free, or clade, approach allows us to examine ecological information within the context of the original research. As examples, known family and genus and species can be placed in a cladogram that indicates relationships, such as those of *Scutellospora pellucida*. In older treatises what we considered a family in 2015 was a genus in 2005. The cladogram does not explicitly denote which level is used but does indicate relationships within the groups. The genus *Glomus* is even more challenging after it was split into many genera, such as a new *Glomus, Rhizophagus, Claroideoglomus*, etc. Most of these have the name *Glomus* in the pre-2010 literature, when Schußler and Walker (647) erected their new structure. An example is the taxa of Glomeromycotina that are common in the American southwest deserts. These were characterized as *Glomus*, which in many cases were as accurate an identification level as could be completed. In denoting something as *Glomus*, I am working with the best taxonomic level that was undertaken at the time the paper was published. I will use the clade concept for the remainder of the book. With this approach, I allow the reader to make an individual decision about the scale of information on which my selection of organism identification is based. Recognize that my use of taxonomic descriptors is limited to that particular clade, and not a more detailed resolution.

Environmental Impacts on Population Differentiation

This has been a controversy and challenge since early efforts to reforest (or forest, where none previously existed) using EM fungi. For example, Goss (296) in Nebraska, Briscoe (140) in Puerto Rico, and Anon (82) in Rhodesia (Zimbabwe) inoculated trees with exotic fungi to initiate EM and establish plantation forests. Inoculated taxa have been distributed worldwide. An example is *Pisolithus* spp. Don Marx and colleagues (485) demonstrated that *Pisolithus tinctorius* (Pers.) dramatically improved the growth and survival of conifers in reclamation efforts. They planted pine trees inoculated with an isolate, *Pt Marx 270*, around the world, demonstrating its ability to establish reclamation forests. Today, we find *Pisolithus* spp. in temperate and even some tropical environments worldwide. I have found *Pisolithus* in mine spoils in Spain, a forest established in the grasslands of Nebraska, USA, far from any source EM fungal populations, and mine spoils in Pennsylvania. Is it native or inoculated? The fungus has many names, dead-man's foot and dyeball, and even horse dung fungus because of the difficulties many students (including some of my own) have in differentiating road apples (horse feces) from *Pisolithus* sporocarps. More recently, many distinct species of *Pisolithus* have been distinguished, especially following molecular identification techniques. And there are distinct differences in host range. In the 1980s, in the very early stages of molecular analyses, we looked at the size of the ITS fragment lengths of sporocarps from around campus and field stations in southern California. The Monterrey pines had one genotype, Australian eucalyptus another, and both had different ITS sequences from native Fagaceae (both *Quercus*, oaks, and *Chrysolepis*, chinquapin (41)). These are quite different taxa. The bluegum eucalyptus EM likely originated from the soil brought in with eucalyptus seedling transplants in the 1860s as part of California's homestead act; the Monterrey pines were likely inoculated with *P. tinctorius*, possibly even Don Marx's isolate; and native *Pisolithus* taxa were associated with native oaks and chinquapin. Morphologically, I could not differentiate any of these from the *Pisolithus* sp. on the coast of Cape Cod or the mine spoils of Cartegena, Spain or even in the *Chrysolepis* stands in the San Jacinto mountains of southern California. It was unclear whether outplanted or colonizing trees had inoculated/exotic versus native *Pisolithus*, and whether having *Pt Marx270* conferred an improvement in fitness upon trees worldwide. As Jack Harley told me in 1979 during a NACOM presentation by Don Marx, the study of *Pt* was a project in need of a genetics study. That is still the case. The population differentiation and evolutionary impacts in structuring populations of both fungus and plant would make a very interesting research dissertation, if not a lifelong research program. One needs to document all the locations of outplanting!

The study of heavy metals has been an especially useful model for studying population differentiation of mycorrhizal fungi. I will discuss the mechanisms for reducing phytotoxicity in Chapter 7, but here I emphasize population differentiation. Gildon and Tinker (285) found a heavy-metal tolerant clade of *Glomus mosseae* by isolating a tolerant strain from a heavy metal site that ameliorated phytotoxity of Zn. But differentiating a resistant population from a tolerant one is a challenge (see (268)). In a remarkable study, Colpaert and colleagues (172) clearly differentiated strains (base on genetic fingerprints) of *Suillus luteus* from polluted (Zn and Cd) sites that had differing mechanisms of tolerance, that were lost upon long-term culturing; a key characteristic of genetic differentiation.

Taxa of *Suillus* have proven to be quite useful in understanding population variation and segregation in EM fungi. Population differentiation in EM fungi is ongoing, just as one would expect for all organisms. In a nicely documented system, Branco and colleagues (137) studied *Suillus brevipes* in populations of pines ranging from seashore *Pinus muricata* and *P. contorta* to *P. contorta* in the Sierra Nevada mountains of central California. The coastal and montane populations are separated by 300 km across the dry Central Valley. Using single nucleotide polymorphisms (SNPs), they were able to demonstrate that the fungal populations were separated by environment, but not by host. The isolates separated genetically (or by DNA) by coastal versus montane populations, including salt-tolerance, likely a key attribute for any coastal population. Moreover, they estimated that the split occurred approximately 25,000 generations ago. In our studies on EM lifespans, we estimated that most EM hyphal turnover averaged annually (e.g., (59)) in another California conifer ecosystem. If we estimate a 25,000-year split, and averaging a growing season per generation, this would take us back to the Quaternary marked by extensive glaciation in the Sierras with lowered coastal sea levels (to the Fallon Islands) and extensive freshwater lakes in the Central Valley, with interglacial rising sea levels and estuarine formation in the Central Valley. The continuous climate shifting provided the opportunity for extensive habitat heterogeneity, many intermittent populations, and the potential for genetic recombination and separation. Pines extended across the region. Since the last ice age ended, about 20,000 years ago for this region, the Central Valley dried, separating the two pine forest regions (501). This fungal species also showed continental-wide differentiation and adaptation among populations (136). Interestingly, with these documented differences, there is a potential for a classical Clausen et al. (169) reciprocal transplant study that includes fungal symbionts with their host pines.

Suillus luteus is distributed globally with the intentional transplant of European and North American pines for wood products. *S. luteus* has become a favored food specialty in Argentina and can be commonly found throughout the mountains and foothills. It is also common in South Africa, Australia, and New Zealand. All were comprised of a monophyletic clade (574) comprised of admixtures of North American and European populations.

AM population differentiation has proven far harder to document. Single nucleotide polymorphisms (SNPs) of AM fungi are not understood adequately to begin teasing apart population variation and selection. Going back to our experimental system host growth response study as an example, and teasing out data from one of our experiments in the early 1990s, it appears that taxon differences may matter. Using the replicated, reciprocal transplant study described in Chapter 4, we studied reciprocally transplanted fungi and populations of a single subspecies of plant between a southern California sagebrush site and a central Nevada sagebrush site. In contrasting the plant responses of both plant populations to the two distinct taxa, *Acaulospora elegans*, and *Scutellospora calospora*, we found that the plants performed better with the home fungus, regardless of the location or plant population between the two genera. Alternatively, when we compared the responses of both plant populations to two populations (from the two sites) of the same fungal taxon, *Glomus deserticola*, the plant growth performances were not different between fungal populations of the same taxon (25; 767). Spore analyses and antisera studies suggested that the fungi survived well between the different sites, but the ability to distinguish taxa using molecular sequencing techniques was not yet possible. These data would suggest that the isolates from the two populations of one clade, *G. deserticola*, separated by more than 600 km, were more similar in host response than distinctly different clades (an *Acaulospora* versus a *Scutellospora*). Unfortunately, we still do not know what is the individual upon which natural selection acts – an individual nucleus, a hyphal tip, an intact clone, or a genetically "identical" taxonomic entity. As molecular differentiation becomes better characterized, I would expect a major revision in our understanding of AM fungal population dynamics.

Population Change and r and K Selection

To tease apart characteristics of a taxon, such as r (maximum rate of increase) or K (carrying capacity of the environment) values, we need a consistent measurement of growth and mortality to assess population change. What is our measure of population change? For AM, the most common measure is *percent infection* (or colonization in the literature – which can be confused with dispersal to a new site – see Glossary). However, percent infection assumes a linear growth between plant and fungus under a standard growth condition. It further assumes knowledge of the rooting structure susceptible to infection. In AM, infection occurs just behind the region of elongation, then ages into arbuscule and vesicle sections, then into secondary and suberized roots (Chapter 3). There is only a narrow region of infection. For example, many reports show very low frequencies of arbuscules – largely because the region of the root where arbuscules form is quite constrained, not because of any characteristic of the environment. Arbuscules are also phenologically constrained to seasons of moist soil and rapid plant growth. Also, in AM, root growth can be rapid, such as under conditions of temporary nutrient addition. This would increase the root length: fungal biomass, showing a lower percent infection, even though the total number of infections per plant has not changed. Plants with slow root growth show a high percent infection compared with rapid root growth, despite an equal number of infections (38). Nevertheless, the percentage of the root length infected is a relatively easy measurement to make, and so will remain the primary measurement employed.

Visualizing in situ infections is becoming possible. We (47) used autofluorescence in a blue–light fluorescence microscope to track individual infections in thin glass plates. A new multispectral camera system (691) has been integrated into a minirhizotron system using a violet and blue light spectral signal to view AM. This approach holds promise for tracking mycorrhizae non-destructively.

Sporocarp and Spore Counts

These have been used to indicate mycorrhizal activity. These measures have been criticized in that they often do not correlate with percent infection or numbers of root tips. However, there is no reason that this correlation should be expected. First, both percent infection and percent short root tips are highly variable, depending upon root growth as much as mycorrhizal fungal activity. Second, there is a distinct phenology of sporocarp and spore production. Spores are formed when the fungus obtains adequate C reserves and the environment is appropriate. In temperate environments, that occurs at the end of the growing season, in tropical climates, with periodic drying, although this is often unpredictable. Finally, spores should really be viewed like seeds of plants – they are reproductive units. As such, they may, in fact, be highly appropriate for determining fitness – maybe better than infection!

Extramatrical Hyphal Lengths

These can also be used to characterize mycorrhizal activity. Measuring hyphal lengths has been in ecosystem toolkits for decades. After spending months in the laboratory, hours at a time, tracing the hyphae between stained then unstained root infections and spores, I confirmed what others had reported; AM fungal hyphae have a relatively distinct morphology – knobby, aseptate hyphae, which could be measured using standard hyphal extraction length techniques (62). Hyphal lengths in sagebrush-grassland ranged from 2 m cm^{-3} of soil in sagebrush (62) to 54 m cm^{-3} in grasses (43). Miller et al. (503) even reported that in tallgrass prairie, that length could exceed 100 m cm^{-3}.

EM and other mycorrhizal types remain problematic for morphological approaches to measuring hyphal lengths. Because EM evolved independently so many times, and the fungi forming orchid, ericoid, or other mycorrhizal types are so widely interspersed among fungal taxa, currently we cannot morphologically differentiate EM from saprotrophic hyphae *in situ*. I know of two ways *in situ* to track individual EM fungal hyphae using a minirhizotron. The first is to observe the hyphae radiating from an EM into the soil. The second is to track a hypha. In mycorrhizal fungi, the absorbing hyphae often radiate from the root outward, then die back to that source. Runner hyphae, those initiating new infections, will violate this pattern, so caution is needed.

Kough et al. (429) and Friese and Allen (265) used fluorescent antiserum to differentiate taxa to evaluate growth of differing taxa. Many researchers have distinguished morphotypes and associated these with distinct taxa based on connections, growth in culture, restriction fragment length polymorphisms (RFLPs), or sequencing. However, many morphotypes look similar among taxa, making tracking difficult.

In some cases, clones can be tracked from the points of origin *in situ*, using high-resolution minirhizotrons (58). We know that both production and turnover can occur within a single day (346). Some hyphae persist for a growing season (725). Because the lifespan of any single hypha can range from a single day (265) to a growing season (59), understanding r and K could have crucial impacts on measuring community dynamics.

Bioassays

Bioassays have been used to characterize the inoculum of a site, particularly to contrast disturbed and undisturbed sites. "Inoculum potential" is an approach that has been used to determine mycorrhizal activity. A bioassay is undertaken in the glasshouse using a field-collected soil sample, and the percent infection of a standard plant (often maize, *Zea mays*) after a standard time, used to indicate "inoculum potential" of the field. This is a good approach but represents inoculum density more than potential function. "Inoculum potential" is a well-defined term comprised of four elements: host and fungal genetics (compatibility), plant–microbial activity (largely phenological), edaphic factors, and inoculum density. In reality, an "inoculum potential" really measures an indicator of inoculum density active at the time when soil is assayed. The use of bioassays is a useful piece of information but limited.

Quantitative Polymerase Chain Reaction (qPCR)

qPCR or Realtime PCR is becoming a useful procedure for not only pulling out species richness, but also quantitatively distinguishing amongst taxa. Landeweert et al. (441) quantified EM fungal mycelium by taxon and Voříšková et al. (755) quantified AM fungi, again by taxa, using qPCR to assess quantitative changes. These approaches should revolutionize how we measure mycorrhizal fungal population and community dynamics.

Measuring Fungal Fitness

This is a special challenge (see (578)). What we do know about fungal populations includes their differentiation, growth, mortality, and persistence, which are the elements of mycorrhizal demography. There are several hundred papers on the impacts of mycorrhizae on the fitness of host plants, and even plant-feeding insects. But there is a remaining gap on quantitative estimates of fitness of EM, AM, or any other type of

mycorrhizal fungi. The effects of mycorrhizae on the fitness of mycorrhizal fungi are >1 if the mycorrhiza persists. (Mycorrhizal fungi are
dependent upon the C from the host.) But how is the fitness affected by
host composition, environment, and perturbation? AM fungal taxa are
widely distributed. Some EM fungi, like *Suillus luteus*, and *Pisolithus spp.*
have been spread by humans globally, and expanded from the initial
inoculations, so must have a high degree of fitness. Basidiomycetous and
Ascomycetous EM fungal sporocarps produce thousands to millions of
spores, and presumably few survive. But, just as in plants like
Sequoiadendron giganteum, a giant tree producing millions of tiny seeds
across its lifetime of several thousand years, we don't know the actual
fitness. For mycorrhizal fungi, quantitatively, we do not have good
information. To my knowledge, there is no study of fitness and
mycorrhizal fungi.

Measuring Plant Fitness

Like studies of plant fitness broadly, measuring plant fitness is largely
implied rather than measured. For host plants, our three divisions of
mycotrophy imply fitness. Nonmycotrophic plants have either no
response or, for some species, a negative growth response to attempted
infections by mycorrhizal fungi. Plants, such as annuals in the
Amaranthaceae, show reductions in seedling survival, indicating a reduction in fitness. For example, Yost and Fox (795) found reduced survival
of nonmycotrophic sugarbeets (*Beta vulgaris*) upon inoculation with AM
fungi. In the glasshouse, after one month, inoculation reduced seedling
survival by 50–70 percent in *Salsola kali*, a nonmycotrophic plant (26).
Field inoculation showed a reduction in plant density by up to 70 percent
over four growing seasons (13). Here, the AM fungi were clearly parasites. While most *S. kali* roots were uninfected, there was a small amount
of infection (47), and the few spores in the *S. kali* rhizosphere extracted
C for sporulation based on spore $\delta^{13}C$ measurements (44).

Direct measurement of fitness in long term studies for both facultative
and obligately mycotrophic species in the field simply do not exist as it is
virtually impossible to maintain a site without mycorrhizae. But in the
facultatively mycotrophic grass, the density over that period was
unaffected by direct inoculation, but plant drought tolerance was generally improved (12). *Atriplex canescens*, a perennial Amaranthaceae, had
improved growth and survival in the field with AM fungal inoculation
(6; 781). On the Mount St. Helens volcano, facultative mycotrophic

plants (*Lupinus lepidus*, *Epilobium angustifolia*, *Anaphalis margaritaceae*) initially established as individuals, then patches, and spread across the open plains of pumice and rocks. AM fungi followed, dispersed by small mammals. These plants appear to have little fitness increase, but the associated fungi clearly required the plants to reproduce and spread. Alder (*Alnus*), cottonwoods (*Populus*), and willow (*Salix*) were able to initially establish without mycorrhizae, but soon became EM (wind-dispersed inoculum, such as *Thelephora* and *Inocybe*), then AM (animal dispersed), and then EM, formed by truffle fungi spread by animals (*Melanogaster*) (69). Conifers, that appear to be obligately mycorrhizal required EM fungi for establishment (33) that had been wind-dispersed. Björkman (123) found that *Monotropa* absorbed ^{14}C from labeled conifer trees and, upon trenching, died, demonstrating that mycorrhizae were essential to these plants. We found that mycoheterotrophic plants did not re-establish on Mount St. Helens until a new layer of litter and EM fungi formed on top of the volcanic ash. The same pattern existed on Exit Glacier, a retreating glacier with exposed new glacial outwash.

Quantitative Modeling of Population Parameters

Quantitative modeling of population parameters such as r and K is a critical step in better understanding how mycorrhizae act in the field. Previously, I suggested that instead of percent infection, the relative numbers of infection units per unit soil volume or per unit soil surface area might be a useful measure of mycorrhizal activity (38). However, this value is challenging to obtain. To generate it, you need to know the total and rate of growth of infectable root length, the number of penetrations per unit infected root, and then the rate at which change occurs. But, in analyzing mycorrhizal infections through time, I observed that the numbers of infections increased continuously. However, the percent of infection varied wildly over a 25-day period, especially in the rapidly growing Indian ricegrass (*Achnatherum hymenoides*), when compared with the slower-growing shrub (*Artemisia tridentata*). However, using total numbers of root segments, I modeled infection using a mutualism model derived from the Lotka-Volterra predator/prey model, where: $N_{t+dt} = Nt + B - D$, where N_{t+dt} is the population at time t + dt, B is the number of new root tips between time t and time t + dt, and D is the number of deaths between time t and time t + dt, where the rate of root growth can be re-arranged to: $N_t = N_0 e^{rt}$, where r is the rate of new root tip formation.

Then we can model the number (or length) of mycorrhizal root infections using my equation for modeling predation (38), although in this case a mutualist, where: $M_t = M_0 e^{rt}$, M is the mycorrhizal quantity, and ρ is the rate of mycorrhizal increase.

In that test case, with roots growing into new substrate, for *A. hymenoides*, $r = 0.71/$day and *A. tridentata* was 0.20, a far higher root growth rate for the grass. But for new mycorrhizal infections that were limited by inoculum, ρ for *A. hymenoides* was 0.201 per day and for *A. tridentata* was 0.202 per day. These values are virtually identical. Here, the limiting factor was the density of inoculum, consistent between the two sets of plants. Percent infection varied from nearly 0 to 70 percent, with daily high variation and no consistency.

But that expansion is not infinite. Eventually the numbers of mycorrhizae, whether infection units, or percent infection, reaches a saturation point beyond which there is no space for new mycorrhizae. I equated that value with K, where: $dN/dt = rN[(K - N)/K]$, where K is the carrying capacity of the environment. When studying the characteristics of a population at a site, or comparing sites, knowing K is as useful as understanding the growth rate (r).

This approach was effective in helping to understand the competitive uptake between two grass species with a competing shrub (152), where the percent infection was not different, but the two grasses had both different numbers of infection units per root length and different root lengths. More detail will be discussed later to develop our understanding of competition and mycorrhizal communities (Chapter 6).

For EM, the proportion of short root tips occupied by EM is the most common measure. This approach may be more useful simply because short roots rarely continue to grow upon EM formation. They remain until they die and disappear. The EM approach (percentage of infectable tips occupied) may also be a more useful approach for AM – the proportion of fine root tips infected. But that has not been a standard measure.

Measuring population parameters *in situ* for any organism is challenging. To my knowledge, no one has attempted to measure population parameters like fitness, r, or K values of a mycorrhizal fungus. In assessing fitness, population biologists often measure seed production as an index. There are many papers reporting spore numbers in very different environments that can be used to construct fitness estimates. But, like seeds, spore mortality is extremely high, from predation and exposure. One important caveat, r and K have been used both as attributes of a population (e.g., (134)) and as characteristics of taxa among members of a community (r-selected species versus K selected species, e.g., (566)). I will use them in both ways, but distinguish, as both are useful concepts. Here, I focus on r and K as population attributes. I will address r- and K-selected taxa in Chapter 6.

As discussed in Chapter 2, cores for sampling never allows for a repeat of the same point through time. Upon coring, one is always sampling a new patch. Automated minirhizotrons (Chapter 2) provide a new technology to evaluate changing mycorrhizal activity dynamically. The rainforest of the La Selva Biological Station, on the transition of the northern slopes of the Barva Volcano to the coastal plain of the Caribbean and dominated by AM plants, provides a useful comparative system. In tracking root and mycorrhizal dynamics at the La Selva Biological Station, in a Costa Rican rainforest, we found that in the control locations, a mature rainforest with minimal seasonality (692), the soil was saturated with both roots and mycorrhizal fungal hyphae. There was little seasonal variation and every new root became infected immediately (Plate 4). Alternatively, leaf cutter ants create forest gaps, pruning new plants and even roots. In these gaps, when nests are abandoned, new roots rapidly colonize the open space where the rate of new root growth is approximately 1 mm per day, and 13 new root branches were growing into the gap, producing 13 mm of new roots per day. However, only five of those roots became mycorrhizal during the seven-day study period. Root growth far outpaced new infections by the fungal hyphal network. As discussed in Chapter 2, the roots are positively geotrophic, whereas the hyphae are not. The new patch rapidly fills with new roots, but there is a lag in filling with mycorrhizae, of several days. These gaps provide opportunities for new root segments for fungal establishment for many new fungi. Analyzing undisturbed (closed or full sites) perturbations (that create open patches) provide unique opportunities to assess r and K.

Estimating ρ, the intrinsic growth rate of a mycorrhiza, requires measuring both root and fungal growth. Here is where a measurement strategy, such as I advocated in 2001, is useful. Understanding r and ρ is especially useful for studying a newly colonized location. That can mean colonizing a newly disturbed environment, such as a gopher mound, an abandoned ant mound, a hurricane blowdown, a volcano, or a retreating glacier. Estimating K, where a population approaches stability, could be useful in generating a maximum number of infection units or even percent infection. One way is to look at sites with little disturbance, where there is stability in inoculum, and with minimal seasonality, where roots and fungi have occupied nearly all space feasible and new growth is dependent upon turnover.

Using the La Selva data (692), the standing crop of roots was 230 g m^{-2}, changing up to 30 percent between samples, but with minimal temporal change. AM hyphal standing crop averaged 50 g m^{-2}, with minimal spatial or temporal variation. Every possible new root was rapidly infected with AM. However, the taxonomic variation is unknown. Importantly, even though the soil volume was fully explored, roots had a short lifespan, averaging 150 days. What this means is that roots die and are replaced in any patch approximately 2.4 times per year. Hyphae turn over even faster, 12.1 times per year (692). Even though the system is saturated, there is a high rate of both root and mycorrhizal replacement, and opportunities for new and replacement taxa. In this case, $\rho = r$, or ρ, representing new mycorrhizal infections, is equal to r, or new root tips formed, and $K \rightarrow 100$ percent of infectable root sections because of rapid hyphal turnover.

Seasonal environments hold opportunities for differentiating K, r, and ρ, as there are always periods of root mortality and mortality of infections and spores. In a seasonal, tropical forest, during the dry season, root and fungal mortality is high, opening resources for new mycorrhizal formation. In places like the Chamela Biological Station, on the coast of Jalisco, Mexico, percent infection and spore counts vary by plant species and soil (21). Here seasonal variation is quite high, with openings amongst the roots of dominant trees. But in these systems, understory species, including cacti adapted to these dry conditions, sustained infection and spores through the dry season. As even more extreme environments are studied, including the grasslands of northern Colorado (679), and the Beartooth Plateau, Montana, an alpine location (18), the same pattern is apparent. In all of these systems what I do not know is how the inoculum density shifted relatively to new root growth. In the spring, is the low infection due to reduced inoculum, high rates of root growth, or both?

There are also numerous reports examining percent infection and spore numbers in a mature environment, where the composition of host plants is not changing but seasonality does. An example comes from the C$_4$ shrub, *Atriplex gardneri* (32), located in a shrub–grassland steppe in southwestern Wyoming, at an altitude of 2,100 m. Fine root growth stops during the summer drought and root infection is largely undetectable. This led many to postulate that, like annuals in the Amaranthaceae, perennial Amaranthaceae are nonmycotrophic. However, during winter, snow cover insulates the soil and it remains unfrozen, and decomposition proceeds, mineralizing nutrients such as N and P. Once snow melts and photosynthesis ramps up, in early spring, new roots form AM and absorbing structures. As the dry season commences, arbuscules rapidly disappear, vesicles and spore formation remain active, and by the middle of the dry season ($\psi_s < -2$ MPa) only spores remain. Upon analysis of the spores for $\delta^{13}C$, they clearly received their C from the C_4 *A. gardneri*, and not the surrounding C_3 shrubs and grasses (44). What this means is that seasonality itself creates growth, mortality, and regrowth of mycorrhizae, making any assessment of infection (ρ) as much a function of r as of K.

This seasonal functionality appears to be especially critical for mycorrhizal fungi of plants with C_4 photosynthesis. A remarkable pattern of responsiveness is the interaction between plants with C_4 photosynthesis and mycorrhizae. Most C_4 grasses have a strong responsiveness to AM, such as maize (*Zea mays*), sorghum (e.g., *Sorghum bicolor*), and big bluestem (*Andropogon hallii* and *A. gerardii*) in comparison with the lesser responsiveness of C_3 grasses such as wheat and species of *Agropyron* and *Stipa* (325; 326; 350). C_4 plants have a higher photosynthetic efficiency and higher temperature optimum. These grasses start growth later in the season, compared with their cohabiting C_3 grasses, but grow faster and later into the warmer, drier summer. Their mycorrhizal fungal activity also shows a phenology that expresses fungal interaction to both species of these interspersed plants (50). At the Arapahoe prairie in the Sandhills of western Nebraska, the AM fungi appear to tap both plants sequentially, potentially increasing populations of the fungus. Activity of *Glomus fasciculatum* will start after the roots of the C_3 *A. smithii* are initiated in the early spring, forming arbuscules in March, with a peak arbuscular infection and a hyphal network in association in June. *B. gracilis* begins arbuscule formation in June, peaking in August. The fungus forms vesicles all season to store C, then peaks in sporulation in September, as no new C is flowing from either plant and the plants begin to die back. That pattern was also found in a host of intermixed C_3 and C_4 grasses (45).

Experimental C and Nutrient Additions

These additions provide unique opportunity to address how resources might shift both r and K values of both fungi and plants. When nutrients are added, more resources are available for plant growth, thereby increasing K. But, does the fungal activity keep up with the additional growth potential for the fungi? Or, are nutrient levels already optimal stoichiometrically, and therefore the plant simply keeps both the additional

C and nutrients? Or, does the limiting resource simply shift, such as from N to P, allowing adaptation, or shifting the composition of the plants or fungi? Considering this additional complexity, in a short-term experiment, that shift could allow for a composite measure of r (root r and mycorrhizal fungal ρ). Alternatively, in a long-term experiment, any shift would more likely represent a change in K (both root and fungal). An example long-term fertilization is presented in the paper where we derived the nutrient limitation model (Figure 1.1 (719)). In the Hawaiian chain, the limiting resources shift from N in newer islands before N-fixing organisms are present to build up N (300 years old), to P in the oldest (4.1 MY old) islands (as the soils leach), but both N and P are relatively high in the medium-aged islands (20,000 years old) (719). Sites were then fertilized and examined for AM fungal infection, and which clades responded (*Glomus, Acaulospora, Gigaspora, Scutellospora*) using immunofluorescence to examine responses. In the fertile site, fertilization increased AM fungal hyphal length (biomass) in both nutrient limited sites, whereas AM fungal mass declined with fertilization in the fertile site. However, percent infections were complicated, probably by changing root growth. N fertilization reduced infection (due to increased root production?), whereas P fertilization increased infection (no change or reduced root growth). Alternatively, in both the N- and P-limited site, fertilization of either N or P reduced infection (due to increased root biomass?). Thus, one is moving up or down the left portion of the curve in the N- or P-limited sites, but near the top of the curve, additional fertility provides excess for the mycorrhizal range, and over the top (see Figures 1.1 and 4.3). However, the infection, hyphal length, and spore counts were complicated by the shifts in taxa. For example, in the *Glomus* clade, fertilization increased root infection in the fertile plots, but showed no change in the N-limited or P-limited plots. Alternatively, in the *Scutellospora* clade, in both of the two nutrient-limited sites, N fertilization reduced infection. P fertilization increased *Scutellospora* infection in both the N-limited and in the fertile soils, but not in the P-limited sites. Every taxon of plant and of fungus responds independently to increased soil resources.

Together, these analyses provide a unique opportunity to generate an understanding of r, ρ, and K. In a continuous environment, such as the rain forest, r and ρ are very high and, as every new root encounters a mycorrhizal fungus, K->90+ percent infection. But turnover is also very high, with lifespans in days, not months. Thus, the overall values have the potential to stay relatively stable. How these play out among individual populations of plants and fungi remains to be studied. In seasonal environments, r and ρ are both high, but vary among populations. But values vary highly amongst populations, species, seasonal drought, precipitation events, length of growing season, and many other factors. I found that in cases that I attempted to analyze, the rate of increase in root growth (r) and numbers of mycorrhizal infection units (ρ) commonly exceeded the *r*-value differentiating linear versus chaotic behavior (38). Using stoichiometric relationships of fungi and plants, and including multiple species with differing resource acquisitions in our models, we were unable to force the relationship to become stable. In some cases, the fungi grew infinitely, in others the plant did (75). Does this indicate that these parameters are unimportant? On the contrary, I suggest that this merely means that we must consider not only how mycorrhizae function, but how they function within the ecosystems in which they exist. The constraints may well lie outside the mutualism, as we will discuss in Chapter 6. But at this stage, suffice it to say that chaotic systems are more the norm than unusual and that constraints outside the parameters of interest in complex systems are rarely surprising.

Summary

- A critical problem is identifying an individual for population assessment. Is an individual a single cell, or an interconnected clone, or an entire clone, with many disconnections? This difference is crucial and not well distinguished either by mycologists or plant biologists.
- Within the concept of an individual, for the fungi, a major issue is whether an individual cell is homokaryotic (same nuclei) or heterokaryotic (different nuclei). A population of cells may be both, but the mechanisms creating them and the frequency are controversial and may be the biggest question across the field of Mycology today.
- How do populations differentiate? Factors like transplanting to new habitats, including continents, and toxicity conditions have been shown to affect these populations. Differentiation of populations is important both ecologically and for application purposes.
- There is a need for quantitative measures of both r and K values to help understand population dynamics as opposed to a static condition.
- For the fungi, what do we measure, individual hyphae, a clone, spore numbers, or sporocarp numbers? Techniques are rapidly changing from individual spore counts to qPCR. For plants, fitness is implied,

but rarely measured. This is especially crucial when assessing the interaction along the symbiosis gradient.

- What is a population, and what is a species? That is also undergoing rapid revision. Here I will use the clade approach. With this approach, I allow the reader to make an individual decision about the scale of information on which my selection of organism identification is based.

- *We need quantitative measures for population dynamics!*

6 · *Community Ecology*

A community has a diverse suite of definitions, as many as there are scientific fields and sociological units. Among humans, there is a place-based variant such as a city defined by political boundaries, or an interest-based variant focused on a group of people defined by their interactions. The same issues plague the field of community ecology, especially when we address a topic such as a mycorrhiza, which is a functional relationship. Early measurements of communities looked for spatial boundaries, the edge of a meadow; a shift in forest type with a physical edge such as a shift in topography; an obvious shift in plant types such as a forest edge. Theophrastus (~44 BCE) noted that most mushrooms were found in forests (which included EM species) but not grasslands (which were almost exclusively AM). Alexander von Humbolt drove some of the earliest developments in community and ecosystem ecology by showing relationships between climate and vegetation both up elevation gradients, such as his beautiful and accurate 1807 *Tableau Physique*, drawings of vegetation up Mount Chimborazo in Ecuador, and his *Geographical Distribution of Plants* – climate consortia of plants – in relation to global climate patterns (371). Those vegetation consortia could also be used to describe mycorrhizal patterns, such as mycorrhizae along altitudinal gradients (e.g., (536; 799)). Mapping vegetation types became a major focus of research in the 1800s (see, for example, (653)). The approach of plant association nomenclature championed by Braun-Blanquet in the early twentieth century is still commonly used in European ecology, also known as Phytosociology. Using this concept, a change in a community exists when the species composition between relevés is greater than a predetermined level (221).

The scale of phytosociology is both useful in measuring a community and in describing any change to that community, but especially challenging to measure for mycorrhizal fungal–plant dynamics. Just as differentiating between an individual versus a population is difficult, so too is identifying what comprises a single community to measure. What is the boundary of a community versus what spills over into the next? Studies since the early twentieth century have largely eyeballed a locale that looked relatively homogeneous and defined that as a community. But Whittaker (776) noted that some boundaries are abrupt, but many are gradual transitions. Parameters of interest were measured within and across that unit, an approach useful in practice today. Most analyses of communities, in either space or time, depend upon this conceptual framework. While useful to measure, conceptually focusing on a spatial definition of an "observable" community is insufficient to extrapolate to dynamics and functional interrelationships.

Because we cannot visualize a community of mycorrhizae, both because we can't see belowground and because much of the community exists in microscopic units, we are forced to sample patches within a defined unit, and usually very small patches. We also assume some degree of homogeneity within a community that may or (more often) may not be appropriate. For example, I began my career studying a short-grass steppe dominated by *Bouteloua gracilis* that exists along the eastern slopes of the Rocky Mountains in North America from just south of the boreal forest in central Canada to the Chihuahuan deserts in central Mexico. But to describe the mycorrhizal fungi (30; 194; 679), we characterized the percent mycorrhizae found within a dozen cores randomly selected per sampling time, 18 cm^2, to a depth of 20 cm, to represent the dynamics of shortgrass steppe, specifically the Central Plains Experimental Range (in Colorado, USA) of 6,280 ha, or 1.6×10^{-13} percent of the surface area of the research station. This context expands to AM communities associated with *B. gracilis* communities ranging from New Mexico (217) to Saskatoon, Canada (353). Both random and spatially designed aggregated samplings and careful statistical assessments are used to estimate the diversity of mycorrhizal plants and mycorrhizal fungi for many different communities across the globe.

Another approach to describing communities is the organism–centered approach proposed by MacMahon and colleagues (472) where a community is identified as organisms that affect each other's fitness. I characterize this tactic as a functional approach toward defining a community as well as an ecosystem. Here, we identify a reference taxon, and try to measure the organisms that influence the fitness of the reference taxon. This approach has the advantage of being functional and more easily linked to ecosystem processes, but the disadvantage of variability in spatial scale, depending on the composition. Trying to measure the impacts of the potential members of a defined community may be difficult empirically, but this approach is theoretically very useful for understanding mechanisms of community interactions.

This functional approach is useful because the variable of interest, the number of individuals existing from one generation to the next, can be measured. Therefore, the approach lends itself to dynamic modeling that scales to understanding critical community processes, including perturbation, succession, and directional change. Once we define our reference organism, then we identify the organisms that affect fitness of our reference organism and then measure their interactions and directional change. Maurer (488) defined a simple linear equation for undertaking this approach, where:

$$dN_i/dt = N_if_i(N_1, N_2 \ldots N_s),$$

where f_i is the effect of species i on the numbers of our reference organism, where $i = 1, 2, 3 \ldots s$.

Of course, this conceptually simple model becomes extremely complex as the dynamic interactions of all members of the community are quantified reciprocally (they are not equal), and as we add the indirect reciprocal dynamics. For example, the AM fungus *Gl. fasciculatum* is indirectly affected by a parasitic nematode, *Tylenchus sp.*, and a parasitic fungus, *F. oxysporum*, on a common host, *B. gracilis*. In considering the soil community, the equation rapidly blows up into a Taylor series expansion across equilibrium densities, resulting in a system of linear differential equations written in matrix form. These also can be damped or exacerbated by the environment and driven by reproduction to exhibit chaotic dynamics.

While more difficult to measure and quantify, this fitness-based approach as a conceptual framework allows us to characterize multiple mycorrhizal fungus and host dynamics of the partners in a far stronger way, allowing some predictive power in the face of change and across large spatial areas (Figure 6.1). Ultimately both approaches, presence/absence of taxa and fitness change, are useful. Here, I will focus on the fitness approach to understand the mechanisms leading to composition and then switch to the spatial approach to measure community organization.

Fitness and Limiting Resources: Defining Mechanisms Delineating Community Interactions

We have already discussed the role of limiting resources in regulating the plant–fungal symbiosis (Chapter 4). We can readily expand that mechanism for both the plants and the fungi comprising a given community.

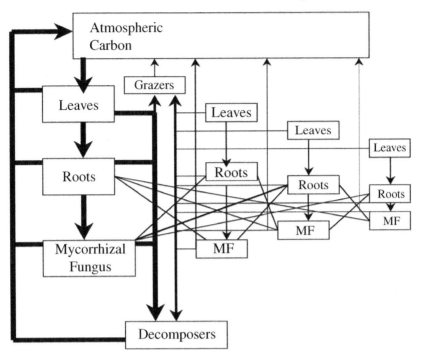

Figure 6.1 Conceptual expansion of connectivity between a plant and its mycorrhizal fungus, with a second fungus, then both fungi connecting to a second plant, connecting to a third fungus, adding a third plant, ad infinitum. Under this scenario, the mycorrhizal relationship becomes a community in reality.

Competition Is a Critical Mechanism Regulating Species Interactions

Competition, thus, becomes crucial to understanding communities. Using the value of r (growth rate) of a population, we can see how individual plants and fungi react to resources as they become available. Growth could be seasonal, such as in a temperate or arctic habitat, or following a disturbance where new resources are available for exploitation. Under optimal conditions, plant roots will access much of the limiting resource, increase C fixation, and create more offspring; but only to a point. The addition of the mycorrhizal fungus increases C fixation of individuals by accessing more resources than the host alone. The increased C can go into plant size or into reproduction, creating more plant biomass beyond that of the single plant alone within the space available. As plant biomass increases, the more limiting a resource becomes and the greater the need for the fungal partner (Figures 1.1 and 4.3). Plant density therefore acts to reduce the resource availability to growth dependent on the most limiting resources available, Liebig's (1840) "law of the minimum" (456). This limit becomes the K-defined level of sustainable availability. Mycorrhizal fungi increase the resource availability to infected plants, allowing the K value to be higher than without infection, but again to the constrained limited level. K is reached either at some point during the growing season in a temperate environment but is really apparent in a tropical ecosystem. Here, where positive growth conditions are continuous, populations are fluctuating around K, unless there is a disturbance. The K value provides a value of the standing crop as turnover must approximately equal mortality and turnover at some defined time frame.

It is important to reiterate the difference between resources and conditions (710). Resources are requirements that can be depleted. Most nutrients are resources such as N or P. Light can be a resource; as the canopy density increases, light is "consumed." Water can be a resource and competition for water can be fierce. Conditions occur and limit growth and reproduction but are not consumed. Temperature is a good example. Climate is dictated by the physical characteristics of the environment and, while it can be altered slightly by plants, is not "consumed" by one or the other.

Importantly, under few conditions, is there a single plant species or a single fungal taxon occupying a location. So, the two or more species respond to the limit of the most limiting resource through competition. If both plant species are mycotrophic and genetically capable of forming the mycorrhiza with the fungus present, then competition takes on the response of that mycorrhizal symbiosis. Fitness of the two plants and seed output are rarely measured directly. But growth is commonly used as a surrogate and is affected by mycorrhizal fungi. It is important to remember that, when assessing K, the limiting resource can shift in time and in space, making the ultimate K value challenging to assess.

Limiting resources, such as water, nutrients, and light, are studied at their most basic perspective using economic models of supply and demand. We rarely consider the role of light in mycorrhizal competition, but certainly light levels affect C fixation, hence allocation of C to mycorrhizae for soil resources (e.g., (68; 368)). Here, a limiting resource by roots is captured at a cost in C. As mycorrhizal fungal hyphae are smaller, that C cost is reduced compared to the large size of the fine root, causing the plant to shift C allocation to the fungus. As resources become more limited, additional fungi can extract novel forms in the supply of resources, but again costing more in C allocation to multiple fungi. The increased demand for resources incurs constraints in any new supply due to the cost in C and the ability of the plant to increase C fixation.

There are three forms of resource supply and acquisition relevant to the study of mycorrhizae. The first is bulk flow. Some resources are supplied through bulk flow, including HNO_3^-, organic PO_4^- (phosphate), Ca^{2+}, and K^+. As water is transpired, these resources flow with the water. When water uptake is increased, more of these resources become available. Under most growing conditions, the majority of plant water is taken up by the large surface area of the root system. Thus, when water is available (where $\psi_{leaf} > -1.5$ MPa), most of these resources are acquired by the plant directly. Early mycorrhizal research emphasized the role of increasing water throughput as the mechanism for nutrient uptake (see (678)) as this mechanism was presumed to be the predominant way in which nutrients were acquired. In this situation, competition results largely from the limited water. Mycorrhizae increase water throughput, especially under drought stress (Chapter 4).

However, cycling of N from organic N to nitrate and from organic P to solution P is slow compared to the rate of photosynthesis. The HNO_3^- in the soil solution is rarely adequate for plants, and organic P breakdown is slow when dependent upon plant-produced phosphatases. This means that a critical limiting concentration of nutrients (N or P) is rapidly reached, creating competition between plants for the nutrients, either directly or expressed by gaining mycorrhizae (Chapter 1). The second mechanism of resource acquisition is to search out pockets of limiting resources. The cornerstone of early twentieth century research in the value of mycorrhizae lies in the study of the role of mycorrhizae in the uptake of bound nutrients. HPO_4^- and NH_4^+ bind to soil particles and in soil organic matter. These resources must be extracted by a nearby root or hypha. In most ecosystems, there are far more fungal hyphae reaching into every pore and crossing every spec of organic matter that can extract these nutrients. The classical work of Hatch (334) for EM, and Mosse (519) and Gerdemann (280) for AM demonstrated the role of hyphae in direct uptake by hyphae and transfer to the host. Both water and nutrient uptake by plant roots and mycorrhizae create depletion zones (Chapter 4). By creating a depletion zone, one plant reduces a resource available to another. The result is competition between the two plants for the resource and, in many cases, for the mycorrhizal fungus that improves access to that resource.

Further, Tinker and colleagues (e.g., (184; 556)) examined the kinetics of uptake, again demonstrating the crucial role of mycorrhizal fungal hyphae in extracting HPO_4^- molecules bound to soil particles. NH_4^+ is even more interesting. Early N uptake models were based on completing the entire N cycle before a plant can acquire an N molecule (which I will discuss in detail in Chapter 7). In these models, plants took up N only after it was released as HNO_3^- by microbial decomposition (organic N to NH_4^+ to HNO_3^-). However, both EM and AM fungal hyphae can take up and transport NH_4^+ directly. Moreover, ericoid, EM, and AM mycorrhizal fungal hyphae can take up and transport organic N. Thus, mycorrhizae also increase both P and N uptake by accessing and short-circuiting the early twentieth century-assumed pathway, making N and P resources more rapidly available to plants than predicted by nutrient cycling models.

A potential third form is elemental transformation from unavailable to available forms. No data exist (to my knowledge) to measure competition via mycorrhizae for elements like Fe. But Fe can certainly limit production as only Fe^{2+} is used in plant cells, and it is a critical resource for electron transport in all organisms. Most Fe is in a Fe^{+3} form. Bacteria and mycorrhizal fungi release chelating agents into the soil. Chelating agents bind Fe^{3+}, facilitating transport into the cells and releasing it as the available form Fe^{2+}. Many elements including Fe have these complex microbial dynamics affecting uptake but the roles in potential competitive interactions are unknown.

Demonstrating Competition

Demonstrating competition is experimentally challenging. The bulk of studies on the role of competition in mycorrhizal symbioses have occurred with AM, as comparative non-mycorrhizal pot cultures are more readily maintained. AM fungal spores are large compared to EM, with less dispersal to contaminate experiments. Fitter (241) initially demonstrated that AM can alter the outcome of competition between plants. Many others demonstrated competitive interactions between plants that were altered by forming mycorrhizae. Miller and Allen (505) examined the crowding coefficients of mycorrhizae and competition between plants in publications through 1990. Mycorrhizae interact with the resource base to determine the outcomes of competition. In some cases, adding a resource such as P could shift the crowding coefficients. For example, using data from Hetrick et al. (349), Miller and Allen (505) calculated that in low P soils, the C_4 grass *Andropogon gerardii* shifted its crowding coefficient from 248 (monoculture M/NM) to 92 (mixed culture M/NM), whereas its C_3 competitor *Koelaria pyranidata* went from 0.84 to 0.14. When P was added, *A. gerardii* went down to

2.5 (monoculture M/NM) from 3.1 (mixed culture M/NM), and the *K. pyranidata* had virtually the same response (0.81 to 0.18). Mechanistically, the shifts were heavily dependent upon the root characteristics and mycorrhizal density. Sliding up the resource scale could alter the mycorrhizal density (Figure 1.1) or the relative growth rates (Figure 4.3). But that would depend upon both the soil richness to begin with and the structure of both roots and mycorrhiza.

In some cases, if one of the plant species is nonmycotrophic, competition studies comparing two situations, with versus without mycorrhizal fungi, are feasible. In this case, with no mycorrhizal fungi present, the nonmycotrophic plant wins; with the mycorrhizal fungus, the mycotrophic plant wins (11). Crowding coefficients can then be calculated between mycotrophic and nonmycotrophic plants under conditions with no mycorrhizae or with mycorrhizae. Not surprisingly, for the mycotrophic plants, the crowding coefficients increased with mycorrhizae, and decreased for nonmycotrophic plants with mycorrhizae.

EM competition studies are more difficult due to the difficulties in controlling spontaneous inoculation from air-borne spores. It is difficult to keep plants nonmycorrhizal. Perry et al. (561) undertook one of these challenging experiments using the two interactive trees, *Pseudotsuga menziesii* and *Pinus ponderosa*, grown for a year. The outcomes are complex for the individual inoculations and I encourage the interested reader to look up the original paper. But for our purposes here, I examine two treatments, a nonmycorrhizal treatment using pasturized soil (nonmycorrhizal initially, but with a *Thelephora terrestris* contaminant) and a whole soil inoculum with multiple species of EM fungi. For the non-inoculated plants (*Thelephora*-contaminated), interspecific competition was greater than intraspecific competition for both plant species. But, when plants were inoculated, intraspecific competition was greater for *P. menziesii* than interspecific competition. For *P. ponderosa*, the fungi were still taking a lot of C (reduced biomass), but interspecific competition was almost non-existent. When measuring ^{32}P uptake, possibly a better indicator of initial competition than growth, without inoculation, *P. ponderosa* took up more ^{32}P, but when mycorrhizal, *P. menziesii* took up more ^{32}P. Despite the challenges in undertaking this study, it appears that EM can also alter competitive outcomes.

Determining Mechanisms of Competition

This is difficult, especially in studying mycorrhizae in the field. Caldwell et al. (152) examined resource acquisition, in this case P, to understand competition between plants, where all three species of plants were mycotrophic and formed AM. In this case, the competition was between two grasses and a common shrub competitor. The grasses were *Agropyron spicatum* and *A. desertorum*, and the shrub, *Artemisia tridentata*. The two grasses were on either side of the shrub in a field experiment. In soil with the interacting root systems, ^{32}P was placed on the side between *A. tridentata* and *A spicatum*, on the other ^{33}P between *A tridentata* and *A desertorum* (and of course, reversed to eliminate any ^{isotope}P effect). Here, *A. tridentata* extracted more of the P in competition with *A. spicatum* (86 percent) than from the side of *A. desertorum* (14 percent). The two grass species had no significant differences in percent infection, with arbuscules or vesicles. In examining the grasses, rooting density was almost doubled for *A. desertorum* compared with *A. spicatum*, and the total number of infection units was more than doubled per centimeter root length, leading to a four-fold increase in total P uptake capacity in one of the grass species, due to the structure of the AM.

In a field study, in a soil with high available nutrients plus high mycorrhizal activity, we shifted our focus to another limiting resource, available water (12; 13). At the site of study, both P and N levels were high but, instead of temperature regulating the end of the growing season, drought was the limiting factor. As drought intensified, in the mycorrhizal condition (soil on a reclaimed stripmine, where stored topsoil with little inoculum-nonmycorrhizal treatment- was compared with plants in plots with added fresh topsoil as inoculum-mycorrhizal treatment), the test plant, *Agropyron smithii*, was able to extract and transpire more water at key periods later into drought periods and phenologically extend growth longer. Together, these mean that more C was fixed in the mycorrhizal than nonmycorrhizal plants. Furthermore, this response was aggravated by the presence of *Salsola kali*, a nonmycotrophic plant. The best condition for *A. smithii* was to be mycorrhizal with no competing annuals; the worst was to be nonmycorrhizal, competing with *S. kali*.

But what happens when a fungus connects both plants? Grime et al. (303) grew an entire community in a mesocosm and found that AM increased diversity by increasing growth of subordinate species. They found that both grazing and mycorrhizae increased the plant diversity within the mesocosm. Plants had differential sensitivity to grazing. By reducing the canopy cover with grazing, more radiation infiltrated, allowing species to co-exist. Mycorrhizae, alternatively, resulted in the survival and growth of subdominant plants. They postulated that a common mycelial network enhanced the growth of these subdominants by allowing for the access of C and nutrient capture.

The Role of a Common Mycorrhizal Network (CMN)

The CMN has driven controversial ways to view the role of mycorrhizae in plant competition. Woods and Brock (789) labeled an AM maple (*Acer*) stump with ^{45}Ca, finding that tracer dispersed among surrounding trees. They postulated that forest trees were interconnected by mycorrhizal hyphae, driving community-scale interactions. Interestingly, the surrounding vegetation was either AM (e.g., *Liriodendron*) or mostly EM, but with an ability to form AM (*Quercus*). Reid and Woods (605) further demonstrated interplant transfer of ^{14}C via a hyphal network in a laboratory growth system, using EM *Pinus taeda* with *Thelephora terrestris* and *Pisolithus tinctorius*. Thus, both EM and AM systems had label transfers. These works set off a firestorm in describing the role of mycorrhizae in plant–community interactions. Chiarello et al. (162) and Whittingham and Read (779) showed ^{32}P, labeled in one plant, would be transported between plants which shared a mycorrhizal mycelial network and, further, the growth of the received plant was improved (252). Finlay and Read (238) found that ^{32}P exchange was unidirectional, but the transport occurred throughout the mycelial network. Alternatively, Ritz and Newman (615) argued that the label was simply exchanged, without a net flow, and further that exchange was a function of loss from the donor plant following root mortality and subsequent uptake (616). By contrast, the Caldwell et al. (152) experiments did not indicate any significant retransport, simply that mycorrhizal fungi picked up ^{32}P or ^{33}P locally and exchanged it for C locally. In another field experiment, He et al. (338) labeled needles of *Pinus sabiniana* with ^{15}N. The objective was to assess whether an EM fungal network increased exchange between individual plants or included isotope leakage into the soil. Over two weeks, foliage of nearby AM herbaceous annuals was enriched. After four weeks, other EM pines and oaks were enriched, but also AM *Ceanothus* and annuals. They concluded that ^{15}N was transported to the soil and mobilized but was not preferentially exchanged between plant cohorts through EM networks.

These diametric views of the efficiency of CMNs are becoming increasingly nuanced; both extreme outcomes (facilitation versus competition) can be found as more research studies are reported. Clearly CMNs are present, at both the individual taxon and community scales.

Fitter and colleagues (e.g., (243)) found that, while C transferred between C_3 and C_4 species, most of the C transfer was a function of the sink strength of the fungi improving their own fitness, but not a significant impact on plant C budgets or fitness. They further note that different AM fungi have very different ecologies (244), and even the EM fungi of a single host have multiple phylogenic backgrounds. Alternatively, Susan Simard and her colleagues have presented probably the most extensive overviews of the facilitation perspective. Beiler et al. (112) described the linkages of cohorts of multiple Douglas fir trees by fungal genets, and suggested that the stands lived or died as interaction networks, depending on stresses (113). Simard's group even created a teaching game, *Shroomroot* (474), to enhance young student learning about mycorrhizae and plant–plant interconnections. The concept has been extrapolated so far as to imply that all the trees in the forest are interconnected, from movies such as James Cameron's *Avatar* to an outdoor sculpture exhibit by Rick and Laura Brown entitled *One Impulse from a Vernal Wood* at the Daniel Chester French Museum. It is difficult to imagine that mycorrhizae reduce or eliminate plant competition patterns that have been studied for centuries. But certainly, CMNs have important implications for the fungi themselves. By connecting multiple plants through multiple mycelial networks, an entirely new conceptual framework based on network theory is gaining traction (385; 388; 672; 674).

We know far less about competition amongst fungi. How do fungi compete, given their small size relative to plant roots (their C source) and soil pores and organic matter fragments (their nutrient sources)? Using postulated physiological differences between fungi, William Swenson modeled multiple outcomes in host plant growth, resulting in differential community compositions (75; 388). Neuenkamp et al. (529) showed that different AM fungi provided different plant species-specific resources, creating an inoculum-specific competitive response. Tedersoo and Bahram (700) found that different guilds of mycorrhizal fungi have different genes for nutrient transformation and uptake that can be used to infer differential competitive ability, thereby altering community composition. There is also the potential for depletion of soil rhizosphere nutrients. Hobbie and Agerer (355) identified a number of strategies employed by different fungi that could lead to the differential exploitation of nutrient acquisition. They then related the different growth strategies to ^{15}N acquisition, suggesting a potential mechanism for competition in resource acquisition.

Does having multiple species of mycorrhizal fungi control the composition of the plant community? Adding multiple species of ectomycorrhizal fungi reduced ^{15}N signatures of the leaves of oak seedlings, indicating that increased richness in the fungal community promoted fungal-increased host N acquisition (Figure 4.4). van der Heijden and colleagues (341) showed that adding additional taxa of AM fungi increased plant species richness. This suggests that reducing fungal richness through competition (or other means, such as chemicals or grazing) can decrease plant community richness.

We are still left with a problem. For what resources do mycorrhizal fungi compete? Is it C acquisition, which also implies root occupation? In early growth of eucalyptus seedlings, the initial roots were AM. But, as the seedlings grew into saplings, EM fungi colonized, covering and replacing AM on root tips (163). We have seen the same pattern in oaks (590). For EM, multiple observations suggest that when an EM colonizes a short root, it can remain for several years (723). Initial occupation is obviously an important competitive process. But roots can be observed breaking out of an EM, growing, and then forming a new EM, with the same or a different fungus. For AM, because the fungus infects a single point, then grows occupying a segment of the root internally, root occupation could also be a competitive process. But there are many points at which fungi invade the root, in the region of elongation, as a root actively grows through the soil. There is a potential for many fungi to occupy a single root, even if it is confined to a small root section. Competition between fungi appears to occur for available root segments. But different fungi also have different foraging strategies (355). While the literature focuses on these strategies for nutrient uptake, the foraging strategies also play a key role in finding roots to colonize.

A final attribute is that mycorrhizal fungi produce hormones that can alter root production and root branching (66; 67) (Chapter 4). By initiating new root tips, the plant is also producing new regions of elongation available for new mycorrhizal infections, colonized by the same or different taxa. Both EM and AM alter phytohormone balance, presumably increasing the C sources for the fungus itself.

Traits of Resource Extraction

Traits of resource extraction of both plants and fungi provide a framework to better understand and extend these individualized experimental systems. Two models, in particular, provide insights into competitive behavior that involve mycorrhizae. Grime (302) studied traits of individual species of plants. In the hallway outside of his office was a long and continually expanding list of traits, with plant species along the ordinate axis. These ranged from physiological traits (photosynthesis) to structural (elasticity). He organized these into the Grime triangle, where plants could be arranged along the three axes of Competitive (C), Stress-tolerant (S), or Ruderal (R). Using an approach similar to Grime's model, we (75) listed a suite of traits exhibited by different mycorrhizal fungi (Table 6.1).

Chagnon et al. (158) organized traits of AM fungi into and along the C–S–R gradients and proposed that we can further use these characteristics to better create a functional life history classification among the different AM fungal taxa. They further note that these appear to be related to phylogeny and can be used in successional and in ecosystem C dynamics.

The second model is the resource-ratio hypothesis, or R* rule, from Tilman (e.g., (710)). The gist of this model is that plants differentially acquire resources. Plant species 1 takes up resource 1 at lower concentrations, whereas species 2 can take up resource 2 at lower concentrations. In this model, there are resource concentrations where one plant species predominates, resource concentrations where neither survives, and a region wherein both coexist. The hypothesis is that mycorrhizae allow a plant to take up resources at lower concentrations than could be acquired by the plant alone (e.g., (195)).

A critical step is to understand the interactions of two plants, with versus without mycorrhizae, or two mycorrhizal fungi and more fungi adding other individuals of the same plant species and then among species. For example, for N and P, two nutrients for which two species of plants are in competition (Figure 6.2). When a mycorrhizal plant of species 1 has a fungal species with high phosphatase activity, the concentration of available soil P in which the plant can grow declines to below that of optimal growth for species 2. The same process can occur, say if a fungal species that breaks down organic N is added. With both species of fungi present, the habitat available to both plant species, and in which competition can occur, can dramatically increase.

This rather simple, but powerful concept, increases the potential range of habitat for plants by including low nutrient soils, but also alters the region for competition. To expand this simple concept to a very complex system, we have only to add mycorrhizal fungi that have additional traits opening resources available to plants, by individual species

Table 6.1 *Functional Traits of mycorrhizae. Modified from Allen et al. 2003 (75).*

a. Specificity of association with host plant (species, age/development)
b. Biomass and morphological/structural characteristics:
  size and growth of individual fungal genets
  total biomass per soil volume
  intensity of root colonization (percent infection)
  characteristics of the radical phase of fungus: structure and thickness of EM
    mantle, presence and Arum versus Paris mycorrhizal structuring
  characteristics of extraradical organization, including architecture and density,
    cords, rhizomorphs, fans, or mats, responsiveness to organic matter patches,
    hydrophilic or hydrophobic properties
  production and size of sporocarps, spores, or other propagules
c. Other life history traits:
  lifespan of individual genets, roots, extramatrical hyphae
  primary strategy of root colonization
d. Physiological/biochemical characteristics
  enzymatic capabilities – ability to utilize organic nutrients
  organic acid production – ability to utilize and access rock nutrients
  membrane transporters
  stable isotope fractionation
  metabolic rate, nutrient uptake efficiency, and cyanide insensitive respiration
  nutrient immobilization
  saprotrophic capacity
  production of antibiotics, phenolics, or other secondary metabolism compoints
  production of phytohormones
  production of heavy metal-chelating compounds
  ability to degrade toxic organics (biotic or xenobiotic)
  exudation of organic materials to the hydrosphere
e. Traits for functional response-type classification
  i optima and tolerance range for seasonally fluctuating or spatially
    heterogeneous abiotic factors
  ii competitive ability against other mycorrhizal fungi
  iii competitive ability against saprotrophic organisms
  iv vulnerability to natural or anthropogenic disturbance, such as grazing by
    fungivores and other disturbances

(specificity), or by complexes of species (generalist fungi). Furthermore, to really make this model complex, we can add other mycorrhizal fungi with an array of traits (Table 6.1) and the tens of hundreds of species of plants that alter each other's fitness, creating another n–dimensional hyperspace that both defines individual species and simultaneously the complexity of species interactions comprising a community. Our

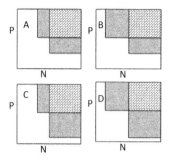

Figure 6.2 R★ patterns for growth of plants in N by P availability both plant species (bricked cross hatching), sp1 (\\), and sp2 (//) with (A) no mycorrhizae, (B) with mycorrhizal fungus 1, (C) with mycorrhizal fungus 2, and (D) with both fungi.

question is what happens in soil when we add one resource preferentially, does it increase or decrease diversity?

For example, we undertook extensive field research including N-fertilization studies at a field site in New Mexico (49). This region is dominated by piñon-juniper (*Pinus edulis-Juniperus monospermum*) vegetation with an interspace mix of C$_3$ (including *Artemisia frigida*) and C$_4$ (including *B. gracilis*) grasses and forbs. The grasses were AM, dominated by small *Glomus* species, such as *Gl aggregatum* and *Gl desertorum*, postulated by the Chagnon et al. (158) model as exhibiting ruderal to stress-tolerant strategies, and the juniper had taxa of Gigasporaceae (competitive to stress-tolerant). Fungi with multiple strategies dominated the EM pine. For example, *Russula alutecea* has a sporocarp δ^{15}N signature of 4.2, or minimal fractionation. This fungus shows a medium smooth exploration strategy, based on fungal architecture (355) or a competitive strategy (in the Grime sense). Alternatively, *Rhizopogon pinyonensis* has a sporocarp δ^{15}N of 9.1, suggesting fractionation (as per Hobbie and Agerer (355)), exhibiting a long-distance exploration strategy, or a ruderal strategy. Both of these mycorrhizae increase the zone of nutrient uptake, R★, for the plants involved. However, when $NH_4^+NO_3^-$-N was added, productivity increased for all plants. The stress-tolerant to ruderal AM fungi (*Gl intraradices*, *Gl fasciculatum*, *Gl desertorum*) were abundant. The AM plants benefited from the stress-tolerance provided by the Glomeraceae. For the pine, initially growth increased but the larger size created a greater demand for water, pushing the availability of water below the acquisition threshold (R★) for the piñon pine. All pines in the experiment died. But the AM juniper thrived with the N fertilization.

I do not view these models (Grime and Tilman) as mutually exclusive, but as different ways to characterize interacting species comprising a community. Indeed, the literature is growing rapidly, documenting the increasing complexity of communities. Large databases such as FUNGuild <https://github.com/UMNFuN/FUNGuild>, which allows for the rapid assessment of traits for taxa that have been described, and mycoDB <https://datadryad.org/stash/dataset/doi:10.5061/dryad .723m1>, detailing which plants respond to which fungi, creates an enormous matrix of host by trait for better modeling the functioning of individual communities.[1] Do organizing mycorrhizal plants or fungi by traits, or determining R★ patterns help in understanding mycorrhizal communities? That question is still open. We sorted AM fungi into guilds based on traits, and experimentally altered water and N availability (766). The respondent species shifted in abundance as expected based on their traits and guilds. Clearly, this is an area needing additional experimental testing, and maybe better criteria to determine boundaries of strategies, traits, guilds, R★, and likely other approaches to organizing plants and fungi of interest.

Spatial Structure

Spatial structure clearly plays a large role in determining functional interactions. The traits organized by Hobbie and Agerer (355) explicitly depend on the spatial structure of the EM fungus. Some, such as *Pisolithus tinctorius* and *Scutellospora calospora*, have hyphal networks that extend long distances connecting many individual plants. Others, such as *Cenococcum geophilum* and *Glomus deserticola*, form a local stress-tolerant mycelial structure. Both groups can connect multiple plants and exchange resources simultaneously. Interplant transfer of both C and nutrients has been demonstrated multiple times. However, the extent of this connection remains controversial. Do big plants, in fact, facilitate the establishment of seedlings? Small plants may well tap into hyphal networks maintained by larger adjacent plants, and locally acquire nutrients from those networks. C is harder to assess. Seedlings acquire small amounts of C from large plants. But, is this enough to alter survival and fitness?

---

[1] Both websites last accessed November 16, 2021.

How long must a hyphal network exist to facilitate establishment? Staddon et al. (677) found that AM hyphae turned over within five to six days using ^{14}C labelling, whereas Treseder et al. (725) found that some AM hyphae could persist for an entire growing season using minirhizotron observations. But, the average AM fine roots survived only 150 days and AM hyphae survive an average of only 30 days in a tropical rain forest (692). EM root tips often appear to survive a growing season in both the Alaskan boreal forest (627) and a Mediterranean mixed forest (59). Rhizomorphs lived 11 months in an arid piñon pine forest (720), but individual hyphae can grow and die within diurnal time scales (346). These data suggest an annual seasonal cycle, but radiocarbon signatures show C formed in EM sporocarps to have been fixed between two and six years previously. Clearly, there is a temporal complexity that varies greatly between vegetation types, mycorrhizal types, fungal and host species, and host communities. The persistence of the network and the ability of the community to support interplant transfer is likely highly variable and needs to be determined for a wide range of communities.

Feedbacks

Feedbacks represent the final regulator to community composition that I will deal with in this chapter. As many researchers have noted, negative feedbacks tend to increase stability in communities whereas positive feedbacks drive one or more species to increase, driving others to extinction and thereby decreasing stability. If this pattern truly predominates, then plant pathogens should be a positive driver for biological diversity, and mutualisms such as mycorrhizae and N_2 fixers should be a negative driver for biological diversity. While the role of negative versus positive feedback in modeling studies support this conclusion, going back to 1974 (491), few experimental approaches do. Is this discrepancy due to limitations of the models or of our understanding of mycorrhizal dynamics?

Many experiments and concepts have been undertaken and described to create complex interactions among multiple species that explain when and why positive feedback can, in fact, lead to stability. These have repeatedly been used to justify the outcomes of increasing mycorrhizal fungal diversity as a mechanism for increasing plant diversity and stability. Klironomos (411) postulated that the range in responses of members of the plant community, from negative to highly positive, coupled with the presence of pathogens, resulted in the observed stability of a mycorrhizal community. Bever (119) showed how, by shifting between plant–fungal preferences, multiple positive feedbacks create negative feedbacks that increase stability of a plant community. All of these are plausible, and not mutually exclusive.

I would argue that these all probably occur, but also that the basic premise of the r and K models and the greenhouse-based micro- to mesocosm experiments are also inadequate. The K is density dependent. It is a single value that is approached, or the population can fluctuate around it, but it is based on limited resources. These models were originally built within the context of a constant environment, and a crucial assumption is that K is approximately constant. In fact, the environment is never constant and therefore K must always vary, often widely. A community is NOT a zero-sum game. A positive feedback to one resource in one patch (one dimension of $R\star$) is different from the positive feedback to the second resource in a second patch (a second dimension of $R\star$), and both patches exist within the same community (up to n dimensions each with $R\star$ values for each species present).

We can build resource feedback interactions as we build complexity. We described the process using stoichiometric relations (75) that we described in Chapter 4. The plant alone will be able to take up N as HNO_3^-, and P as available HPO_4^-. However, adding mycorrhizal fungus 1 allows the plant to access NH_4^+ as well, thereby increasing its production and competitive ability, to a limit of HPO_4^- availability. But the neighboring plant may have fungus 2, which has an alkaline phosphatase increasing its P uptake, to the limit of HNO_3^-. The fungus 2 on plant 2 may cross over, forming a mycorrhiza with plant 1, increasing both overall productivity but also improving the competitive ability of plant 1. Plant 2 then might connect with fungus 3 using organic N, again shifting both overall production but also the competitive interactions. Importantly, there has been no change in the plant, simply the addition of mycorrhizal fungi with different resource extraction abilities.

Mycorrhizal plants interact with an increasing complexity of mycorrhizal fungi, which we can describe using our simple stoichiometry model. For example, we studied shifting resources in a piñon pine-juniper community in New Mexico, USA. As available N was added, the N supply to the plant obviously increased, and the pine (*P. edulis*) grew more rapidly, expanding its leaf area (49). However, another resource is locally depleted, such as HPO_4^-. The tree, with multiple EM fungal symbionts, could have switched from depending upon a *Russula alutacea* (smaller, intense exploration and depletion) to a

Rhizopogon pinyonensis and explore additional, distant patches (based on the isotope ratios and using traits described by Hobbie and Agerer (355)). As leaf N increases and grows, HPO_4^- becomes more depleted, the tree might switch to a *Hysterangium* that produces oxalates, weathering bound inorganic P, such as $Ca_3(PO_4)_2$ (55). If both N and P increase, then demand increases for both those two resources, and water becomes limiting, based both on isotopic ratios and declining root production (49). We see this process emerging, as *Cenococcum geophilum*, a fungus known to be drought tolerant, increases. The EM pine growth collapsed upon the limitation imposed by the third limiting factor (water). But, the oak (*Quercus turbinella*) and AM juniper (*J. monospermum*) persisted right on through the N fertilization and the drought (49).

In fact, the more complex the community composition, the more strategies for P acquisition the plant can depend upon. The same process can benefit the mycorrhizal fungus. A *Scutellospora pellucida* in a seasonal tropical forest will tap a tall *Caesalpinia eriostachys* with leaves in the top of the canopy during the rainy season, but as the soils dry out and leaf drop occurs, the mycelium will tap cacti such as an understory *Opuntia decumbens* that has extremely high concentrations of chlorophyll, or an *Opuntia excelsa* that grows well into the canopy. Both are active and mycorrhizal during the dry season sustaining the fungus until the rainy season and new tree growth recurs (21). Clearly, seasonal environments create at least annual opportunities for new taxon combinations.

Importantly, even the most stable environment has a spatially and temporally patchy soil environment. For example, in an old-growth tropical rain forest, such as the La Selva Biological Station, within the alluvial soils, there are many nests of leaf-cutter ants (*Atta cephalotes*, between 1.8 and 2.3 nests per hectare (233; 560). By harvesting seedlings establishing on the nests, and creating clearings of up to 100 m², these ants create patches within the vegetation. Root and hyphal mortality within the nest footprint is higher than in control soils (692). Individual ant nests persist between 10 and 16 years. Each nest has both patches of soil without roots or mycorrhizae and nutrient-rich patches of nest garbage dumps (692). When they are abandoned, the open patches show a remarkable rate of new root colonization (Plate 5). AM fungi follow roots into these patches. But importantly, the fungi do not recolonize and infect roots as fast as the roots grow. This creates opportunities for both new roots to colonize from different plant patches surrounding the nests, and different AM fungi, colonizing from a neighboring patch. The continuous disturbance is one mechanism creating the opportunity to increase local diversity. Other disturbances include animal diggings, individual tree mortality, and severe storms. We will discuss the importance of patches and succession processes later. But models built on an assumption of a constant environment are inadequate and do not represent the dynamics of real communities. Storms, animal disturbances, and tree turnover represents continuous opportunity for multiple taxa to colonize, maintaining diversity. Stability is useful in modeling, but is it relevant where disturbance is continuous?

Spatial Structure of a Community

Distinguishing a community, whether by eyeballing a patch of relatively homogeneous vegetation, or by identifying mechanisms and boundaries through statistical properties, produces opportunities to measure and compare communities. We often perceive boundaries, such as between a meadow and a forest, and use that boundary to differentiate communities. However, what is a patch, versus what is a community? Subsequently, what is a disturbance creating a patch, and how do we differentiate that from a disturbance large enough to initiate succession? Finally, how do mycorrhizae fit into understanding ecological succession? These are critical questions to address in any study of mycorrhizae and community ecology. What differentiates a disturbance that creates simply a patch within a community, versus one that initiates succession creating a new community? Patches may have small perturbations within a larger community that do not reset the community, but allow for the opportunity for new connections, and even invasions. Alternatively, succession disturbs a large enough unit of surface area to reset the community living on that patch of land. It can be either a minor perturbation, leaving soils largely intact – secondary succession – or a disturbance severe enough to eliminate soil structure and organisms – primary succession. I will address these questions to the level of the patch in this chapter, addressing succession in Chapter 8.

Describing a Community Means Defining the Boundaries

Defining a community is initially spatial. The community is usually identified subjectively, then assessed using statistical tests for homogeneity based on species similarity in numbers, coverage, or biomass. Change in the distribution of organisms comprising a community is to be expected. The fungal species composition found at the base of a shrub differ from those at the edge of the canopy, which, in turn, differs from the interspace (e.g., (62; 416)). In assessing the EM fungi of *Pinus edulis* across two harsh sites in Arizona, Gehring and colleagues (278) found that one or a few EM fungi dominated nearly every tree, but that the fungal communities of nearly each individual plant were different. Are these part of the same community? Do the different fungi compete to occupy each individual tree? Is occupation simply a function of who arrives first or micro-gradients that we do not understand? To date, host plant communities define the mycorrhizal fungal community boundaries almost exclusively, not measurements of interaction effects on fitness. So how can we use measures of plant communities to help us interpret community structures of mycorrhizal fungi, then on the overall community?

First, what is the edge of a community? Edges, whether at the below–aboveground interface or between two vegetation types, are intrinsically crucial. In some cases, it is obvious that one moves from one plant community into the next; in others, more of a gradient exists (777). For the fungal community, it is far more challenging. Once we move belowground, it is extremely challenging to know when a boundary has been crossed. To define the community boundary sampling the fungi using a grid or transect becomes a first step. As samples are analyzed along the transect a rarefaction curve is generated. As the curve levels off, then you have defined the community. In a classical rarefaction curve taxon numbers increase as the area or number of samples across a grid (or transect) are taken, approaching an asymptote as the taxa within that community show no additional new species. When there are jumps in the curve, a "broken-stick" model, then a boundary is reached (75).

Nor does space alone create boundaries. Time and edaphic conditions may as well. Taniguchi et al. (695) found that, through the dry season, EM fungi found on black oak (*Quercus kelloggii*) roots declined. But with a small precipitation event, richness (the number of species) of EM fungi rapidly increased. With added water, even more were found. Thus, multiple seasonal sampling and even sampling of events are needed to describe the taxa comprising the community. Are the drought and watered communities different? As most of the fungi are the same, with a few additions, we assumed that both samplings are of the same community. Going further, when examining the communities of *P. ponderosa* along a N deposition gradient in California, some EM fungal taxa disappeared as N was added, either by fertilization or by N deposition, but others appeared that were not found in the control plots (661). All taxa were present, so why were they detected differently? Determining where boundaries exist, and what comprises a mycorrhizal fungal community remains a challenge. Only when boundaries are identified can they be analyzed. How do we measure communities?

α Diversity

α Diversity is the diversity of a single community. Once a community has been delineated, the second challenge is to characterize it. Three measurements are used by ecologists to quantify species in a community: richness (S), or the number of taxa; evenness (J), or the relative frequencies across the taxa; and overall diversity, which integrates both elements. The latter is most often assessed as Shannon–Wiener diversity (H') or as Simpson's diversity (D). Richness (S) is simply how many species are detected within a unit area using a particular sampling technique (such as sporocarps, morphotypes, or operational taxon units (OTU) based on sequence similarities). In past studies, community composition was based on the richness of sporocarps or, for AM fungal communities, morphologically distinct chlamydospores. Commonly, between 10 and 40 taxa of EM fungi were identified per community, and 5 to 10 AM fungal taxa. Sequencing has demonstrated that these are underestimates, and richness of AM communities may be more similar to EM communities, often with 60 or more taxa in temporal climates (716) or even several hundred in tropical communities (514). Although the newer technologies continue to increase the numbers of OTU's per community, the first step still remains to S. At our study site, we examined the characteristics of the community of EM fungi associated with Q. *kelloggii* (695). Using 12 composite samples per sampling period, we found an $S = 123$ (OTUs). The next step is to assess diversity, which incorporates both richness and evenness. Using Simpson's Index of diversity (D) where:

$$D = 1 - \left(\sum n(n-1)/N(N-1)\right),$$

n is the number of individuals of each species and N is the total number of individuals. Here, $D = 4.2$, or the Shannon Index, H', where:

$$H' = -\sum_{i=1}^{R} (p_i{}^* \ln p_i),$$

and p_i = proportion of individuals belonging to the ith species. Here, $H' = 2.0$, and evenness, J, where:

$$J = H'/\ln S.$$

Here, $J = 0.9$.

We found that drought reduced EM fungal diversity, but that diversity then rapidly increased with the addition of water. Overall, these values allow investigators to experimentally examine community dynamics and, from there, compare different communities and better understand how taxa are organized across regions, perturbations, or other topics of interest or γ diversity.

γ Diversity

γ Diversity allows us to identify larger scale biogeographic patterns in the structure of communities. Once we have described multiple single communities, we can then compare and contrast multiple communities. For example, we can simply count those species present in each of two communities and calculate Sorensen's index of similarity (SI), where:

$$SI = 2w/(a + b),$$

where w = species in common, a = species unique to community a, and b = species unique to community b. This number can be converted to an index of dissimilarity = 1 − SI. To contrast multiple communities SI values can be converted to γ diversity using multivariate ordinations, such as a Bray-Curtis ordination, to display how multiple communities relate to each other, and provide scores that can be correlated with environmental parameters.

How are these analyses undertaken? Here, I step through an example of a simple Bray-Curtis ordination, as this technique was devised before the availability of computer programs (139). As John Curtis was one of the major professors of my major professor, Martha Christensen, I originally learned ordination from her analysis of the fungal communities (164). This can provide insight as to how ordinations can be constructed. I provide an example from an AM spore dataset across the range of sagebrush in western North America (16).

Artemisia tridentata subspecies *tridentata* (sagebrush) radiates out from the Great Basin, in western North America, which began opening about 30 MYa. This dominant plant concentrates between the Wasatch Mountains in Utah and the Sierra Nevada in California. One question that emerged is, as this plant fingers out northwest into British Columbia

(Canada), southwest into Baja California (Mexico), northeast into North Dakota (USA), and southeast into New Mexico (USA); did it take its mycorrhizal fungi (AM) with it, or did it adjust to the AM fungi already present in those regions? To study this question, we undertook spore collections, isolations, and identification of Glomeromycotina taxa from 99 sites across the range (16). From that analysis, I created an ordination of site by Glomeromycotina taxa (Figure 6.3).

Here, the richness per site, with the exception of the center of the range, was mostly 10 to 15 taxa. Across all sites, there were 48 distinct morpho-taxa. Importantly, H' was approximately 2.5 and evenness slightly less than 1 (16). Altogether, this tells us that as *A. tridentata* radiated out, it probably did not carry the baggage of past AM fungi with it, most likely it adapted to those already located in the invaded environment.

As computer technology improved, and sequencing became both less expensive and easier, more sophisticated analyses that incorporate numbers of individuals, either counts of spores or sporocarps, or reads of identified OTU's, have proven of value. Modern variants include analyses such as Canonical Correspondence Analysis (CCA), de-trended correspondence analysis (DCA), Non-Metric Multidimensional Analysis (NMDS), and others. Together, these provide powerful tools to help characterize, describe, and compare communities and study mechanisms correlated with shifting composition.

As examples, Krüger et al. (433) used NMDS of OTUs to study AM fungal taxonomic diversity and community composition as a function of age of restoration of a site. Interestingly, as sites aged, the variation in composition was greater in older than younger sites, suggesting that all of the factors that change with succession, including plant communities, spatial distribution, and soil chemistry contributed to the increasing diversity of the fungi. Morgan and Egerton-Warburton (514) looked at OTU numbers across two seasonal tropical forest sites and found soil chemistry to be a strong determinant to sample richness. Urcelay et al. (731) compared the compositions of AM fungal taxa as different plant functional elements were removed using PCA and compositional com-parisons. Here, a rather surprising outcome was that the AM fungal community was rather resilient, persisting well after plant functional groups were removed. This outcome was supported by a recent study based on molecular sequencing of shrub versus grass roots. Phillips et al. (565) found that the sites of two chaparral communities had a similar pool of AM fungi, but NMDS analysis of the OTUs indicated that the

A

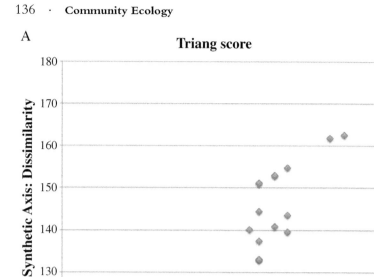

B

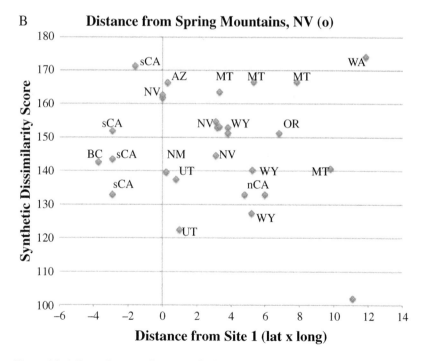

Figure 6.3 A Bray-Curtis ordination of Glomeromycotina (from a morphology-based taxonomy (636)) associated with the widespread plant Artemisia tridentata

fungi dominating within the roots differed between the native shrubs and the invasive grasses within a site, but only differed slightly between sites.

Egerton-Warburton and colleagues (217) expanded the scale of the AM fungal communities to compare across North American grasslands. They used NMDS to study the relationship between N and P, for AM fungi across the North American grasslands, ranging from shortgrass steppe in New Mexico and Colorado, to tall grass prairies in Kansas and Minnesota. While grass species shifted across precipitation and temperature gradients, the AM fungal spore community composition was driven by fertilization, N, P, N:P ratio, along with annual net primary production and plant species composition.

Using an interesting and increasingly common approach, Rosinger and colleagues (623) evaluated the results from Europe-wide comparisons of EM fungal communities. They compared the EM fungi of sites dominated by *Fagus sylvatica*, *Picea abies*, or *Pinus sylvestris*. Here, they defined the community as the fungi associated with a single tree species at a particular location defined in their database. The overall Simpson's diversity (D) was 7.76 for *F. sylvatica*, 6.81 for *P. abies*, and 7.57 for *P. sylvestris*. Given a total richness of 253 for *P. abies*, 84 of which were distinguished as taxa, means that less than half could be identified at the genus or species level. The richness was surprisingly similar (269, 177, and 218 for the three tree species). The average taxon richness per site was 18, with an evenness of 0.016, 0.015, and 0.017, respectively. The evenness value is very low, indicating that each location had a few very dominant taxa that were different from those at the other sites. This

---

Figure 6.3 (cont.) subspecies tridentata. Here, a simple matrix of similarity, then dissimilarity was created for each site. Each step is described in Bray and Curtis (139). Sites were organized along both Axis 1 (from the southernmost site, northern Baja California, Mexico, to the northern most site in Washington, USA, and the Synthetic Axis of Dissimilarity (A). There was a clear geographical pattern. So, from the triangulation scores, I used the distance from site 1 (the Spring Mountains, Nevada, USA, located in the center of the range of A. tridentata tridentata, within the Great Basin, USA, against the Synthetic Dissimilarity Score. (B) The distance from site 1 south- and westward across southern California (sCA sites) into Baja California to the left of site 1, and northward from other Great Basin locations in Arizona (AZ), New Mexico (NM), and Utah (UT), and then northward into northern California (nCA), northern Nevada (NV), Wyoming (WY), Oregon (OR), Montana (MT), and Washington (WA). The plant subspecies remained the same, but as this plant dispersed into new regions, we postulate that it adapted to both the edaphic conditions and the local AM fungi.

pattern was not as extreme, but not dissimilar from that found by Gehring et al. (278) for individual trees across a volcanic soil, cinder-cone environment. But these differences allowed them to hone in on trends in community composition across different types of gradients. Two interesting patterns emerge. First, they reported that diversity does not change with N deposition. This is really a rather startling conclusion, in direct contradiction of earlier studies. For example, Arnolds (87) and Lilleskov et al. (459) found that, across shorter N input gradients, EM fungal diversity declined. But Rosinger and colleagues (623) found that there were interfering factors. Two bell-shaped curves emerged to Simpson's D values. First, stand age showed a peak at approximately a century. They postulated that this represented a time period when both early-stage- and late-stage-EM fungi overlapped. Second Simpson's D peaked at a pH of about 6.5. But, adding NO_3^- reduces soil pH; the two factors are not necessarily independent. In European forests the critical load for N is 10 to 20 kg ha^{-1} (436), complicating the deposition interpretation. So, what are the mechanisms for change across the continent? Was the shift in composition random or a turnover in taxa as climate changed? Did N deposition cause a random shift in the composition of the community? For teasing out these dynamics, we need to understand β diversity. This simple parameter would allow a reader to assess the causal mechanisms for change across the gradient, whether linear or curvilinear.

β Diversity

β Diversity, or the turnover in taxa as we move between communities across landscape or region, or other gradient allows us to begin to understand patterns that will become crucial to understanding ecosystem characteristics and predict ecosystem change (Chapter 7). Occasionally, distinct boundaries are present but, more commonly, communities grade from one to the next (778). Here, I show the use of the indices discussed for α and γ diversity to help us understand how to use them to help describe community dynamics. This example comes from our work using a N deposition gradient in southern California, coupled with paired N-fertilization plots. Using old-growth (>100 years) *Pinus ponderosa*, 50 km apart across an N-deposition gradient within the San Bernardino mountains in southern California (data from (661)), in addition to the deposition gradient of low and high deposition, both sites were fertilized with N to understand how added N would affect the EM fungi. EM were analyzed using the ITS region of the nuclear rRNA (275). Interesting responses were observed across this gradient and treatments. The low deposition site had the highest richness, which was reduced with each N addition. D also declined, although H' changed only at the high deposition plus fertilization. J increased with N (Table 6.2).

What changed was the turnover, or β diversity. Using Sorenson's index of dissimilarity where: β = 1 − SI (where SI is Sorenson' index of similarity). With each N addition, the turnover was more than half of the taxa shifted. A number of taxa dropped out, and additional replacement occurred. Using a CCA (661), the impact of increasing N is obvious. Some diversity is lost but, more importantly, what changed was the composition of the communities as we move across space or a gradient such as N deposition. This is crucial. Rosinger et al. (623) found, using a CCA, that NO_3^- deposition had a significant impact on fungal community composition, along with stand age, pH, and temperature. Hence, the impacts of perturbations such as N-deposition may be less on overall diversity, but on the composition and the traits exhibited by the individual taxa of mycorrhizal fungi, than on diversity *per se*.

Island Biogeography

Island biogeography provides another approach to assessing turnover and dispersal (e.g., (35; 54; 287; 557; 558)). This approach may work especially well with plants that are widely distributed but in clumps and fungi that are wind-dispersed (35). In southern California, live oaks (*Quercus agrifolia* and *Q. engelmannii*) are distributed in patches where groundwater is near the surface, or perched pockets of water are located, creating patches in a range of sizes across the landscape. My lab collected sporocarps and assessed root tip morphologies for four years in the 1990s on old-growth (>100 years) oaks nested watersheds within a matrix of chaparral and grasslands (54). SI indices indicated that many of the same fungal species occupied both of the live oaks, *Q. agrifolia* and *engelmannii*. Species increment curves showed a broken stick model, but a continuous increase becoming asymptotic of approximately 130 taxa, across 20 "islands."

Island biogeography looks at species turnover as a function of the area of the island, where $S = ca^z$, where c is a constant, a is the area of the island or patch, and z is the rate of change in richness with island size. The z-value is especially useful as it is independent of size, and allows

Table 6.2 Community indices of the EM fungi of Pinus ponderosa across the San Bernardino Mountains of southern California. These sites represent a nearly background old-growth stand, from a low-deposition stand to a high N deposition stand (approximately 30 kgN ha^{-1} y^{-1}). At both sites, N fertilizers were added at 150 kg N ha^{-1} y^{-1} (from 661).

Site and treatment index	Low deposition control	Low deposition + fertilization	High deposition	High deposition + fertilization
Richness	69	54	53	49
D	1.64	1.60	1.44	1.46
H'	0.91	0.91	0.99	0.88
J	0.78	0.81	0.86	0.88
β versus control	–	0.57	0.57	0.60

comparisons of different communities (622). We studied live oak (*Quercus agrifolia*) islands in southern California. Within a single canyon, using root-morphotypes and RFLP patterns, our z-value was 0.57 (r^2 of 0.96 for the species area curve). Across Camp Pendleton, a large Marine-Corps base with a natural area footprint of 52 thousand hectares, in southern California, the z-value based on four years of sporocarp collection was estimated at 0.58 (r^2 of 0.975 for the species area curve). This is a higher value than the z-value of 0.23 reported by Peay et al. (558), who sequenced root tips from soil cores and fruit bodies for a single season on "tree islands" of *Pinus muricata* near the California coast.

This type of analysis leads to several interesting questions of the mechanisms controlling diversity, because it allows for predictions that can be tested. As background, z-values for species of true islands range from 0.06 to 1.31. Further, z-values for true islands tend to be higher than for isolated patches within terrestrial landscapes (468). Peay et al. (557) expanded their study assessing richness, resulting in 107 species of EM fungi. Here, the size of the island did not appear to be a significant determinant. But distance from the forest edge or "mainland" was critical in determining EM fungal taxon richness. The species compositions indicated radiation from the forest edge. They postulated that dispersal and genetic barriers constrained the diversity of fungal partners. In the oak study, the size of the patch was a major determinant. Could the *P. muricata* "islands" be more like expanding (or contracting) terrestrial patches, whereas the *Q. agrifolia* "islands" were like true islands? *Q. agrifolia*, being a "live" or evergreen oak, requires access to a patchily-distributed, year-round water source (590). These water-access patches include groundwater or perched water tables that are patchily distributed around the landscape. Alternatively, *P. muricata* stands are comprised of seedlings that are always trying to establish beyond the edge of the main forest.

A second question also emerges. In comparing hillside oaks (*Q. agrifolia*) with those in the valley, the greater richness of the EM fungi with *Q agrifolia* in the valley was attributed to a lowered drought stress. However, might the diversity also be a function of the widely dispersed hillside trees versus the larger "island" of trees in the valley (590)? Also, projecting richness to an oak island of 100 HA of land (about as large as a patch as I have observed), we would project that there could be 500 taxa of EM fungi. Could this be accurate? It is rare that an oak "island" would be larger than this size. What then would be the turnover between islands?

The EM fungi of subalpine tree "islands" of *Pinus albicaulis* and *P. contorta* dispersing from the forest "continent" edge of Yosemite National Park, California, were analyzed (287). Out of 238,895 sequences, 96 OTUs were found, with 10.7 OTUs per tree. Both "island" size and distance from the forest edge showed expected island biogeography patterns; the number of OTUs increased with patch area and decreased with distance. But these trees were evaluated as individuals of various sizes. The forest showed a richness of 49 OTUs, considerably less than the 96 OTUs identified from the tree islands where even the nearest individual tree "island" was only 8 OTUs. Although they developed interesting analyses of EM fungal taxa, and relationships between individual tree size and distance, what happens when these individuals become organized into island patches? And, how do stands of pines and oaks compare with continental diversity, such as across Douglas Fir (*Pseudotsuga menziesii*), comprised of large stands across western North America? How do these values compare with new stands planted in other continents for forest harvest? These are all questions that could be addressed once we have additional analyses using the various basic models for assessing biodiversity of mycorrhizal fungi.

Architecture

The architecture of vegetation is also important to understanding mycorrhizal communities, but rarely considered in describing mycorrhizae at a stand level. There are two distinct types of architecture that we need to consider, and for which there is very little information. The first is from the soil surface into the tops of tree canopies. The second, as roots and mycorrhizal fungi extend downward through the soil and even into bedrock. Architecture remains a critical factor in addressing biodiversity issues for mycorrhizal fungi.

Dickie et al. (206) noted that EM fungi from different soil layers varied, as different taxa adapted to the marked chemistry changes. Mycorrhizae can also be found deeper into the soil, even in fine cracks in the bedrock. As a broad, weakly-tested generalization, the mycorrhizal fungi found deep in the soil profile are a subset of those found in surface soils (34; 214). In part, this habitat is challenging for mycorrhizal fungi due to the nature of geotropism; plant roots are positively geotrophic, but mycorrhizal fungi are not. But this hypothesis is hardly tested and should be.

As we move up into canopies, epiphytes play critical roles in the ecology of forested communities. We know that mycorrhizal epiphytes exist high up in communities ranging from tropical rainforest canopies (e.g., (524)) to canopies of the tallest trees including the majestic red-woods of California (223). A number of epiphytic plants are in groups that form mycorrhizae, such as Cactaceae and Bromeliaceae (293). However, we do not know the taxonomic composition of their mycor-rhizal fungi. There are AM relationships, EM seedlings can be sometimes observed, orchid mycorrhizae are highly diverse, and dark-septate-looking fungi within roots can often be seen (72). But are their species a subset of those found on plants of the forest floor or are they unique?

Other Factors

Many other factors also affect the richness and diversity of mycorrhizal fungi in a community. In earlier samples, identifiable spores and spor-ocarps formed the basis of sampling efforts. These represent what had been or might become active, but not necessarily the taxa currently active. More recent molecular approaches find different reservoirs of fungal taxa in soil than on active root tips in both EM (e.g., (590)) and AM (630) fungal communities.

Other anomalies manage somehow to persist in patches in commu-nities. *Gnetum* vines (a gymnosperm) can be found in tropical rainforests associated with EM of a *Scleroderma citrinum* complex of fungi in neo-tropical and old-world tropical rainforests. Mycoheterotrophs form link-ages between mycotrophic photosynthetic plants and heterotrophic plants, although their benefits to the photosynthetic host, or even the fungus, are still unclear (e.g., (120; 701)).

We also have virtually no understanding of the fine endophyte, Mucoromycotina fungal communities. This work is just beginning and will hold interesting clues to the early history of mycorrhizae, but also to the composition and functioning of mycorrhizal fungal communities.

Orchids represent a unique situation but hold some clues for mycor-rhizal fungal biodiversity in general. Hadley (307) postulated that orchids were broadly non-specific with their mycorrhizal fungi. However, that is not necessarily a consistent feature. Kartzinel et al. (402) found a wide diversity of fungi, which should not limit re-establishment. However, they suggested that the environment was the constraining factor limiting establishment of individual fungal taxa. Downing et al. (210) found that invasive species of orchids formed mycorrhizae with a broader range of fungi, whereas native orchids were more specialized. This finding has critical ramifications for understanding the role of invasive species (both plant and fungal, as we will address in Chapter 10).

Food Webs and Communities

Food webs are complex interactions of producers and consumers. Mycorrhizae affect food webs both directly, in that both plants and fungi serve as food for animals, bacteria, and even other fungi, and indirectly by providing resources to plants, thereby interacting with plant–animal community dynamics (Plate 6).

The bane of every mushroom collector are the small insect larva that also search out edible (and nonedible) sporocarps. One must get out early in the morning to find newly emerging mushrooms and beat the competition from animals. Yet, the larvae and emerging insects are the base of forest food webs, supporting larger insects, small mammals, and owls, small birds, and hawks. The potential complexity of the web for mycorrhizal sporocarps is enormous. But mycorrhizal fungi are directly involved in both microscopic and macroscopic food webs. Nematodes, mites, collembola, and a host of other invertebrates graze on mycorrhizal fungi. Across a community, the average mycorrhizal fungal lifespan ranges from 10 to 30 days in tropical rainforests, with many surviving less than a day (692), to up to 48 days in a seasonal mixed-conifer forest (59). Invertebrate grazing is probably the most common cause of mycorrhizal fungal turnover (85; 245).

In the Pacific Northwestern United States, one of the iconic endangered species is the spotted owl (*Strix occidentalis caurina*) that preys largely on northern flying squirrels (*Glaucomys sabrinus*) and red tree voles (*Arborimus longicaudus*). The primary food of these rodents, in turn, are EM truffle fungi (487), commonly found in old growth forests (583). EM fungi here form an intricate component of the food web. I continue to wonder if part of the problem of a declining endangered species in southern California, the California spotted owl (*S. occidentalis occidentalis*), is the current decades-long drought, resulting in low numbers of truffle fungal sporocarps, and a potential local extirpation of San Bernardino flying squirrel (*G. sabrinus californicus*) across the Mount San Jacinto area (65). Importantly, the interdependency of rodent-like mammals, EM fungi, and host trees is common in forest food webs around the globe (555).

The complexity of mycorrhizal relationships might also be topsy-turvy. During an unusually wet spring, under a snow pack, numerous dead spores were found, each with a hole that we postulated as nematode grazing (51). Alternatively, Francl and Dropkin (254) found that *Gl fasciculatus* will attack the endoparasitic nematode, *Heterodera glycines*. Is this a case of an AM fungus that feeds on a nematode? This observation needs additional work and confirmation.

Just as importantly, there are many indirect effects of grazing animals, from insects to vertebrates, on energy channels through food webs. For example, AM increase the populations of the insect parasitic wasp, controlling whitefly on tobacco; a case of an influence of AM through three trophic levels (790). We will discuss these patterns in Chapter 7 on ecosystem dynamics. It is important to note that we know little about the composition of the mycorrhizal fungal taxa comprising the change in energy allocations, a critical topic in our quest to better understand ecosystem dynamics.

Time and Community Structure

Most studies describing mycorrhizal communities focus on the composition at a limited point in time that does not encompass variation in composition over time. However, nature never stands still. Every community has a time element. Temporal anomalies can exist in small patches or may result from environmental incongruities, be successional, or be driven by global change. There are distinct properties associated with each of these drivers that determine the community composition. How do we distinguish among these temporal anomalies, the mechanisms regulating each, and their ultimate impacts on community structure and composition? Just as important, each of these can be superimposed on each other. Climate is always changing. Succession may be neverending, as even old growth stands change, a community always exists across environmental gradients and patches always emerge. We will deal with each in turn.

Every community contains patches and some are more distinctive and readily defined than others. In arid ecosystems, plant communities contain shrub "islands of fertility" that are comprised not only of nutrients and organic matter, but also mycorrhizal fungi. There are distinct differences in composition of mycorrhizal species between a shrub crown, under the canopy, and in the interspace (62; 416). Oaks in grassland savannas also support different EM and AM fungi, depending on whether they are isolated, living on perched water tables, or in stands where groundwater can be accessed (590). Patches also exist within an established community that might be barely discernable when looking across a landscape (Plate 5). Sunflecks are variable in time and space. Gaps in leaf distribution occur which, when consistent in time, can result in dramatic differences in soil temperature and moisture, thereby altering root and mycorrhizal fungal activity. A pocket gopher will emerge, creating a patch of soil with a different inoculum level than the surrounding community. In a relatively stable community, the inoculum is reduced creating an opportunity for a less mycorrhizal-responsive plant to invade (422; 423). A badger may dig into a gopher burrow, creating a patch of soil richer in inocula than the surrounding soil (28). These are not trivial disturbances. Gophers are common across temperate North America, and can maintain a meadow community (that is AM) within the matrix of a forest (that is EM) by turning over the soil every 10 to 14 years, constraining tree establishment (64). Lightning might strike a single tree, burning a few square meters, but creating a gap in the forest for reestablishment of mycorrhizal fungi and seedlings. In a tropical seasonal forest, blowdown gaps caused by hurricane winds (21) are often less than 100 m^2. Here, the plants in these gaps were largely either understory species or seedlings of trees similar to the surrounding forest. All were mycorrhizal, with the same taxa as present in the surrounding forest. Even "normal" events shift fungal species compositions. Taniguchi et al. (695) found that as the normal dry season progressed, EM fungal richness declined, but increased again with a monsoonal precipitation event. What was particularly interesting was that the predominant taxa emerging were the infrequent tails of the species increment curve. Are such disturbance events especially important to maintenance of rare species?

As the perturbation grows in size, the temporal event becomes less frequent. Large disturbances are more likely to disturb, either partially or completely, the existing community and initiate succession. Historically, the study of succession has been treated as a community process. Odum's (539) classical paper, in *Science*, changed our view as a species-change process to rethinking of succession as the interactions of plant community composition and ecosystem processes. Much of my career has focused on mycorrhizae and succession, from the perspective of both plant–plant and plant–fungal species interactions and how those influence water, nutrient, and carbon cycling. Because succession is such a major ecological process, I will defer the discussion of succession and communities to a special chapter on succession, Chapter 8. As described by Odum, I will also treat the process of succession as both a community and an ecosystem process, consisting of both interdependent characters and independent processes.

Integration: Mycorrhizal Community Diversity Patterns

Are there generalities that characterize mycorrhizal communities? I will try to organize contrasts in AM and EM communities that might be useful hypotheses and then look to see if there are other generalities that we can draw out of the literature.

Richness patterns emerge that require further testing and refinements but might be useful to understand community patterns. Firstly, EM communities tend to have higher richness than AM communities. This pattern was especially observed in older sampling based on morphology. Newer sequencing techniques increase the richness in both, but especially in AM fungal communities. More studies are needed, but a couple of observations stand out. In looking at temperate region communities, 150 to 300 distinct taxa are often found in EM fungal communities of limited plant diversity, often oaks or pines, and 40 to 100 in AM fungal communities in highly diverse plant communities. But EM fungi have evolved multiple times, whereas Glomeromycotina AM communities are likely monophyletic. (Note: We know virtually nothing about Mucoromycotina AM fungal communities). Interestingly, if we look at single clades, maybe they are not that different? For example, in southern California, in a *P. ponderosa* community, where we found 145 EM fungal taxa, we found 19 Boletaceae and 15 Thelephoraceae operational taxonomic units (OTUs). In a coastal-sage grassland AM community, we found 84 Glomeromycotina taxa, but 21 Glomeraceae OTUs, a number relatively similar at the family level (766). As our understanding of basic diversity emerges, a better understanding of richness patterns should also emerge.

Michener (499) reported that diversity in bees peaked in temperate compared with tropical regions. In part this might occur because hyper-diversity of tropical plant communities constrained specialization by the bees. A similar argument has been used to address mycorrhizal fungal diversity. Is there really an inverse relationship between plant and mycorrhizal fungal richness? In part, as we better understand AM fungi, does the richness pattern remain? In the late twentieth century, still only about 150 species of AM fungi had been described based on morphology, several with global distributions (16). Using small subunit rRNA genes and pyrosequencing, Öpik et al. (542; 543) estimated 282 virtual taxa using a 97 percent similarity across the globe. While sequencing increases diversity of AM fungi, it is not a particularly dramatic increase (746). Certainly, nothing has been found on the order of the 5,400 species found forming EM symbioses (510). Two-thirds of the taxa reflected a climate zone or a continental limit. The tropics do not stand out as having high AM diversity despite the known hyper-diverse plant communities.

The diversity of EM versus AM fungi at the scale of a single, bounded community may not be dramatically different. While the richness of AM fungal communities is higher than older morphological surveys suggest, the differences are insufficient to alter the hypothesis. Egerton-Warburton et al. (217) found 63 taxa of AM fungi in temperate grasslands based on spore morphology, with 54 in mesic grassland and 45 in short-grass *B. gracilis* steppes. This compares with the 63 taxa of OTUs found by Hiiesalu et al. (353) in a shortgrass steppe *B. gracilis* community in Saskatchewan, Canada. Varela-Cervero et al. (735) reported 71 OTUs in a Mediterranean community, with 47 from spores and 47 obtained in roots, and Weber et al. (766) reported 84 AM OTUs in another Mediterranean community. This is slightly lower than the 80 to 150 taxa generally reported for EM fungal communities. In South American temperate *Nothofagus* forests, 72 morphotaxa of EM fungi were found (537), a range often reported for temperate EM fungal communities. A value of 2,000 taxa is often quoted as indicative of EM fungal taxa found in association with Douglas fir across its range, a nearly mono-dominant tree in many communities of western North America (16), but that is distributed across half the continent.

Why is the relative biodiversity of a single community relatively similar between AM fungal and EM fungal communities, whereas the overall biodiversity of EM fungi is much higher than AM fungi? Morton et al. (517) assert that because of the ancient lineages of AM fungi, at large scales, there are simply historic limits to biodiversity and distribution. Another hypothesis (not mutually exclusive) might also be at work. Eighty percent of all plants form AM, but only 30 percent form EM (with some forming both). AM communities include hyperdiverse plant communities in grasslands and tropical rain forests, whereas EM communities tend to exist in monodominant to low richness tree communities. We do not have good estimates for β-diversity or z values, but I would hypothesize that EM fungal communities exist as "islands" within the larger matrix of the AM world. Especially in the tropics, these might be islands of *Quercus* (291; 660), conifers such as *Abies-Pinus-Quercus* patches (149), or *Dicymbe* (179; 345) in Neotropical forests, Dipterocarpaceae in southeast Asia (e.g., (670)), or Leguminosae in Africa (e.g., (7)). To test carefully, we need better information on the various measures of community structure, α, β, and γ diversity, and turnover, such as are measured using island biogeography tools.

Plate 1 Types and basic structures of mycorrhizae.

Plate 2 Evolution of mycorrhizae from the Devonian Period. Shown is (A) Marchantia sp., a modern liverwort in the Los Gigantes mountains of Argentina, (B) the Marchantia thallus, with a arbuscule mycorrhizae in the rhizoids, and hyphae of both coarse AM (Glomeromycotina) and fine endophyte (presumed Mucoromycotina), and (C) Sigillaria sp. fossil root with emerging rootlets from Nova Scotia. *Evolution of mycorrhizae in the Cretaceous Period.* Shown is the rain forest with the primitive conifer Gnetum (D) and its EM root, specifically Scleroderma sp. (E). Also emerging are orchids (F) and their unique mycorrhizae (see Plate 1H), as well as both EM mycoheterotrophs such as orchids and Ericaceae, and AM mycoheterotrophs including Arachnitis uniflora (G). Nonmycorrhizal Cluster roots emerging with plants in extreme nutrient deficient conditions (H). *Mycorrhizae change from the Miocene Epoch to the Anthropocene Epoch.* With the reduction in CO_2, C_4 photosynthesis emerged and nonmycotrophy expanded. Shown is a nonmycotrophic Salsola kali (J) in the deserts of Central Asia and (K) the browning response showing Salsola tragus rejecting AM fungi (44; 47), to the (L) widespread distribution of the nonmycotrophic Papaveraceae in naturally nutrient-rich soils surrounded by distant grain fields with high levels of agricultural fertilization (see discussion in Chapter 4).

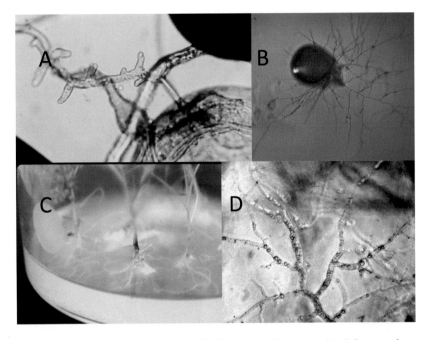

Plate 3 What comprises an "individual" of an arbuscular mycorrhizal fungus of Glomeromycotina? Are these both haploid, diploid or polyploid, homokaryotic or heterokaryotic? Shown are (A) a germinating hypha emerging from a germ shield (arrow) of a *Scutellospora sp.* and branching, (B) the multiple germinating hyphae from a single spore of *Glomus (=Rhizophagus) fasciculatus*, (C) The establishment of a new mycorrhiza from surface-sterilized germinating spores onto *Bouteloua gracilis* without other organisms (68), and (D) nuclei scattered along aseptate hypha exploring the substrate in the field. Photo courtesy of Mayzlish–Gati and Allen.

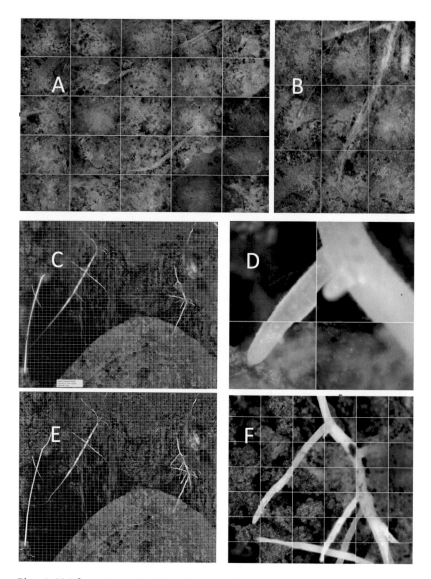

Plate 4 AM formation at La Selva. Showing Control (A and B) where every root is infected nearly immediately, versus roots invading an abandoned leaf-cutting ant nest, where AM hyphal cannot keep up with growing roots. Images (C) and (D) are comprised of growing into the abandoned nest (500052 scan), followed by (E) and (F) (500059 scan), seven days later.

Plate 5 Disturbances that create gaps may locally alter mycorrhizae but have minimal impact on the surrounding community and are rapidly reinvaded. Shown is (A) a lightning strike that burns a patch approximately 10 x 10 m, (B) a gopher bioturbation where gophers feed on roots and mycorrhizal fungi, (C) a bison wallow creating a small disturbance patch, (D) a gap created by a leaf–cutter ant nest in a tropical rainforest. The sunfleck light is on the forest floor, (E) light penetrating the gap created by ants to the forest floor (the PVC structures protect the sensors), and (F) root regrowth (using an automated minirhizotron) after abandonment of the ant nest.

Plate 6 Food webs of mycorrhizal fungi. Shown are (A) grazing of a growing hyphal tip by Symphylans, (B) a Cordyceps growing after parasitizing an Elaphomyces, (C) grazing of a Russula sporocarp by ants, (D) moose (Alces alces) grazing, that reduces EM, and (E) insect grazing on *P. pinyonensis*, where the large tree was resistant to grazing and heavily mycorrhizal, and the small one susceptible with less mycorrhizae. See text for further details.

Plate 7 The Serengeti, where mycorrhizae play a critical role in supporting the entire grazing landscape. This view shows the crossroads between the P-rich soils in the south and the P-deficient soils in the north where mycorrhizae are absolutely critical to plant production. The wildlife cycle around the Serengeti following production. Without the P provided by AM fungi, the northern production would be inadequate to sustain the grazing, hence the predator densities (682). Photo from © Rick Johnson Photography, reproduced with permission

Plate 8 Bootstrapping and EM fungi. This site in the San Bernardino Mountains, in southern California, burned in 1952. The first vegetation re-establishing was *Arctostaphylos patula*, forming arbutoid mycorrhizae with many of the same fungi existing in the former forest. Now, pines (*P. ponderosa*, *P. monophyllum*) are establishing in the *A. patula* stands, forming EM with many of those same fungi (see text for details).

Plate 9 Overview of Mount St. Helens research sites through succession time. Shown is (A) the pre-eruption view (from Jim Trappe with permission), (B) a nearby control study forest, (C) the blast zone above Bean Creek looking towards Mount St. Helens (the author in 1982), (D) the 1984 Pumice Plain gopher introduction patch (see text), and (E) the 2018 Pumice Plain site looking towards Mount St. Helens.

Plate 10 Impact of Hurricane Wilma on the El Eden forest.

Plate 11 Mycorrhizae are critical to conservation of ecosystems (A) that support endangered species, such as (B) Geoffrey's Spider Monkey (*Ateles geoffroyi*).

Plate 12 Reserves protecting diversity of mycorrhizal fungi. Some are intentionally designed for the fungi, such as (A) the Luci National Reserve in the Czech Republic with Boletus rhodopurpureus, a rare mycorrhizal taxa with Quercus, others such as (B) Boletus (=Baorangia) bicolor in Crawford Notch, New Hampshire, USA, with widely disjunct populations, and (C) Boletus edulis at Lake Dennison, Massachusetts, USA, the "same" species across Europe and North America.

Why does this matter? Currently, we are in the early stages of a human-caused mass extinction (Chapter 9). If we broadly have a high diversity of EM fungi, but that diversity is distributed in small, individually low (or moderate) diverse patches or islands, then each patch has novel EM taxa and every large disturbance contributes to a measurable loss in global biodiversity of one of the groups of organisms supporting plant communities. I will discuss the implications of this loss in Chapter 9, but they could be critical for the future of the planet.

Summary

- Mycorrhizae do not exist in isolation, but in communities comprised of the mycorrhizal partners and interacting organisms. The first challenge to measuring the interactions comprising a community is defining what it is, and what are the boundaries.
- Two concepts of communities are useful here: the first based on fitness, and the second on space. One definition of a community is the organisms that affect each other's fitness. The second definition is a spatially similar group of organisms. These two approaches are not mutually exclusive but do measure the community differently.
- Identifying members of a community based on fitness is conceptually straightforward, but very difficult to actually identify and measure. Measurement of community processes can be undertaken, such as competition and facilitation. Relationships can be arrayed along Grime's Competition-Stress-Ruderal triangle based on traits and on locations. Facilitation between plant and fungus is often measured. Competition is challenging and studies are rare. They can be either response related using crowding coefficients, or mechanistic, looking at $R^\star$ relationships.
- Because it is difficult to see belowground spatial boundaries, sampling of spatial structure for mycorrhizal fungi, using sporocarps, spores, cores, and sequencing are challenging both spatially and temporally, but need to be well described. Architecturally, mycorrhizae exist from several meters deep to the tops of the canopy. Spatially and temporally species increment, or rarefaction curves facilitate identification of community boundaries.
- Once the boundaries are derived, measures such as α diversity of a single community, and γ diversity (landscape diversity) are useful. α-diversity, measuring richness, evenness, and diversity indexes are useful

to help describe and understand a community. For γ diversity, similarity and the use of ordinations provide good information. Between α and γ diversity, β diversity is the measure of taxon turnover. Turnover is important in scaling from a landscape to regional diversity and biodiversity broadly. One useful approach is to use island biogeography. The z-value is a useful index but needs additional research. The role of patches and. ultimately, succession creates disequilibria, but as disturbance is both a normal, and an abnormal process, the characteristics of the site are important. Because island biogeography theory assumes that richness is in equilibrium, this attribute should be defined.

- Food webs are an element critical to understanding communities. Who consumes whom links autotrophic plants to C exchange, decomposers, and animals. Every element of biodiversity has a distinct and crucial role in community foodwebs.

7 · Ecosystem Dynamics

Mycorrhizal fungi are a critical linkage between the soil, containing water and inorganic and organic nutrients, and the plant itself, the photosynthetic unit that converts solar energy to complex C compounds upon which the terrestrial world depends. Sir Arthur G. Tansley (696) first defined the term ecosystem, noting that organisms cannot be separated from their environment, thereby forming a biological–chemical–physical system, or an ecosystem within a relatively stable but dynamic equilibrium. This, the origin of the term ecosystem, breaks down Clements' concept of biome, where climate is the overriding regulator of the community, into a view of the multiplicity of interacting biotic, chemical, and physical components that allow for the existence of life.

By building upon Tansley's definition of ecosystem, we can study how mycorrhizae affect and are affected by their ecosystem. But how do we approach this? An ecosystem is an integrated group of processes. By definition, then, there must be at least two components, a primary producer, and a decomposer. Starting as atmospheric CO_2, Carbon (C) is fixed by the primary producer. The C, H, and O are fixed into carbohydrates, using catalytic enzymes comprised of N (especially the abundant RUBP carboxylase) and P as the energy storage and transfer compound, as well as a host of other elements for various processes. The complex compounds comprised of C, N, P, O, and H (in addition to other elements) are then mineralized by decomposers and recycled. The simplest examples of ecosystems have no mycorrhizae, such as the endolithic community of cyanobacteria and saprotrophic bacteria within silica rocks of Antarctica (262), a type of enclosed community sought on Mars as the best chance for an ecosystem beyond the confines of Earth. As the area increases, the complexity of ecosystem processes also increases along with the diversity of organisms comprising the occupying community. The Antarctic Dry Valleys utilize carbohydrates remaining from ancient fixation or from atmospheric deposition, supporting a simple decomposer food web of bacteria, yeasts, saprotrophic nematodes, and a single predatory nematode (157). As life on Earth continually evolved for over 4 billion years, the communities involved and the cycling patterns have grown ever more complex, including mycorrhizae, the symbiosis that connects primary production and nutrient cycling across most terrestrial ecosystems.

As such, mycorrhizal fungi are an integral part of all aspects of terrestrial ecosystem functioning and comprise a large module of carbon (the medium of energy exchange) and nutrient dynamics and budgets. Mycorrhizal fungi are often the single largest consumer measured in most terrestrial ecosystems and, by regulating nutrient and water allocations, determine the overall rates of net primary production (NPP) and nutrient cycling. The literature on ecosystem dynamics of mycorrhizae, while dramatically improving, remains underwhelming in comparison with quantitatively lesser processes such as grazing. A cursory glance at publication numbers using Web of Science in 2020 shows 511 papers on mycorrhizae and primary production where mycorrhizae consume between 10 and 40 percent of the NPP, versus 1,131 papers on cattle and NPP where cattle consume between 1 and 5 percent. The study of mycorrhizae and ecosystem dynamics is becoming much more exacting and exciting with a plethora of new tools, from isotopic signatures, to the identification of genes for specific ecosystem processes, to new models that tease apart how elements are transformed to global models incorporating mycorrhizae into global atmospheric composition and change.

By the mid-twentieth century, EM were known to be intimately involved in forest ecosystem processes. Frank (258) noted that mycorrhizal fungal hyphae permeated the forest floor, especially concentrating at the litter–humus interface, thereby playing a critical role in the N nutrition of a forest. Stahl (678) postulated that mycorrhizae increase water throughput from soil to leaves in forests. Rommel (619) showed that mycorrhizal fungi were dependent upon direct C fluxes from the host using a trenching approach. Hatch (334) proposed that the hyphae of mycorrhizal fungi extend from roots across the forest to acquire nutrients, and Kramer and Wilber (430) demonstrated uptake of ^{32}P by mycorrhizal fungal hyphae and transfer to the host tree. Melin and colleagues (reviewed in 1953 (497)) found that EM fungi would short-circuit the N cycle by shuffling amino acids to the host tree from the forest floor litter. Went and Stark (770) went on to propose that mycorrhizal fungi formed a large proportion of the soil fungal biomass, involved in recycling nutrients from decaying organic matter to active plants.

The first estimates of mycorrhizae and NPP in the 1960s to 1980s demonstrated that mycorrhizae were a large state variable in C cycling (247; 248; 318; 676; 714; 753), likely a greater sink in many terrestrial ecosystems than all the animals combined (Figure 7.1).

Just as for a community, how do we define a boundary? In MacMahon et al. (472), the ecosystem boundary for a taxon would be its distribution. That could be small, as a single valley for some plant species, or the entire globe for a great white shark. For actual measurements, some sort of bounded unit – a watershed for the Hubbard Brook LTER site, a stand of several ha for a coniferous forest, a patch of several m for a grassland. For studying mycorrhizae, all are useful. The key for mycorrhizal studies is that the investigator should define the unit, whatever it is, so that comparisons can be understood and calibrated.

Biomes and Mycorrhizae

Mycorrhizae evolved early and multiple times. Importantly, that process included adapting to new habitats that derived from changing landscapes and climates across the last 450 million years of the Earth's history. Today, we are embarking on a journey into the Anthropocene Epoch, whereby humans are modifying even atmospheric CO_2, and thereby the entire range of global ecosystem processes. We have already seen that a large fraction of terrestrial C fixed by plants was allocated to mycorrhizal fungi (Chapter 4) as early as 450 MYa. The many types of mycorrhizae separate themselves based on the host plant types and the different plant-mycorrhiza types separate into different regions of climate space, each of which affected C exchange.

Using the climatological plant distribution concepts of von Humbolt and Bonpland (371), the biome concept of Clements (170), and the ecosystem perspectives of Tansley (696), we can divide the world into biomes that represent different vegetation types. Moser (518) broke the world into higher latitude EM forests and mid- to lower-latitude AM forests. Read (597) modified the diagram adding high latitude ericoid mycorrhizal heath, and linking the mycorrhizal types to N and P cycling. Allen et al. (16) added AM deserts, shrublands, and grasslands, and noted the crucial role of patches of EM monospecific stands within the lower-latitude biomes. Importantly, both deserts and grasslands can be found at both high latitudes and high altitudes within low latitudes (as first mapped by von Humbolt and Bonpland (371)). In general, most of the

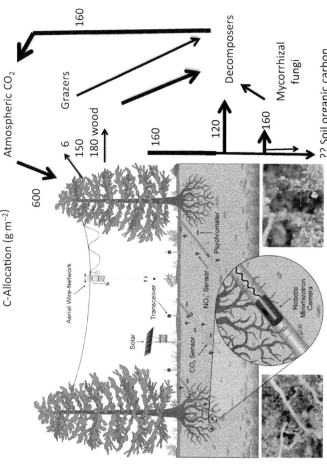

Figure 7.1 A simplified overview of annual flows of C using a Soil Ecosystem Observatory (SEO) measurement system, at the James NRS, in a California Mixed Forest Ecosystem. Sensors and models of C flux were coupled with daily or more frequent measurements of standing crop of roots and fungi; these lead to the study of the dynamics of ecosystems, not just static features (58; 59; 76; 346). Figure drawn by Jason Fisher, with permission

ecosystem-scale EM forests in the neotropics predominate at higher elevations, although not exclusively.

Why is this important? Vegetation types are dependent upon climate characteristics and mycorrhizal types also revolve around soil–climate interactions. Interestingly, ecosystem dynamics appear to shift along biome–mycorrhiza type axes. Vargas et al. (743) undertook an analysis of 50 flux-tower sites globally for estimates of fluxes by plants and climate. They analyzed 236 site-years of data. Using structural equation models, they characterized the relative contributions to forest mycorrhizal type of annual temperature (T), precipitation (P), or photosynthetic photon flux density (PPFD). PPFD was not statistically significant. Using standardized coefficients as estimators of control, interannual variation in T, with a structural coefficient of 0.67, was controlling in EM-dominated vegetation, accounting for 50 percent of the variation in gross primary production (GPP) which, in turn, accounted for 74 to 93 percent of ecosystem respiration (R_e). Alternatively, CO_2 fluxes in AM-dominated ecosystems were regulated by variation in precipitation, with a structural coefficient of 0.52, accounting for 81 percent of the variation in GPP and 93 percent of the variation in R_e.

A number of other studies also differentiated ecosystem functioning of AM and EM vegetation, and estimated implications for broader global processes. Fisher and colleagues (240) developed what they called the FUN model, or the Fixation and Uptake of Nitrogen model, built around the C costs of N exchange to plants. This model includes the role of mycorrhizae *implicitly* (based on forest mycorrhizal type) in C and N cycling, which can be assimilated into the larger Joint U.K. Land Environment Simulator (JULES), a dynamic model that integrates N into general circulation models (GCMs) that form the backbone for predictions of global change. Brzostek et al. (145) developed FUN2.0 that used C costs for N uptake to differentiate AM and EM, and tested these against experimental forested sites. Shi et al. (656) then incorporated AM and EM vegetation types into FUN2.0, where the mycorrhizal types were scaled to global levels, providing a means to link FUN into the Community Land Model (CLM), a vegetation model that feeds into JULES. For breakout of vegetation types, Shi et al. (656) broke the world into 17 plant functional types (PFT's), ranging from unvegetated bare soil, to different forest biomes, grasslands, and agricultural systems that were used to assess N/C, and thus C exchanges between the terrestrial and atmospheric components of the CLM (see also 673; 689). I will focus in the outputs and impacts of such modeling later (Chapter 9), but will

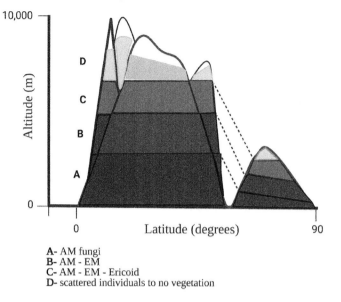

A- AM fungi
B- AM - EM
C- AM - EM - Ericoid
D- scattered individuals to no vegetation

Figure 7.2 Mycorrhizal types shift across both altitude and latitude (see text for details). Figure created by Danielle Stevenson, with permission

build the ecosystem dynamics supporting these models through this chapter.

The concept of biome, based on the von Humbolt and Bonpland (371) model of vegetation belts, and linking with the mycorrhizal-specific ecosystem types as characterized by Read (597) and Allen (36), can be used to build mycorrhiza-centric C, N, and P cycling patterns (Figure 7.2). Here the differing vegetation belts are sorted by climate patterns from sea level at the equator, poleward in latitude and higher in altitude. The tropics at low latitude and altitude are predominantly AM with AM-based nutrient cycling. At an altitude of 1,000 m, or a latitude of 15°, we begin to see an increase in EM forest taxa that continue to increase to an altitude of nearly 4,000 m in the tropics, or at a sea level at a latitude of 60°. Approaching this zone, forests are increasingly domin-ated by an EM overstory and an AM understory. Upon reaching an altitude of 5,000 m a return of AM forb-grass begins. At high altitudes, water becomes highly limiting because the diffusion coefficient dramat-ically increases. Above 5,000 m, plants and mycorrhizae become scarce, but AM, EM, and ericoid mycorrhizae can be found. Alternatively, at low altitudes but high latitudes (~50°), evapotranspiration is low, leaving high soil moisture and a shift towards a predominance of ericoid mycorrhizae. Higher than 55°S or 70°N, the few plants, like at high elevations, form AM, dark septate, and likely Mucoromycotina AM. In each categorization, from Humbolt to the present, it is critical to remember that there are always exceptions that are both compositionally and functionally interesting. Understanding these patterns (663) can play a large role in modeling the impacts of climate change.

In using a biome-based approach, it's important to remember that biomes, like communities, rarely have sharp boundaries, and they are constantly shifting with human disturbance, climate change, and geological activity. Mycorrhizal associations respond directly to these shifts (104; 681). For this chapter, I focus on four predominant mycorrhizal types, nonmycorrhizal (NM), AM, EM, and ericoid (ErM), in which to describe the relevant ecosystem processes. These processes can then be rebuilt into types that can be used for the Plant Functional Type (PFTs) for global change modeling in Chapter 9 (see FRED, Fine-Root Ecology Database, https://roots.ornl.gov/).[1]

Energy: Element Stoichiometry and Ecosystem Currencies

Energy is the basic ingredient of life. For most of terrestrial life, plant leaves take in photons of light and use that energy to create potential energy, splitting $6H_2O$ and synthesizing it with atmospheric $6CO_2$ to glucose, $C_6H_{12}O_6$, releasing $6O_2$. The potential energy is used as kinetic energy respiring C as CO_2 back to the atmosphere. $C_6H_{12}O_6$ is used by the plants, and by all organisms dependent upon that sugar, including mycorrhizal fungi, animals, parasites, and decomposers. We have already described the roles of N, P, and other elements at the level of single organism physiological process (Chapter 4). Here, we think about the integrated process at the level of the stand, or watershed, or patch, or whatever spatial unit one choses. The ecosystem is comprised of the pieces that process the flow of energy and how it is coupled with elements that carry and utilize that energy, in both biotic and abiotic forms.

The challenge is to take the models of Treseder (Figure 1.1) and of Ågren (Figure 4.3) and convert them from units of an individual leaf to units of surface area or volume to scale up to the ecosystem. In ecosystem studies, the surface area of ground cover is often used, generally meters

---

[1] Website last accessed November 16, 2021.

squared or hectares. This value could include the biomass extending upwards to the top of the canopy, including leaves, branches, detritus, animals, aboveground fungal sporocarps, and anything else that is comprised of or utilizes the fixed C. This value includes biological activity that extends well up into the clouds – a topic beyond our scale, but worthy of study and potentially contributing to our understanding of environmental change. The value for net C fixation should (but often does not), include the roots, fungal hyphae, microorganisms, animals, and biomass distributed below the ground surface. Researchers generally focus on the top few centimeters, maybe up to one meter deep. However, mycorrhizal root systems extend several meters below the surface. Unless defined otherwise, mycorrhizae alter C and nutrient cycling of m^2 to ha of the soil surface, affecting C from the top of the canopy downward to the depth of the root tips.

Top Down: C and Mycorrhizal Costs for N and P

A key concept for modeling is to find a currency that can be tracked. In economics, that concept is money, whether dinars, dollars, euros, or other currency. In ecosystems, the key unit is energy, and the currency that exchanges energy is C (540). What controls C exchange between atmospheric CO_2 and biosphere C forms? Leaf photosynthesis, the conversion of CO_2 into potential energy, or $C_6H_{12}O_6$, comprises GPP. What controls the rate of photosynthesis? Fifty percent of global N is estimated to consist of one enzyme, ribulose-1,5 bisphosphate carboxylase, which promotes that conversion. The energy from the conversion is stored as ATP, which is converted to ADP plus inorganic P (P_i). Thus, N, P, and water become crucial regulator valves for C modeling.

To understand the roles of ecosystem components, like any good sleuth, "follow the C." This becomes clear in understanding what regulates respiration of those state variables (roots, mycorrhizal fungi, saprotrophs) acquiring N and P. Vargas and Allen (739; 740) and Vargas et al. (742) examined soil respiration (R_s) across ecosystems, using continuous measurements of soil temperature (T_s) and soil moisture (θ), compared with NASA's moderate-resolution imaging spectroradiometer (MODIS) estimates of R_s and soil CO_2 production (P_s). GPP explained 73percent of the variation in R_s and 30 percent of R_s. Including lags between R_s and T_s and GPP linked the differing vegetation types, phenology, and the control of autotrophic and heterotrophic C fluxes.

Moreover, the response of R_s to T and θ were different between EM and AM forests (743). Thus, to track C, we need to know who is finding the N and P and giving it up; in other words, who regulates the acquisition of the N and P "funding" the C.

The physiological mechanisms of N uptake of EM and AM were discussed in Chapter 4. FUN2.0 was upgraded from FUN1.0 (145) by incorporating C costs to EM for N acquisition, including under high soil N, directly as NO_3^- from the soil solution, moderate N levels as NH_4^+ by mycorrhizae (EM in the model), and when N becomes limiting, as organic N (N_o) through EM. (ErM and N_o uptake has not been integrated into the FUN models (to my knowledge), but represents a probable important next step, for understanding extreme N limitations in peaty histosols and extreme lithosols.) Several researchers then began breaking EM and AM biomes for global modeling purposes (656; 673; 689). For example, Shi et al. (656) followed FUN2.0 by breaking the different pathways spatially among the 17 CLM4.0 Plant Functional Types (PFTs). This provided a global estimate of the role of EM in C transfer between atmosphere, vegetation, and soils. R_s in AM and EM biomes respond to temperature and precipitation differently (743), thereby setting the stage for building global vegetation models based on mycorrhizal type. K. Allen et al. (23) followed FUN2.0 with FUN3.0, incorporating P pools from 45 temperate forest sites and 13 seasonal tropical forest plots by adding in the costs of phosphatases and the high costs of nitrogenase for the C and P cost from the mycorrhizal fungi.

This direction represents a huge transition in integrating mycorrhizae into the global C modeling effort. These pathways can be assessed stoichiometrically and thereby integrated into large-scale C exchanges. Another advantage of these models is that they are modular, allowing them to be successfully integrated into the larger global change models that are becoming so absolutely critical to understanding global change phenomena. A challenge, from my perspective, is that these models still do not incorporate the C exchange of the fungi and soil ecosystem itself, a next step.

Bottom Up Mechanisms: CNP Stoichiometry, Mycorrhizae, and Nutrient Cycling

Another approach to determining whether mycorrhizae are important to C cycling is to look at competing sinks for C. For example, Gehring and

Whitham (279) studied insect grazing in piñon pine. They identified a scale insect (piñon needle scale – *Matsucoccus acalyptus*) that grazed on individual trees. Interspersed were individual trees that were resistant to the herbivore. These trees grew bigger, not only because of the C that was not grazed, but also because some of the C retained by the tree was reallocated to increased EM fungi, improving plant nutrition (see Plate 6). Another study used fencing to exclude grazing moose, which again promoted more EM activity (624). Alternatively, in a grass–AM system (70) clipping effects on AM activity varied as a function of precipitation. Clipping the leaf area reduced the total root length and AM root length in irrigated treatments, but increased AM root length when moisture was limiting. Likely, these data indicate that with reduced leaf area, soil moisture was more available and that, in this high elevation site, radiation intensity was high enough so that plant photosynthesis remained high enough for adequate C allocation to the mycorrhiza. What this tells us is that we need to better understand how mycorrhizae influence nutrient uptake and C allocation across biomes.

Another approach is to follow C from the atmosphere to the plant, to the mycorrhiza, to the soil, and back to the atmosphere. I built a simplified C flow model, tracking C, N, P, and H_2O around our Soil Ecosystem Observatory (SEO, Figure 7.1), comprised of sensors and *in situ* microscopic scale minirhizotrons (58; 76; 346). We measured pools, including atmospheric CO_2, soil CO_2, root and saprotrophic and mycorrhizal fungal standing crop, CHO, H_2O, and NO_3^-. By measuring frequently enough, we quzntified not just standing crop, but also production and turnover, the key dynamic measures that describe ecosystem processes. Isotopic fractionations provide powerful checks on the quantitative measures (49).

To better understand mycorrhizal dynamics, I start by following C then, using C:nutrient relationships, will begin to track how mycorrhizae really provide a keystone role to ecosystem processes.

Carbon Cycling

Carbon cycling means that plants fix CO_2 through photosynthesis into $C_6H_{12}O_6$, where it is exchanged with other organisms and converted into billions of CHO compounds. The first step in understanding C and mycorrhizae is to determine how much C mycorrhizal symbioses add to photosynthesis and net primary production. As we saw in Chapter 4, photosynthesis can increase as much as 40 percent in a mycorrhizal versus a nonmycorrhizal plant. This comprises a large net C sink for atmospheric C, and a C source for soil C processes. At the ecosystem scale, for terrestrial ecosystems, the most common approach to measuring C accumulation is through a measurement technology known as eddy covariance, whereby CO_2 exchange between a canopy and the atmosphere is measured at 10 times per second, or 10 Hz. This speed is necessary to measure CO_2 concentrations in atmospheric flux eddies, along with temperature, windspeed and direction, and water vapor, near the canopy surface, capturing the atmospheric CO_2 concentration (downward flux) and canopy CO_2 concentration (upward flux), the difference being the rate of exchange. Daytime values represent photosynthesis minus respiration, and nighttime values equal respiration. Because this approach has proven so useful, flux towers have been used for nearly four decades in almost all terrestrial ecosystems globally. Flux networks such as AmeriFlux, ChinaFLUX, CARBOEUROPA, Integrated Carbon Observation System – ICOS, and many others, with over 900 sites and over two decades as networks (see (100)). These systems provide an integrated measure of fluxes at many time scales, from seconds to decades. Because these have been used over the many conditions and locations, and because gas exchange is so tightly coupled to leaf N, and leaf fluorescence is dependent upon leaf N, satellite imagery also provides large, spatial-scale C flux information. This forms the basis of the Shi et al. (656) FUN2.0 model. But just as importantly, eddy covariance provides the C inputs and outputs bounding C fluxes within research sites. The James Reserve is a UC Natural Reserve, in a California mixed forest ecosystem dominated by an EM oak-conifer forest with AM meadows (see (59)), and the La Selva Biological Station, an AM tropical rainforest in Costa Rica. I will use these two sites as examples for assessing mycorrhizae and C exchange.

At the James Reserve, the measured net ecosystem exchange (NEE) for 2011 was between –20 and –30 g m^{-2} y^{-1}. (A negative value represents the total C lost from the atmosphere to the biosphere or the total amount of C pulled from the atmosphere by the ecosystem). This value was less than the average anthropogenic C inputs to the atmosphere of about +50 g m^{-2} y^{-1}. That provided an overall picture of the accumulation of C. But of course, 90 percent of the C fixed is respired by the plant before biomass ever accumulates. For that, we need to know the net primary production (NPP). NPP can be determined using direct measurement of growth increments. At the James Reserve, we used a model based on daily temperature, precipitation, soils, and vegetation, called DayCENT (www2.nrel.colostate.edu/projects/daycent/).[2] Using the DayCENT model, we estimated an NPP of 600 g C m^{-2} for 2011, between 3 and 4 percent of which remained as NEE in wood and soil organic matter. Of this, DayCENT estimated that 200 g m^{-2} was allocated to roots. We then turned to the Soil Ecosystem Observatory (SEO) (58) to measure microbial dynamics. The EM fungal hyphal biomass within fine roots was estimated to be 40 percent of the fine root mass (see (319; 753)), or 80 g m^{-2}. The external EM fungal biomass peak standing crop was 19 g C m^{-2} (varying through the year).

By tracking individual hypha, we determined the lifespans, and from that we calculated that there were 3.2 generations per year, a turnover of 30 g C m^{-2}, the external EM fungal hyphae NPP was therefore calculated to be 82.6 g m^{-2}. The EM fungi thus comprised 162.6 g m^{-2}, or 27 percent of the total NPP for the forest.

For this C investment by trees to mycorrhizal fungi, we could also determine fractionation of ^{15}N during the exchange of N from fungus to host tree. Using the Hobbie and Hobbie (358) model (which we will discuss in detail in the N cycling section), we estimated that 43 percent of the N in the tree leaves came from EM fungi, again for an investment of 34 percent (compared with a measured 27 percent) of the host's fixed C. To follow the currency (C), then, an investment of 27 to 34 percent of the trees C for 43 percent of the N seems a pretty good annual return, certainly one I would use for my retirement.

Broadly, most studies suggest that the 20 to 40 percent of NPP is a reasonable range for wildland forest ecosystems. Mycorrhizal fungal hyphae may represent the largest component of microbial biomass (248) and EM mat fungi, such as *Hysterangium*, may occupy up to 17 percent of the soil volume (186). Vogt et al. (753) estimated that EM fungi took 15 percent of the NPP in a Pacific Northwest forest and Fogel and Hunt (248) estimated that 50 percent of organic matter throughput went to the EM fungi. In the James Reserve forest, 27 percent of the NPP was allocated to EM fungi.

AM C budgets are actually far trickier. Much of the challenge comes from the absorbing mycelial branching network, the rapid exchange of C, and the rapid dynamics of hyphal life. EM fungal hyphae, comprised mostly of Basidiomycetes and Ascomycetes, tend to have hyphae that branch but maintain a cylinder form, keeping the hyphal diameter more

---

[2] Website last accessed November 16, 2021.

constant. The AM fungal arterial network, that is readily observed in $100\times$ minirhizotron observations, averages 8 to 10 μm or larger in diameter. But, these fungi also form an absorbing network that drops in diameter with each dichotomous branch down to 2 μm (see discussion of Friese and Allen 1991 (264) model in Chapter 2). This structure makes AM fungi extremely dynamic in nature. Several approaches have been taken to quantify hyphal turnover, mostly based on transfer of labeled-C in laboratory experiments. Staddon et al. (677), using ^{14}C labeling, found that hyphae turned over on the order of 3 to 5 days. Wang et al. (760), using ^{11}C, a rapidly decomposing radioactive C isotope, found that C allocation from host plant to the AM fungal hyphae occurred within a day. In the field, Hernandez and Allen (346) reported diurnal growth and mortality of fungal hyphae. Here, the average individual hyphae persisted for an average of about 30 days. In arid annual grassland, Treseder et al. (725) found that coarse AM fungal hyphae persisted about a growing season.

How these dynamics play out across a stand becomes even more challenging. Overall, Harris and Paul (321) estimated that about 1 percent of the plant's fixed C went to AM fungi, whereas St. John et al. (676) suggested that the amount of C allocated to mycorrhizae could exceed 30 percent of NPP. These very different estimates make determination of AM C allocations extremely challenging. Field studies suggest that in a mixed forest–meadow stand, allocation of C to AM fungi may be less than to EM fungi. For example, in a small AM meadow at the James Reserve, I estimate 0.160 g C m^{-2} (346) hyphae with a 20 day lifespan, for an NPP of 1.3 g C m^{-2}, compared to the surrounding forest with a standing crop of 30 g C m^{-2} of external hyphae for EM fungi and 82 g C m^{-2} NPP (59).

But across ecosystems, what are the values, and how do they vary? Several key parameters are needed. Some are straight forward, such as the length of the growing season and the hyphal standing crops. Lifespan and turnover are challenging to measure. Using isotopic signatures, Staddon et al. (677) estimated a 3–5–day turnover of AM fungal hyphae whereas we measured (using SEO observations) a 20-day lifespan across a complete network (75; 264). In a meadow with a standing crop of 160 mg C m^{-2} and a growing season of 100 days, the resulting annual production (NPP) would equal between 3.2 g C m^{-2} (5 day lifespan) and 0.8 g C m^{-2} (20 day lifespan) in the James Reserve meadow. From the minirhizotron observations, we know that, within this meadow, both production and mortality occurred and varied at daily time scales (346). In the tropical rain forest, lifespans averaged 35 days (692). Far more information is needed to determine turnover rates and C cycling in AM systems. Even a rough estimate shows a range of over two orders of magnitude between ecosystems and all AM ecosystems appear to be less than the NPP of EM ecosystems. *Far more work is needed to understand the C dynamics of AM ecosystems.*

But a key goal is to measure the C costs and N and P gains across enough sites to incorporate into a model hierarchy such as is underway for the FUN model series. These tandem approaches are essential to not only modeling C exchanges but to tease apart alternative management scenarios for future C sustainability.

There are many complexities in C cycling that make a mycorrhiza of interest mechanistically, well beyond the C lost. David Read has argued that EM fungi have enzymes that can break down organic C (668), that allow access to soil organic N. We will address N cycling later in this chapter, but it is important to remember that the impacts of mycorrhizal fungi may be subtle. Mycorrhizal fungi, in searching for N, may alter nutrient cycling by N mining. Here, the lack of N reduces decomposition, a version of immobilization of N within the N cycle. This is similar to green manuring (adding litter with a high C:N ratio) or reducing the numbers of litter grazers, slowing the rate of litter turnover. In this case, the EM fungi first take up N, immobilizing N, but then release it more slowly than a non-EM background N cycle. This is defined as the Gadgil effect (named for Gadgil and Gadgil (269)). It is based on work in an exotic *Pinus radiata* plantation in New Zealand. In some cases, this ability could go to extremes. Chapela et al. (161) studied an exotic introduction of *P. radiata* from California into a native grassland of Ecuador, along with its mycorrhizal fungus *Suillus luteus*. They found a loss of 30 percent of stored soil C within 20 years of conversion to pine plantation. The isotope values indicated that *S. luteus* was utilizing older soil C, in addition to that produced by *P. radiata*. Apparently, the tree did not produce and transport enough C to support *S. luteus*. This fungus was productive beyond the amount of C allocated by *P. radiata* by directly utilizing and thereby reducing soil C.

Once incorporated into mycorrhizal fungal mass, where does the C in mycorrhizal fungi go? There are three distinct pathways for C flow: (1) respired as CO_2; (2) fixed into the fungal mass, much of which is allocated into structural C and metabolic compounds that remain within the cell walls until used by grazers or decomposed; and (3) materials secreted into the soil environment, such as soil enzymes and organic acids. A significant, but understudied fraction is respired. Paul and Kucey (553) estimated that 3 percent of the C was allocated to AM fungal hyphae, and Wang et al. (760) estimated a 20 percent increase in plant C assimilation, with 45 percent remaining in the leaves with a C sink of AM roots being nearly twice that of nonmycorrhizal roots. What this means is that a better understanding of C allocations by the fungi themselves is needed.

The large fraction of CO_2 respired by mycorrhizal fungi play a role beyond simply the C throughput. The soil CO_2 from root respiration and respiring hyphae themselves changes soil chemistry. Knight et al. (418) found that soil CO_2 levels doubled in soils dominated by AM versus nonmycorrhizal plants. This amounted to the equivalent of doubling the carbonic acid to soil, reducing soil pH, and weathering P, from unavailable $Ca_3(PO_4)_2$ to available HPO_4^--P. This mechanism also applies to other cation-dominated ecosystems, for example in soils with $AlPO_4$ and $FePO_4$ (395). As we will discuss in detail shortly in the P section, the added weathered P is critical to P-limited ecosystems. The respired CO_2 may also be critical in sequestering C. In arid lands, in soil solution, HCO_3^- will bind to free Ca, forming $CaCO_3$. This $CaCO_3$ is leached downward until the moisture evaporates, leaving a $CaCO_3$ layer, also known as caliche. There is more C stored in soils and geological layers (e.g., caliche, limestone, marble, dolomite) than is present in the atmosphere. At a global scale, small changes to this inorganic C could make a large difference to the global C budget. Schlesinger (638) and Schlesinger and colleagues (e.g., (637)) studied isotopes of $CaCO_3$ in deserts of the southwestern US, and concluded that a large amount of the CO_2 came from microbial respiration, and a dominant microbial biomass component in these deserts is comprised of AM fungi. Finally, there may be as much dissolved HCO_3^- moving through soil profiles in the tropical rainforest as organic C, providing a large fraction of the C in tropical rivers (Harmon and Allen, unpublished data).

That portion of the C not respired is allocated to mycorrhizal fungi and directly fixed into fungal biomass. Soil animals graze some of the biomass C (245; 414). This pathway occurs especially in EM systems. For AM, grazers will clip hyphae, probably reducing the absorbing networks. AM fungal hyphae are less palatable than EM or saprotropic fungi. This pathway probably plays a large role in increasing turnover and reducing N immobilization. Klironomos et al. (415) and Allen et al. (61) found that mites grazed a large fraction of fungal biomass in the field, reducing the net accumulation of soil C under elevated atmospheric CO_2. The structural material itself is important, as fungal walls are largely chitin. Chitin comprises up to 60 percent of the cell wall material. While it is a relatively simple compound, containing desirable N (230), there is an extensive literature that indicates that chitin is a layered molecule that resists rapid decomposition (292). (Remember, that chitin is comprised of linked glucose molecule chains that become a tough fiber and resist decomposition, compared to numerous individual glucose molecules). Melanin is also common in many mycorrhizal fungi, and melanin compounds are highly resistant to decay (231).

Finally, there is a third way by which C gets into the soil environment. C is lost either by being secreted into the soil environment in the form of soil enzymes, organic acids, or as other proteins.

Mycorrhizal fungi also produce large amounts of enzymes that are released into the soil to extract nutrients from organic nutrient pools. These will be discussed shortly in our nutrient cycling sections, but enzymes from glucosidases, to peroxidases by EM fungi, to phosphatases (by both EM and AM fungi) are constantly secreted and then utilized as C sources by other microorganisms. EM fungi that form mats are well known for producing large quantities of organic acids. The production of oxalic acid by mat-forming fungi is well documented. Cromack et al. (186) showed that the whitish mats of *Hysterangium crassum* accumulated in surface soils. *H separabile* also produced high levels of oxalates in chaparral ecosystems (55). Understanding the rates of oxalate production is challenging, because the methods for measuring oxalates in soil solution are difficult, at best (457), oxalates are difficult to extract and some exist in crystals along the hyphal surfaces (e.g., (55)), and many microorganisms utilize oxalates as a C source breaking down into CO_2 (515). While oxalates appear to be continually released into soil (153; 251; 419), they do not accumulate, indicating that the turnover is somewhat similar to production rates. A background standing crop value of 4 mg kg^{-1}, for instance, does not help us to determine the quantities of organic acids produced and turned over. However, the quantities are high enough to have a dramatic effect on HPO_4^- and on soil cations.

Soil aggregation has long been an indicator of healthy soils in agricultural ecosystems. Aggregates are clumps of soil particles, organic matter, and microbes that are bound together. By clumping these materials, they create a diversity of soil conditions at a sub-centimeter scale that includes patches ranging from anoxic conditions, within the middle, to large gaps with O_2 and water between. Sutton and Shepard (690) found that AM fungal hyphae bound sand particles into aggregates. In the late 1980s and early 1990s, Julie Jastrow and R. Michael Miller showed that the formation of soil macroaggregates was associated with AM hyphae in prairie and restored prairie soils (see (502)). Wright and Upadhyaya (792) identified a novel protein that was abundant in soil aggregates, a protein of AM fungi that they named glomalin. Extensive subsequent research has indicated that AM fungal hyphae are a major contributor to stable soil organic matter. Of interest, EM associations appear not to support soil aggregation (Figure 7.3). Soil aggregation and glomalin are not common in EM forests or in NM successional patches or croplands. But they are commonly encountered in AM-dominated ecosystems ranging from prairies to tropical rainforests. Key elements in soil aggregates are proteins that bind those soil particles together. The glomalin identified by Wright and Upadhyaya appears to be one of those proteins. This compound (or class of compounds) is interesting in that it forms a slime that helps bind particles and organisms together into aggregates. Aggregates with glomalin both immobilize C, but also bind many cations such as Fe, Cu, Cd, and Pb. These aggregates increase C sequestration (612). Glomalin is a heat-shock glycoprotein that is largely associated with AM fungi. Whether this compound (or class of compounds) is produced only directly by AM fungi, or by AM fungi and other bacteria in the mycorrhizal microbiome (211; 475), remains a research question as of this writing. But the role in soil aggregation and in C sequestration looms large. Glomalin comprises up to 25 percent of the recalcitrant organic matter. How stable this material is, again, remains contradictory in the literature. In the tropics, estimates ranging from 2 years (467) to 25 years (613) can be found.

Finally, it is important to recognize that there is an important spatial aspect to how and where mycorrhizae deliver C that is critical to ecosystem C modeling. Mycorrhizal fungi grow with roots deep into soils and rock fragments, both in extracting resources and in sequestering C. In desert soils, Virginia et al. (750) found AM down 4 meters just above the seasonally fluctuating water table. In granitic rock outcroppings Egerton-Warburton et al. (214) found EM following cracks in granite up to 3 meters deep in soils. Bornyasz et al. (131) reported roots growing through earthquake faults down 3 meters into bedrock. They found both EM and AM hyphae winding into granite where water was available. Graham et al. (297) noted that development of cracks large enough for hyphal penetration could occur within about 80,000 years of exposure following deglaciation, and roots could invade after 120,000 years, as measured by cosmogenic ^{36}Cl surface exposure dating.

Figure 7.3 Soil aggregation within a mixed forest stand. Two adjacent patches were sampled about 2 meters apart on the same soil and with the same climatic conditions. Shown is (A) the soil surface from a patch of EM eastern hemlock (*Tsuga canadensis*) and eastern white pine (*Pinus strobus*), and (B) the EM roots from that patch, (C) the adjacent patch of AM red maple (*Acer rubrum*) and white snakeroot (*Eupatorium rugosum*), and (D) the AM roots with soil aggregates. From the Berkshire Mountains in western Massachusetts, USA

Jongmans et al. (391) found that rock etching was associated with EM fungi deep into rock substrates, and Balogh-Brunstad et al. (101) found that EM fungi stimulated silicate dissolution, that was then associated with biofilm formation and etching. While these C distributions do not appear to be large, on a surface soil m^2 basis, the large volumes deep in the soil and the likely long-term nature of turnover make this a potentially important process in C sequestration.

Knowledge of the Water Cycle across the Landscape and at Depth

This is critical to understanding how and where mycorrhizae function. I previously discussed the importance and mechanisms whereby mycorrhizae influence water uptake and throughput in Chapter 4. Here, I focus on where not enough water is present, or too much water is present to influence mycorrhizal functioning. In general, mycorrhizae probably play a minor role in nutrient dynamics in wetlands. Because the O_2 concentration is low, anoxic conditions are detrimental to the growth and metabolism of mycorrhizal fungi. At the other extreme, if water is too limiting, as in extreme arid lands and on rock surfaces that do not hold water and dry out on a daily basis, mycorrhizae may also be of lesser importance. However, unlike bacteria, hyphae can transport water across air gaps, thereby becoming critical for plant production in arid lands. Ericoid mycorrhizae are commonly found in bogs (597). Under extreme wet conditions, AM are present in floating mats, and may well exist in aerenchyma roots where O_2 is pumped into roots under water (e.g., (125)). In tropical rainforest, we found AM activity to persist through even very wet periods, when soils were saturated. One mechanism is redox reactions that sustain O_2 (310). For example, denitrification converts NO_3 to NO_2 to N_2O to N_2, thereby releasing three molecules of O. The coupling of root pipes and hyphal micropipes in these soils with redox reactions may well provide O_2 to sustain mycorrhizal activity under soil saturation. At the other extreme, in dryland ecosystems, AM and EM hyphae form bridges across dry soil pores that can carry water even from ultramicropores (42).

Mycorrhizal fungi may also depend upon the plant for water. During drought periods, deep groundwater that has been tapped by deep roots is lifted to the soil surface during the day when transpiration is rapid. At night, some of that water moves laterally through roots and then into hyphae via hydraulic redistribution (588). Some of the water is exuded out of the hyphae and into soil. That water can acquire added $^{15}NH_4^+$

(218). In the morning, as the stomata open and transpiration starts back up, those water droplets, with the dissolved N, are reabsorbed by the mycorrhizal fungi (590). Individual trees and the mycorrhizal fungal network are sustained by this water over the course of the dry season (408; 589).

We can envision how the role of mycorrhizae in water flux alters the physiology of individual plants (Chapter 4), but what about water provisioning for the ecosystem? In evergreen forests of southern California, trees and their mycorrhizal fungi are dependent upon deep groundwater to survive through the dry season. Nearly 33 percent of the total water transpired by the forest comes from weathered bedrock, generating 61 percent (3.5 mm day^{-1} of the total 5.7 mm day^{-1}) of the respired soil CO_2 as fine roots and mycorrhizae (408). Without hydraulic lift, the perennial plant communities and their EM as well as AM would not persist.

Again, at the individual plant scale, in *B. gracilis* plants, AM increased water throughput by 68 percent, 51 percent of which was due to a reduction in the gas-phase resistance (74) or the increased stomatal conductance of water vapor transport. This means that more water flows through a plant, allowing greater CO_2 uptake. For example, the increase in transpiration rate of AM versus NM plants ranges from 10 to 20 percent (Figure 4.4). However, estimating the differences due to mycorrhizae at the stand scale is virtually impossible as diurnal measurements shift by the minute, dependent upon sunlight, leaf area, temperature, and even wind.

Nitrogen Cycling

Nitrogen cycling was the original focus of research into mycorrhizal functioning. In 1885, Frank postulated that mycorrhizae logically would be responsible for N uptake from the forest floor. Carbon and N are intimately linked, as RuBP Carboxylase, the enzyme catalyzing conversion of CO_2 to $C_6H_{12}O_6$, comprises nearly 50 percent of the global biological N and is directly correlated with CO_2 fixation (e.g., (235)). The role of mycorrhizae in N uptake is the basis for Fisher and colleagues' (240) FUN model. Understanding the relationships between mycorrhizae and N dynamics is crucial for understanding the functioning of mycorrhizae in ecosystems.

N cycling is complex and dynamic. The N cycle is presented in every general ecology and plant and soils textbook. The usual focus is that N_2 is

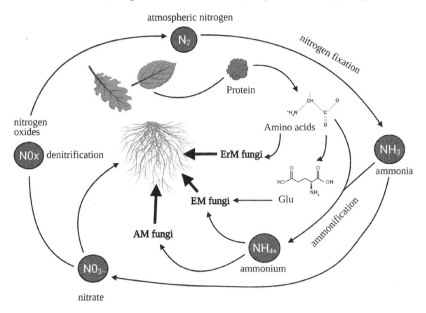

Figure 7.4 The N cycle. Shown are the locations within the cycle that are short-circuited by mycorrhizal fungi. Ammonification is the step between amino acids and NH_4. Nitrification is the step that converts NH_4 to NO_2, then NO_3. Denitrification is the step that converts NO_3N_2, losing N to the atmosphere. Figure drawn by Danielle Stevenson, reproduced with permission

fixed by either lightning (into HNO_3^-) or biological fixation into NH_3 by dinitrogen-fixing bacteria, either in symbiosis or free-living. That N is assimilated into amino acids, which are released into soil upon death of microbes and their plant hosts. When amino acids decompose, the process of ammonification releases NH_4^+ into soil directly. The NH_4^+ may be taken up by an organism or be nitrified into NO_3^-. When fertilizer is applied, it is usually either some form of HNO_3^- or urea, which is rapidly converted to HNO_3^- and then the HNO_3^- goes into the soil solution and is taken up by the plant by mass flow. Upon uptake, the plant must convert HNO_3^- to NH_4^+ and then glutamine before it can be used to construct other amino acids and proteins, an energy expensive process (Chapter 4).

Part of the reason that HNO_3^- is readily taken up is that this form of N is soluble and moves with mass flow of water. Alternatively, NH_4^+ binds to soil particles in the vicinity, making it challenging for a large (relative to soil pores) root to grab. Both AM and EM fungal hyphae, on the other hand, can sneak between soil particles and through small soil pores, absorbing NH_4^+ molecules. Ericoid and some EM fungal hyphae secrete proteolytic enzymes that break down proteins into amino acids and take those up directly. Thus, probably the most critical function of these mycorrhizae is to short-circuit the N cycle (Figure 7.4). In every experimental addition of N, the first group of organisms to acquire that N are the microbes, including mycorrhizal fungi (see (494), as an example). The mycorrhizal fungi then exchange that N for C. This short-circuiting of the N cycle has been observed for every mycorrhizal type, in most types of terrestrial ecosystems.

Ericoid mycorrhizal fungi almost completely short-circuit the N cycle by secreting ligninases, cellulases, and proteolytic enzymes that obtain N directly from dead biomass (598). This process is crucial in ecosystems such as boreal bogs, where low ET rates leave high water tables and N would be rapidly denitrified if the full cycle were to occur. The same process is also critical in dry ecosystems like the Sand Plains of western Australia where ericoid mycorrhizae absorb organic N before it is leached or lost via ammonia volatilization in the high temperatures. Ericoid mycorrhizal dynamics have yet to be included in global models, even though ericoid-dominant ecosystems are globally extensive in tundra, taiga, and some semi-deserts.

AM fungi take up and transport NH_4^+ (e.g., (794)). How much N moves in short-circuited pathways remains controversial. Hodge et al. (360) found that three times as much N was taken up in AM compared with NM patches. They concluded that hyphae were present where N was released, or scavenging, into highly organic patches. Hyphae are known to grow preferentially into organic-rich patches (676). Further, Whiteside et al. (773) demonstrated that AM fungi were capable of taking up organic N in a boreal forest, but I know of no data showing that AM fungi secrete proteolytic enzymes into their environment. Also, other decomposing organisms, the microbiome of the hypha-sphere, would supply the fungus with organic or mineral N. The one interesting process beyond simply having hyphae in the right locations (including growing into micropores and ultramicropores containing NH_4^+) is the Fenton chemistry process described by Frey (260).

EM ecosystems appear to be far more complex and interesting. The most obvious mechanism is that EM fungi, just like AM fungi, have the ability to grow into organic matter fragments and into micropores (although not ultramicropores), to extract NH_4^+. Gebauer and Dietrich (276) found considerable variation in isotopic $^{15}N{:}^{14}N$ ratios

Box 7.1 *Mycorrhizae and Isotopic Fractionation*

Isotopic fractionation is conceptually simple, although the instrumentation and output can be quite complex. Conceptually, the lighter isotope tends to be preferentially transported through membranes or undergo chemical reactions. For example, as ^{15}N is heavier than ^{14}N, ^{14}N is preferentially transported to the plant resulting in a depleted (low) $\delta^{15}N$ signature, whereas the fungus retains ^{15}N. This fractionation results in an enriched (high) $\delta^{15}N$ signature of the sporocarp. Depleted forms, such as NH_4^+, may result from a more rapid N cycle, where the ^{15}N in organic forms remain in the soil. Other signatures are similar. For example, ^{13}C is heavier than ^{12}C, and reactions such as lipid synthesis (a common C form in AM fungi) discriminate against ^{13}C. For water, ^1H to ^2H, also called D, for Deuterium, is commonly referenced as δD. ^1H and ^{16}O diffuse faster than ^2H (or D) and ^{18}O, making it possible to differentiate water uptake zones between water in surface soils compared with deep groundwater. However, while isotopic ratios are a powerful tool, a single value results from multiple processes, making interpretation challenging. There are many publications and, indeed, many courses to learn isotopic analysis concepts and techniques. For example, the ITCE: Inter-university training for continental-scale ecology <https://itce.utah.edu/courses.html> has been a valuable resource for many of my students.[3]

in different trees, soils, understory plants, and heterotrophic organisms, suggesting considerable variation in N cycling patterns (see Box 7.1). Taylor et al. (697) found that the $\delta^{15}N$ signatures of sporocarps were similar within genera but, more importantly, there was a strong and significant enrichment in the $\delta^{15}N$ signature of EM compared with saprotrophic fungi. $\delta^{13}C$ of EM fungal sporocarps were significantly depleted in $\delta^{13}C$, although the difference was not large (a mean of − 25.8 versus −23.4 in EM versus saprotrophic fungal sporocarps, respectively).

Frey (260) outlined four mechanisms whereby EM fungi affect organic matter decomposition. These mechanisms, while altering C, appear to be most important in releasing N into the environment for uptake. These

---

[3] Website last accessed November 16, 2021.

are (1) enzymatic degradation, including the production of cellulases, hemicellulases, and ligninases, (2) priming, or the production of simple C exudates and necromass that "primes" saprotrophic activity, (3) the Gadgil effect, or the absorption of N needed by saprotrophic microorganisms inhibiting decomposition, and (4) Fenton chemistry, or the production of organic acids that bind with Fe_2O_3 under acidic soil conditions.

Enzymatic Degradation

Enzymatic degradation is relatively straightforward. Cellulases, hemicellulases, ligninases, proteases, and other enzymes are exuded by EM and ErM hyphae into the soil solution. These enzymes then cleave their various carbohydrates and proteins, releasing amino acids. EM and ErM fungi can take up many of the amino acids directly and transport them to the host plant. This is the short cycle most common in EM forest stands. Alternatively, the released enzymes can mineralize amino acids to NH_4^+ into the soil solution. Hence some N continues through the long cycle (Figure 7.5).

A challenge to understanding N uptake is that the forms readily available in soils for plant uptake, NO_3^- and NH_4^+, are both toxic in moderate-to-high concentrations. EM short-cycles N uptake by especially accessing NH_4^+. Plants and mycorrhizal fungi get around this problem by converting NH_4^+ to glutamine (gln) via glutamine synthetase using glutamate (glu) and energy. This reaction occurs in both the EM fungus and host plant. Interestingly, this conversion may fractionate against ^{15}N. The preferential ^{14}N-gln appears to be the form transported from EM fungus to host plant, resulting in a low $\delta^{15}N$ signature (higher ^{14}N) in the host. The greater ^{15}N-NH_4^+ remains in the fungus, and concentrates in the hyphae and sporocarps, resulting in a high $\delta^{15}N$ signature for the N retained in the fungus. Many researchers have used the $\delta^{15}N$ signature to help distinguish saprotrophic from EM fungi. This may occur, as the interspersed saprotrophs are not transporting low $\delta^{15}N$ to a host, but rather retaining the ^{14}N within the hyphae and sporocarps. For example, Hasselquist et al. (330; 333) found high sporocarp production with the high amount of litter produced after Hurricane Wilma in the Yucatán Peninsula, and were able to distinguish EM sporocarps using $\delta^{15}N$ signatures (all >5) compared with saprotrophic sporocarps (all <4). In the mixed California forest, the $\delta^{15}N$ signatures of saprotrophic sporocarps ranged from 1 in *Agaricus silvicola* to 2 in *Helvella griseolaba*

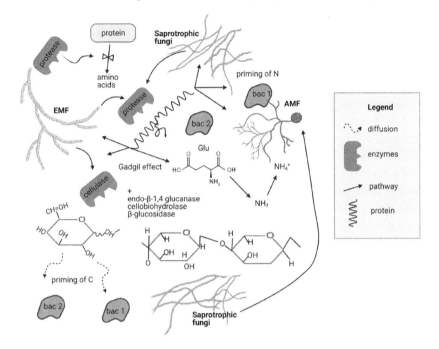

Figure 7.5 Acquiring N within a competitive mycorrhizosphere. Within the soil solution, multiple organisms are simultaneously releasing enzymes that chew up complex organic compounds, releasing simple C and N compounds that can be acquired. The trick is to create the greatest diffusion gradient towards oneself. Figure drawn by Danielle Stevenson, reproduced with permission

and *H. lacunosa*, compared with 4 in *Russula veternosa* to 16 in *Cortinarius duracinus*. Hobbie et al. (357) codified this fractionation using the Nitrogen Isotope Fluxes in Terrestrial Ecosystems (NIFTE) model to compare the signatures from the soil, the enriched $\delta^{15}N$ of the sporocarp, and the depleted $\delta^{15}N$ signature of the host plant leaves. This model simulates the N transformations to predict depletions or enrichments in ^{15}N between the soil and plant (357). In a mature piñon–juniper forest, there was a clear $\delta^{15}N$ fractionation signature (−3.3) in piñon needles, compared with −0.13 to 3.7 in saprotrophic fungi sporocarps and 4.2 to 9.5 in EM fungi (49).

Does the same fractionation occur in AM? Klink et al. (410) found that isolated intraradical hyphae of *Glomus* (=*Rhizophagus*) *irregularis* had an enriched (>5‰) $\delta^{15}N$ signature compared to plant tissue. Schweiger (649) reported that under conditions of N deficiency, roots had a depleted foliar $\delta^{15}N$ signal, whereas the roots, with a high proportion of fungal tissue, were ^{15}N enriched. (Unfortunately, AM spores were too small to measure N isotopic ratios.) But AM juniper leaves may have exhibited a possible fractionation signature of −1.43 (49). In another study (154), the $\delta^{15}N$ signature of AM plants (*Artemisia tridentata*, *Platanus racemosa*) had signatures similar to those of EM *Quercus agrifolia*. Both EM and AM fungi transfer NH_4^+ (see Chapter 4), and NH_4^+ has a depleted $\delta^{15}N$ signature compared to the larger organic N pool in soil (276). Alternatively, organic N in the soil pool is enriched and, if that pool is tapped (360), that would enrich the hyphae and the host tissue. In another approach, Whiteside et al. (773) used quantum dots to demonstrate the uptake of organic N by AM fungi.

Hobbie and Hobbie (358) followed up the Hobbie et al. (357) NIFTE model, by developing a model for NPP of EM fungi using the enrichment and depletion signatures. By analyzing the ^{15}N for soil, plant, and fungal sporocarps, and estimating the C and N needed for plant production, based on C:N stoichiometry, estimates of C and N allocations between plant and fungus were modeled. They estimated that 0.9 g $C\ m^{-2}\ y^{-1}$, or 60 to 86 percent of the total N uptake in plants, came from the mycorrhizal fungi, and only 14 to 39 percent directly from soil. To get this fungal N, plants exchanged 15 to 31 g $C\ m^{-2}\ y^{-1}$ to the EM fungi, or 8 to 17 percent of the photosynthesis. (The fungi retained 0.7–1.6 g $N\ m^{-2}\ y^{-1}$, more than the N in the plant). This approach provided the first clear effort to determine N and C allocations without extensive soil coring and provided reasonably good estimates. In studying C and N allocations in the arid land pinõn–juniper stands, the EM fungi needed approximately 20 percent of the NPP-C to acquire the N necessary to support that production (49). Adding fertilizer N (as NH_4NO_3) reduced (or eliminated) EM fungal sporocarp development by providing enough N for plant growth. Of course, the EM provided other resources, especially P and water that were simultaneously eliminated to the detriment of the plants. In a mixed conifer–hardwood stand using direct measurement of root and hyphal production coupled with NPP modeling using DayCENT, the allocation of NPP-C to EM fungi was estimated at 27 percent, compared to an estimated 34 percent using the Hobbie and Hobbie N fractionation model (59). These approaches indicate that coupling of isotope signatures with ecosystem measurements and models might provide useful estimate of the allocation of both N and NPP-C between mycorrhizal fungi and host plants across ecosystems.

The Mechanisms of Priming and the Gadgil Effect

These mechanismsremain more conceptual than quantified. Frey (260) presented a comprehensive assessment of the mechanisms and literature documenting the extent of these two competing processes. In effect, both priming and the Gadgil effect result from the most basic mode of resource acquisition by fungi, that is the release of enzymes into the soil solution, breakdown of complex C–N compounds, and uptake of the simpler amino acids and inorganic N compounds. The relative amounts transformed depend upon the limiting factors to microbial growth. In essence, mycorrhizal fungi, like microorganisms broadly, exude enzymes into their growth medium. These enzymes break down proteins, complex carbohydrates, and other plant residue and microbial and animal necromass, by cleaving the more complex compounds into simpler forms that are available for uptake. But, as the simpler compounds dissolve into the soil solution, they always diffuse towards regions of lower concentration. This could be to attach to organic particles and soil ions, such as NH_4^+, or to diffuse toward the organism that is creating the greatest gradient by taking up simple compounds (Figure 7.5).

The separation of "normal" transformation and uptake from priming and the Gadgil effect is a matter of degree; the pathways are the same, but the quantities depend upon the local environment. A key issue for "normal" uptake following decomposition, and for the priming and the Gadgil effect, is that diffusion and concentration gradients are dynamic and dependent upon the distribution of water droplets, soil pores, organic matter residues, temperature, and the dispersion of all of the soil organisms. In soil, a large fraction of the soil organisms are saprotrophs, largely fungi and bacteria that produce a broad range of decomposition enzymes. These organisms then take up any available N (amino acids, NH_4^+, NO_3^-), C (such as glucose), or other nutrient, creating a depletion zone, whereby more nutrients diffuse into that gap. Since the fungi and bacteria are so widely distributed in soil, microbes are usually the first to grab any released or added sugar or N source. It is only as saprotrophs are consumed by grazers, or die and their necromass becomes subject to enzymatic action, that the nutrients and sugars are re-released. This process generally takes days to weeks (e.g., (494)). In extreme cases, such as when new plant material is added (green manuring), the soil microbes can immobilize almost all free N, reducing the N uptake by plants.

Mycorrhizal fungal hyphae, being far smaller than roots and root hairs, are present at the same spatial scales as the soil saprotrophs, and active at those scales. Enzyme production is taxon–dependent. AM fungi appear not to release proteases into the environment. EM fungi, that have independently evolved multiple times largely from brown-rot fungi (Chapter 3), produce proteases and cellulases (along with cellobiohydrolase and b-glucosidase) that release amino acids and sugars into the soil solution. Ericoid mycorrhizal fungi and orchid mycorrhizal fungi may even be white-rot fungi that degrade lignins, releasing amino acids and NH_3 into the soil solution (which in acid conditions bind the free H+ to create NH_4^+). While AM fungi access organic N, the reason that AM fungi are effective at taking up NH_4^+ appears to be not because they are producing proteases in the soil, but that their hyphae rapidly take up NH_4^+ as it is released, and create a tiny depletion zone, absorbing even more of the NH_4^+ as it is released. In the case of EM, they enzymatically breakdown proteins and, because they are nearby, rapidly take up the available amino acids or NH_4^+. Thus, mycorrhizal fungi rapidly access any available N.

The Gadgil–effect occurs when there is an excess of EM fungal hyphae. This phenomenon was described in an exotic forest in which the exotic plant litter inhibited native saprotrophic organisms (270). The exotic EM fungi were capable of grabbing so much N that saprotrophic microbes become N-limited to the point of reducing growth. EM fungi then transport the N to the host in exchange for C. Under these conditions mycorrhizal fungi reduce decomposition, which created a feedback loop ultimately reducing N availability.

"Priming" is the polar opposite. Here, the exudation by mycorrhizal fungi and a large input of mycorrhizal fungal necromass can actually stimulate decomposition, by releasing both N and C for the soil saprotrophs (e.g., (260)).

Patterns of N and C allocations in forests reflect a wide range of complex mechanisms for N and C allocation, re-allocation, and secondary re-allocation. Mycoheterotrophic plants, in particular, have isotopic signatures that reflect uptake and fractionation of N by mycorrhizal fungi and acquisition of C from trees, then reallocation to the mycoheterotrophic plants (e.g., (284; 800)). Evidence also exists for a small amount of complex exchange between mature EM trees and seedlings of both C and N, especially appearing in seedlings attached to larger mycelial networks (see Chapter 6).

Organic Acid Production by Mycorrhizal Fungi

Organic Acid production by mycorrhizal fungi may catalyze a broad range of nutrient uptake, including N, but also P and Fe. The process of

Fenton chemistry, or the production of organic acids that bind with Fe_2O_3 under acidic soil conditions (260), has an enormous but underappreciated role in soil nutrient dynamics. Just as mycorrhizal fungi grow into every possible crack and crevice, their mere presence triggers many interesting chemical reactions. EM fungi, and AM fungi where triggered (we do not know the stimulus), produce organic acids, including oxalic acid, citric acid, and others. Lignocellulose depolymerization is triggered by brown-rot and EM fungi. Many mycorrhizal fungi produce catechols, which are oxidized derivatives of aromatic compounds that can be siderophores (see Chapter 4) that improve Fe uptake by solubilizing Fe^{3+} and releasing it as Fe^{2+} (Fe^{2+} is the soluble form that is important for physiological processes, but is unstable in most soils). Mycorrhizal fungi produce oxalic acid ($C_2O_4{}^{2-}$), which preferentially binds the Fe in $Fe(OH)_3$ in soil to become ferric (Fe^{3+})-oxalate (395), or may bind Fe^{2+}, creating ferrous oxalate. Ferrous oxalate is soluble in water as a dihydrate ($Fe^{2+}C_2O_4*2H_2O$), whereas ferric oxalate ($Fe^{3+}{}_2(C_2O_4)_3*6H_2O$) is only slightly soluble in water. Oxalic acid protects hyphae by reducing the rate of Fe^{2+}-oxalate oxidation, inhibiting reduction of Fe^{3+} by siderophores and diffusing into organic matter. Away from the hyphae, oxalates are decomposed to CO_2 by a range of fungi and bacteria (515), thereby releasing Fe^{+2} (or Fe^{3+}). Within the organic matter, peroxides and Fe^{+2} react producing Fe^{+3} plus hydroxyl radicals degrading C–N complexes, and facilitating protein degradation.

Other Elemental Cycling

Other elemental cycling, including P, Fe, and Al, are also directly affected by organic acids and more broadly by mycorrhizae. Mat-forming EM fungi are well-known for producing high concentrations of organic acids, including oxalic acid. Kermit Cromack and colleagues (e.g., (186; 298; 476)), Fox and Comerford (250; 251), and Allen et al. (55) all found oxalic acids, which, when combined with cations such as Ca, Fe, and Al, become oxalates, forming recognizable crystals that can be distinguished morphologically, and on the elemental composition binding to those crystals, using energy dispersive spectrometry. The dominant form of $PO_4{}^{-2}$ in soil is typically bound to cations, especially Ca, Fe, and Al. Both EM and AM fungi exude oxalic acids, with the oxalates then preferentially binding the cations, releasing $PO_4{}^{-2}$, which after binding with an H+, becomes available for uptake (see details in (395)). Because Ca-, Fe-, and $AlPO_4$ are widely distributed and segregate between soil types and biomes, and as these organic acids are also players in Fenton chemistry, the role of mycorrhizae and organic acids in both N and P mineralization and uptake should not be underestimated.

In addition, organic acids released by mycorrhizal fungi are the same as those produced by lichens weathering rock. Hyphal acids, coupled with osmotic forces, decompose and push their way into bedrock. In soft carbonates, the rock can be dissolved and Ca taken up directly by the fungus, leaving behind root and hyphal channels in the rock. In quartzite and in granite, hyphae force their way into microcracks, opening routes for nutrient and water exchange (e.g., (644)).

Just as for N cycling, release of enzymes by mycorrhizal fungi plays a large role in P cycling. Many plants produce acid phosphatases. Both EM and AM mycorrhizal fungi also produce and release alkaline phosphatases (e.g., (73; 83; 95)). Importantly, just as for proteases released into the environment, phosphatases, upon release, do not direct the flow direction of mineralized HPO_4^-. If an adjacent saprotrophic fungus or other mycorrhizal fungus, or even plant, creates a greater depletion zone, the released HPO_4^- will diffuse to that organism. Thus, the overall cycling is often a function of the simultaneous exudation of phosphatases from multiple sources, and the uptake of the released HPO_4^- to multiple sinks.

One additional note, many forms of organic P are soluble and may move by diffusion and by mass flow. These compounds are then subject to degradation by enzymes, which creates a secondary pattern of P dispersion through soil. Unfortunately, I know of no analyses that examine the interaction of organic P dispersion and implications for uptake by mycorrhizal fungi.

Finally, to make nutrient cycling even more complex, we need to better understand the dynamics of organic chemistry in elemental cycling. One example goes back to glomalin, produced by AM fungi. This glycoprotein binds cations. By one estimate, 9 percent of the mass of glomalin in the aggregate may be comprised of Fe, and immobilized heavy metals can include Al, Cu, Pb, Mg, and Cd (448; 752; 763).

Limiting Factors: Ecosystem Cycling

There is a long history of research on the mechanisms whereby mycorrhizae regulate individual ecosystem processes, such as C, N, and P cycling. However, our goal is to put these all together. Liebig's law of the minimum dictates that growth is regulated not by the overall resource availability, but by that of the limiting resource. This limitation can be scaled for an individual plant, or stand NPP, all of which regulate rate of cycling and ecosystem functioning.

At the global scale, Mediterranean ecosystems provide a unique perspective on the comparative dynamics of mycorrhizae. Mediterranean ecosystems are hailed as the ultimate example of convergent evolution because climates are similar in dispersed regions across the globe, consisting of warm, moist winter/spring growing seasons, and hot, dry summers. In these environments, distinctly different phylogenies of plants show remarkably similar leaf morphologies and above-ground plant architectures. This convergence is taught in every plant ecology text. However, soils in these regions are dramatically different leading to very different mycorrhizal outcomes (16). In the Mediterranean-climate regions of Australia and South Africa, most of the soils are ancient, and what little P there is is often bound as $AlPO_4$. Here, in the most extreme conditions, many non-mycotrophic plants show a special adaptation to the most extremely low P soils by secreting organic acids, especially oxalic acid, binding with Al to form Al-oxalates, then taking up the released P. When P is added at high to moderate levels many of these species show P toxicity (see (437)). AM and some EM plants predominate under less-extreme conditions but effectively are essential for other plants to survive, by increasing P uptake (668). At the other extreme, in California and the west coast of South America, soils have only built up recently, geologically. Most of these soils are rarely P-limited. Total P levels are in the range of a few g kg^{-1}. Although often bound as $CaPO_4$, these become available with CO_2 and H_2O, or organic acids. Generally, the limiting resource is water, and secondarily N. Nonmycotrophic plants in the Amaranthaceae and Brassicaceae readily invade newly exposed soils (which we will discuss in greater detail in Chapter 8 on succession) that are relatively rich in nutrients. AM and EM (and arbutoid EM) plants flourish, facilitating N and P acquisition. The Mediterranean Basin itself has adequate P, but that P is often bound into $(Ca)_3(PO_4)_2$ in the widespread calcareous soils. Every type of mycorrhiza can be found, many producing organic acids that facilitate P and N acquisition. Thus, while there is obvious aboveground convergence, the same plants show diverging mycorrhizal and root structural characteristics (36).

But just as temperature cycles diurnally and precipitation patterns shift seasonally, no single resource is consistently limiting over time. Seastedt and Knapp (650) proposed their "Transient Maxima Hypothesis," in which resource 1 initially limits production, which shifts to resource 2 with changing conditions, be that seasonality or perturbation by pressures such as grazing. In virtually all ecosystems, limiting resources change with conditions, temperature, and precipitation with nutrients, N, P, Fe, K, Ca, etc. This limitation can vary with climatological characteristics ranging from El Niño-Southern Oscillation events, to diurnal cycles, to sun flecks. We have already identified how mycorrhizae affect the physiology of individual plants, and how these mechanisms, in turn, scale to ecosystems. But there are processes ranging from single stand to biome spatial scales, and from minute time frames to decadal temporal scales that provide additional insights into ecosystem dynamics.

An example is the work on the Serengeti, initiated by Sam McNaughton and colleagues and continuing to the present by Nancy Johnson and her colleagues (see (496; 581)). They showed that AM fungi are regulated by rainfall events and the edaphic conditions of soils (84). As everyone knows who watches nature shows on television, massive herds of ungulates track grass productivity, which tracks rainfall. Here, rainfall moves seasonally between the south and north. Soil P is high in the south and low in the north where there was less historic volcanic activity (581; 626). Because of the heavy grazing, N turns over rapidly all across the range. Stevens et al. (682), followed up using data inputs of AM, climate, grazing, and soil nutrients into a spatially explicit ecosystem model. They found that in the P-sufficient south, productivity following the rainfalls was sufficient to sustain the grazers, hence the predators. However, in the P-deficient north, productivity without AM would be inadequate to support the high populations of grazers. Without the production in both regions, the seasonal cycling of the grazing populations would crash (Plate 7).

Because of the climatological and edaphic factors, the Transient Maxima Hypothesis is especially appropriate, as rainfall shifts seasonally, promoting plant production, but grazing follows, resulting in a reduction of photosynthetic surface area (leaves), but rapidly turning over N. Total AM activity, while higher in the south (with higher P), had the greatest impact on plant production in the north, where P was extremely limiting. In effect, these outcomes indicate that AM increase P, important to production in both the higher (total) P south, during high precipitation, and *essential* to the lower P north portion of the grazers' rotational range.

Mycorrhizae and NPP: Putting the Pieces Together

Constructing estimates of the mycorrhizal contributions to NPP remains a primary challenge for mycorrhizal research. Here, I will construct two estimates of contributions of mycorrhizal fungi to ecosystem processes. I will focus on two extremely different ecosystems where mycorrhizae and NPP were studied in detail *in the field*, recognizing that similar estimates have (and should continue) been undertaken for a number of study locations. Then I briefly describe a comparative study undertaken across multiple biomes. In the end, these should be compared with the top-down modeling estimates from the FUN modeling. It is only with a synthesis of these approaches that we can design ecosystems to provide for conservation and agricultural sustainability (Chapter 10).

Svartbergets Experimental Forest, Sweden: A Boreal EM Forest

One of the first study systems *in the field* undertaking a broad ecosystem analysis was undertaken by Peter Högberg and his colleagues. Much of the history leading to this work and the research outcomes are reviewed in their 2017 review (365). One of the remarkable outcomes from their overall research was the observation that no one of the studied mechanisms and processes occurred in isolation, especially under complex field conditions. To understand any one factor, C and N cycling, base saturation, plants, decomposers, and mycorrhizal fungi, and many others need to be simultaneously understood. They first noted that the small size of the inorganic N pools was inadequate to support tree production, supporting classical early work by Melin and Nilsson (497). From that beginning, the work by numerous ecosystem scientists began showing the short-circuiting of the N cycle and the mechanisms for obtaining N for both fungal and plant NPP. Högberg et al. (see (365) and references) note that boreal forests tend to be strongly N-limited, but that there are hotspots within the landscape. They describe a ratcheting whereby plant N uptake is connected to belowground C allocation, which determines either immobilization or N leaching. Sinks for N and C between plants, EM fungi, saprotrophs, and soils are not necessarily even directionally readily predictable. Among the detailed analyses (331), unmanipulated sites had a high number of EM fungal sporocarps (60 to 80 per 100 m^2), and high species richness (12 to 14 species). The fractionation of $^{14}N{:}^{15}N$ between fungi and trees was consistent with the mechanisms described here and by Högberg et al. (363; 364). The isotopic signatures indicated that trees were heavily dependent upon EM fungi for N, requiring considerable C transfers (362; 366).

One means of better understanding the overall C dynamics was to compare shade versus sun plants. Understory plants, especially seedlings and saplings, are often C limited, in part due to the lowered solar radiation intensity, to levels where photosynthetic rates are reduced. Hasselquist et al. (332) found that, when shaded, photosynthesis was reduced by 59 percent, EM sporocarp biomass by 41 percent, EM fungal biomass by 22 percent, and R_s by 31 percent. In essence, this means that the greatest loss was in C available to the ecosystem (nearly 60 percent), whereas the EM fungal mass declined far less (22 to 41 percent). Soil respiration also dropped, but again, not as much as decline in C inputs. With ^{15}N labeling, EM tips accumulated 65 percent more N in control than shaded plots and double in control than shaded plots. These data show a high dependency of plants on EM for N, and that the EM fungi are reduced as C fixation is reduced. The relative gain by the EM remains higher than the overall C accumulation by the plant, indicating that several changes are occurring that remain unknown. Is there a reduced fraction of C going to wood production and proportionally more allocated belowground (234)? Are EM fungi shifting towards mining soil C to persist (as per (161))? As boreal forests represent a large fraction of the sequestered global C, answering these questions is a high concern (see Chapter 9). Furthermore, clearly the fertilization of N, by either direct application or air pollution, dramatically alters the exchange dynamics between EM fungi and host plant.

La Selva Biological Reserve, Costa Rica: A Tropical AM Forest

The second model system is the La Selva tropical rainforest in eastern Costa Rica. This ecosystem is well described in publications over the past five decades. It is diverse biologically, ecologically, and biogeochemically (for example, see (234)). But two features make it of special interest. First, the plant communities are predominantly AM, with intrusions of EM individual trees in the Myrtaceae and *Gnetum*, and orchids scattered throughout the forest. So, here, we focus on AM as the base of ecosystem dynamics. Second, the site is located at 10°N with temperatures that vary more daily than annually, with nearly continuous and high rainfall of about 4 meters annually, and with a short but moderate dry season (when it only generally rains once or twice per day). Like the James Reserve ecosystem, there is a large primary forest with little historic human use. We began constructing our Soil Ecosystem Observatory (SEO) in 2010, with elements running until 2018.

Table 7.1 *Net Primary Production allocations to AM in a tropical rainforest, in g C m^{-2} for the La Selva Biological Station[a]. Here the Net Fixation was 3,100 g m^{-2} y^{-1} and Net Respiration 2,650 (see text for details).*

Standing Crops	value in g m^{-2} y^{-1}	lifespan days^{-1}	generations y^{-1}
Aboveground	17,000		
Fine Root	250	150	2.4
AM Fungal Hyphae	17[b]	40	9.1
NPP estimates			
extramatrical AM fungal hyphae	154		
root	600		
leaf	750		

[a] Values from Loescher et al. (464), Clark et al. (168), and Swanson et al. (692).
[b] Standing crop was calculated using the following data: 46 percent of the hyphae were AM based on a phospholipid-fatty acid (PLFA) composition of 16:1w5c versus 18:2wc for fungi, and assuming 30 percent dry mass and 50 percent C.

Using data my research group and others have generated over the years, I constructed example budgets using a meter-squared and annual basis (Table 7.1). From eddy covariance estimates, net CO_2 fixation averages 3,100 g, with a net respiration of 2,650 g, resulting in a NEE of 450 g. Using a variety of techniques, respiration can be broken into canopy respiration (R_c) of 1,831 g, and soil respiration (R_s) of 1,134 g, for a net respiration of 2,965 g. Using $R_c + R_s$ as the measure of net respiration yields a NEE of 135 g. It is important to remember that each of these numbers are developed from different methods and done at different times, and, while close, will not directly add up. Aboveground standing crop averages 17 kg, with a leaf standing crop of 225 g. Belowground, we have a far less detailed view of the ecosystem. Soil coring data shows that there is a standing crop of about 120 g (224). Net annual primary production of leaves is 750 g, and wood is 500 g. We have no published estimates of belowground NPP. Also of note, the dissolved organic C, moving from soil into waterways, amounts to 5 g annually. Glomalin, an AMF-produced glycoprotein, averaged 145 g m^{-2}, accounting for 3.2 percent of soil C and 5 percent of soil N in the surface soil (467).

From this base, we designed an SEO based on our SEO from the James Reserve (58; 76; 628). We undertook daily images first from three locations in the mature forest, combined with measurements at three depths for CO_2, temperature (T), and soil moisture (θ), taken at five-minute intervals. These measurements were undertaken for three years (2010 through 2012). Using the RootView program (Rhizosystems Inc.) to measure root production, there was a standing crop C of 71 g m^{-2} in roots. By tracking the fates of individual roots, the average fine root lifespan averaged 60 days. AM fungal hyphal length ranged from 1 to 15 mm mm^{-3}. Because measurements were made on a daily basis, I was able to model the daily hyphal standing crop, using a negative binomial distribution, where:

$$(\ln \text{ hyphal length}) = -0.14 - (4.68\theta_8) + (0.12T_8) - (0.05R_s).$$

Here, $n = 248$, $p < 0.0001$, $r^2 = 0.22$, where: $\theta_8 =$ soil moisture at 8 cm, $T_8 =$ temperature at 8cm, $R_s =$ soil respiration.

Because individual hyphae could be tracked, I used a mark–recapture approach to compute the lifespans of the population of roots and hyphae (409; 692).

Based on the assumption that the AM fungal C is part of autotrophic C respiration, this estimate of 754 g m^{-2} y^{-1} belowground NPP is approximately equal to the *rate* of the aboveground NPP. The *external* AM fungal hyphae accounted for approximately 10 percent of the total NPP. What does this mean to the overall C dynamics of this ecosystem? In these wet tropics, a high amount of N is cycled through the ecosystem. N_2 fixation is high, with a large proportion of leguminous trees, N_2 fixation from lichens and free-living organisms in the canopy, and nearly continuous high soil moisture facilitating free-living N_2 fixation. But N cycling is also quite rapid (548). However, although decomposition is rapid, with 1 to 15 mm of AM fungal hyphae penetrating nearly every millimeter cubed of soil, grabbing NH_4^+ before it can be denitrified, these hyphae would make a large contribution to the productivity of the forest (360). P is then presumed to be the primary limiting element, a resource especially regulated by AM. In part, this is due to the high need for P in N_2 fixation and overall basic plant physiology.

Direct access and uptake of HPO_4^- by AM fungal hyphae must be high to provide plants with needed P resources. Behind every new root tip, AM fungi infect the roots and produce an extensive network of absorbing hyphae (15 m g^{-1}) extending into the soil. It is also important to remember that, in wet soil, much of the respired CO_2 from AM roots and AM fungal hyphae becomes HCO_3^-, which in turn reduces soil pH, weathers $AlPO_4$, and releases HPO_4^- for uptake. AM fungi may also acquire organic P during decomposition through release of phosphatases and phytases. Because of the high soil moisture and even surface standing water in these clay soils, soil CO_2 reached up to 50,000 ppm (compared with the atmosphere of 400 ppm (232)). Much of this CO_2 dissolves into the water as H_2CO_3, and HCO_3^-, also reducing pH, and releasing HPO_4^- from $FePO_4$ and $AlPO_4$ for uptake (395). Also, Lovelock and colleagues (467) found that the glomalin in soil accounted for 3.2 percent of soil C and 5 percent of soil N in the surface layers. Soil aggregation in the top 20 cm comprises the majority of the soil (Figure 7.3), and soil aggregation is tightly coupled to soil glomalin. The production of glomalin by the AM fungal hyphae would reduce the rate of cycling of N, P, and cations, thereby reducing leaching and retaining these elements within the ecosystem to sustain soil fertility.

Finally, every soil has variable distribution of resources that must be taken into account. But new patches of resources open continuously due to local disturbance, such as the creation and abandonment of leaf-cutter ant nests. These nest patches had higher available N and P. Total fungal hyphal length was 35 mm mm^{-3} and root length 3.55 mm mm^{-3} outside ant nests (692). In the nests, turnover rates increased dramatically as root and fungal lifespans dropped from 32 days in the control (between nests) to 10 days in the nest, and root lifespans dropped from 152 to 27 days. The high density of new roots and hyphae growing into gaps created by disturbance agents, such as leaf-cutter ants (692), provides new hyphal networks to extract P when and where it is released.

Comparative Ecosystems

In 1997, a group of us (Kurt Pregitzer, Ron Hendricks, Roger Ruess, and myself) initiated a comparative study where we used the same methods across biomes to measure mycorrhizal and belowground dynamics at sites ranging from forested ecosystems in central Alaska, south to central New Mexico, east to Michigan, and then south to Florida. The study encompassed both coniferous and deciduous trees, with both EM and AM dominant trees of both mycorrhizal types. The root length of each of the first three orders of root (finest to third order) varied across all species and root length varied. As a perturbation, N fertilizer (NH_4NO_3) was added to all sites at 10g m^{-2}. Fertilization had few consistent effects on root morphological characteristics, or root C:N ratios, specific root length, or length per unit N. Percentage AM infections were highly variable between species and dates, but virtually all first-order EM short-root root tips were mycorrhizal. We found no consistent differences in root characteristics between AM and EM roots. First-order roots had the highest specific root length and N concentrations and mycorrhizal fungal activity in all species (576). Root respiration was tightly coupled to root N concentration (150). Interestingly, although root respiration (R_r) was linear with N concentration (corrected for temperature), AM roots showed the two extremes with the highest rate of R_r and N concentration, with the highest being the AM angiosperms and the lowest the AM gymnosperms. Q_{10} (temperature versus respiration) relationships did not vary among sites or species. But AM and EM stands appear to have different N cycling rates, with more open and rapid immobilization and turnover in the AM forests. AM and EM forests differed in that when labeled glycine was added, it was rapidly mobilized indicating a C limitation for soil microbes in the EM forests compared with AM forests (494).

This study comparing AM and EM forests across latitudes indicates that mycorrhiza–root dynamics behave similarly at the macrosystem (regional to continental) scale (*sensu* (77; 339)). This bodes well for models like FUN3.0 focused on quantifying and characterizing biosphere–atmospheric exchanges, where:

$$C_{growth} = C_{NPP} - C_{acqP} - C_{aqN} \quad (23).$$

However, the variation across all of the measurements shows considerable disparity, despite their statistical significance. This suggests a strong need to couple ecosystem measurements with fungal community composition. It is highly likely that within the macrosystem constraints, community members change with N, T, θ, and each mycorrhizal taxon, functional group, or guild may, or may not, respond even down to individual events (695).

Summary

- Sir Arthur G. Tansley defined an ecosystem as the organisms together with their special environment forming a physical system. From this perspective mycorrhizae are an interface between plants and C inputs, and soils and nutrient and water inputs.
- Mycorrhizae differ by Biomes. Broad patterns are used for global-scale modeling. Although broad patterns exist between AM, EM, and

ericoid mycorrhizae across latitude and altitude gradients, many patches comprise exceptions.

- Stoichiometric relationships based on physiological interactions (Chapter 4) can be used to help identify ecosystem constraints and flows.
- Broadly, 20 to 40 percent of net primary production is allocated to mycorrhizal fungi in ectomycorrhizal ecosystems, and 5 to 10 percent in AM ecosystems. All ecosystems need additional research.
- Mechanisms whereby mycorrhizae exchange carbon and nutrients between plants and soils include secretion of enzymes, input of H^+ ions altering pH, direct input of organic acids, and facilitation of soil aggregation. These are especially important in short-circuiting the N and P cycling.
- Because limiting resources change in space and time, understanding the shifting transient maxima is important to describing ecosystem dynamics.
- Future approaches to understanding mycorrhizal dynamics of ecosystems should link ecosystem measurements with isotopic ($\delta^{15}N$, $\delta^{13}C$, $\delta^{14}C$) information from sporocarps, root tips, and soils, molecular sequencing of mycorrhizal composition and functional genes, and frequent direct observations for hyphal dynamics and turnover. In order to manage ecosystems for optimal production and for global carbon, putting ecosystem and community information together is becoming essential.
- Altogether, although more work is needed, these preliminary forays into mycorrhizal functioning of extremely different ecosystems demonstrate that mycorrhizae are very much a critical element of all terrestrial ecosystems.

8 · *Mycorrhizae and Succession*

During the nineteenth century, the Selkirk Settlement of Canada and the Homestead Act in the United States led to some of the most dramatic and widespread destruction of native ecosystems across a short time period in history. Soils across the Great Plains, from the Mississippi River to the Rocky Mountains, from the Chihuahuan desert of Mexico to the Boreal Forests of Canada, an area of around 4 million kilometers squared, were nearly all turned over for agriculture, from prairies with dense grasslands to riparian regions with extensive forest cover and deep roots, within the decades from approximately 1820 to 1890. By the 1890s, a protracted drought led to a collapse of agriculture in the United States, Canada, the Ukraine (the Selk'nam genocide), and elsewhere. The young field of ecology was just beginning, documenting the community ecology of recovery at lakeshores (182), glaciers (175), and from the abandonment of highly disturbed agricultural lands (170). Frederic Clements proposed a theoretical framework to describe "succession," a term used as early as the nineteenth century by Thoreau (708). Clements' model of nudation, colonization, ecesis, and reaction (see (170; 472)) provided a framework. But he also proposed that succession leads to a climax vegetation based on the climate. Following up, Tansley (696) began to view succession as an intercept of physical–chemical–biological processes and factors, termed an "ecosystem." Linkages between plant community composition, nutrient limitations, and microbes became of ever-greater interest. Because of the clear interactions between taxonomic composition and ecosystem processes, it becomes impossible to think of compositional change through time ignoring ecosystem change, and vice-versa. Across the twentieth century, ecologists, concerned with recovery, began trying to intimately link these two conceptual frameworks.

Successional patterns and mycorrhizae have been topics of interest in mycorrhizal research for nearly a century. Stahl (678) noted that a number of weedy plants that invaded disturbed areas (e.g.,

Amaranthaceae and Brassicaceae) were nonmycotrophic and replaced by mycotrophic plants. Dominik (209) examined the roots of plants at the forefront of coastal dunes on the Baltic and found that they were nonmycorrhizal. As organic matter accumulated with succession, mycorrhizal activity increased. Frydman (267) even followed this work up by recording the mycorrhizal status of plants reestablishing on the Warsaw ghetto ruins after World War II. Nicolson (532) first truly tied mycorrhizae to succession, observing that, in sand dunes, the first colonizing species were the nonmycotrophic *Salsola kali*, followed by AM grasses, then EM trees. Schramm (645) noted that mycorrhizal activity was nonexistent in slag heaps from coal mining, but were naturally colonized as plants established.

In 1969, Eugene Odum published his classical paper clearly linking community with ecosystem processes occurring during succession (539). He outlined a suite of characteristics from early · to late succession, including the shift from open- to closed-cycling of nutrients with succession, and the importance of symbioses. From this conceptual base, research on the role of mycorrhizae in succession exploded. The role of mycorrhizae in succession and vegetation recovery was observed across many regions of the globe, from forests to grasslands and deserts (16).

A problem that arises as we start to describe the role of mycorrhizae in succession is what is a disturbance that initiates succession? A disturbance can range from an anthill or gopher mound, to a landslide, a volcano, or a retreating glacier. They can be natural or anthropogenic, such as abandoned agriculture or a mine spoil. If small, or if the soil is not disturbed, we can view the perturbation as a patch. *Every ecosystem has disturbances.* If the soil remains largely intact, then that simply becomes a patch. I view a small animal disturbance, such as an ant nest, the blow-down of a single tree, or even a small fire as creating patches, but not necessarily initiating succession. I've already discussed examples of patches in Chapters 6 and 7. So *here I define succession as occurring after a disturbance is large enough and severe enough so that soils are disturbed, plants are killed, and mycorrhizal activity is disrupted.*

Early- and late-seral plants are well known and described for almost every terrestrial habitat. Mycorrhizal fungi also show successional patterns. Early- and late-stage EM fungi have been well described (e.g., (196)). Succession can be initiated at the patch scale, such as a landslide or vehicle disturbance patch, or it can even be regional. The Great Droughts in the Plains of western United States in the 1890s and 1930s caused so much mortality that succession was initiated across large

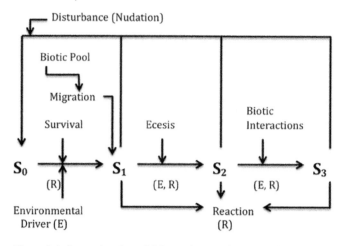

Figure 8.1 Successional model for evaluating the roles of different processes, derived after Clements (170). S refers to the successional state of the system. The Disturbance Event is termed Nudation. S_0 is following disturbance, before organisms are present, S_1 is the total pool of surviving and invading organisms before they start growing, S_2 is after organisms establish (ecesis) and is largely a function of soils and climate, and S_3 is after the biotic interactions such as competition. E refers to the environmental driver, and R to reaction of organisms to these actions. It is important to remember that the States (S_0 to S_3) do not equate to years. From Allen et al. 2018 (69).

regions. In the Colorado and Wyoming plains, both disturbances resulted in the establishment of nonmycotrophic early seral plants such as *Salsola kali*. Mycorrhizal fungi reinvaded from patches supporting late-seral mycorrhizal grasses (10).

David Janos (377) proposed a model of succession based upon his observations from the La Selva Tropical Forest, using the mycorrhizal categories proposed by Stahl (678). Here, the first colonizers of a site were nonmycotrophic plants that grew independent of mycorrhizal fungi. As the fungi themselves invaded, first facultative, then obligately mycotrophic plants established. Mycorrhizal fungi invade a site in patches, carried by animals feeding from a neighboring mycorrhizal patch, or by wind, from patches where eddies can reach either the sporocarps or dry soil wherein spores can be entrained (Figure 8.1). Because facultatively mycorrhizal plants could grow and survive without the fungus, they provide resources that attract animals and vertical structure that extracts energy from eddies resulting in the deposition of the spores and soils that were in the wind. Spores and spore mimics were observed migrating up to 1.5 km from a source area by wind (56).

Animals disperse spores. For example, rodents dispersed spores in Amazonian rainforests (379), in Pacific northwest forests (487), across the Pumice Plain on Mount St. Helens (63; 69), in exotic forests (538), and in arid forests in Australia (166). The deer mouse, *Peromyscus maniculatus*, moved spores up to a kilometer into sterile soil material (764); and birds dispersed spores to new islands over several kilometers (180). (For a comprehensive update of bird dispersal, see (555).) Once inoculum is established on facultatively mycotrophic plants, then obligately mycorrhizal plants can establish using the mycelial networks sustained by the C from facultative plants.

Large Infrequent Disturbances: Mechanisms of Succession

More recently, the dynamics of mycorrhizae have been treated as a key suite of interdependent processes for understanding recovery of what are now termed Large Infrequent Disturbances (LIDs). However, even LIDs can vary in size and intensity, making generalizations challenging. These range from disturbances that directly impact largely aboveground plant community composition and structure, to gigantic LIDs that destroy entire swaths of vegetation and soils, even bedrock. We should envision that a LID will usually encompass several hectares in size. The disturbance agent can occur rapidly (minutes to days) such as fires or floods, or persist over long time periods, to centuries, such as grazing or shifting cultivation. Recovery of the community is the process of succession and has been a focal area of study in mycorrhizal ecology for the past half century.

Where disturbance is large enough to initiate succession, a large array of initial conditions can exist that would make the simple Janos model quite complicated, by pushing conditions only partially back through the successional model, say to the facultative portion of the curve if soil inoculum persists through the disturbance. Moreover, because there are multiple mycorrhizal types, the pattern may be very different in different environments; a legacy effect (64; 69). For example, Allen and Allen (14) derived different patterns of succession depending upon the biome, whether the initial condition was early or later seral, which influenced the role of mycorrhizae in individual plant competition. Each of these factors was regulated by nutrient status and effective precipitation.

Before we discuss mycorrhizae and succession, it is crucial to understand the basic types of succession and the patterns of succession. *Primary succession* occurs from a severe disturbance in which soils and organisms have been eliminated from the site of disturbance. Examples include volcanoes, retreating glaciers, severe landslides, and severe anthropogenic disturbances such as strip-mining and severe perturbation where the soil has been removed and replaced with subsoil or rock. *Secondary succession* occurs from a large disturbance that destroys the aboveground portion of the community and disrupts the soil but does not necessarily remove all substrate materials and organisms. Examples include large fires. Importantly, a single disturbance can result in both types of succession. For example, in our strip mine study, some areas had subsoil only, whereas other areas had respread topsoil providing predisturbance nutrients and organisms.

The environment also dictates the pattern of succession. The best-studied successional communities are those in which early colonizing plants are replaced by mid- and then late-establishing vegetation, or *relay floristics*. These tend to occur in mesic biomes where temperature and moisture can support many species. Suites of plants replace other suites of plants as the soils increase in nutrients and/or microbial symbioses. In more extreme environments, where severe environmental conditions constrain the diversity of species that can survive, an *initial floristics* succession may occur. Here, the plants and microbes that initially occupy a site are limited to the same ones that persist potentially at least until the climate changes.

The model (Figure 8.1) originally proposed by Clements (170), modified by MacMahon et al. (472) and our larger group (69), provides a useful framework to describe succession and how mycorrhizae regulate the patterns of succession (see (36; 69) for more detail). This model incorporates Odum's (539) ecosystem characteristics, including entropy, symbiosis, and open-to-closed nutrient cycling, but are Clementsian in view of the regulating processes, as both the fungal and plant communities reflect community dynamics (as discussed in Chapter 6) and all the physical–chemical–biotic elements are interactive (as defined by Tansley (696)).

Nudation

The first step is to define the nudation process, or the disturbance event. That determines the initial conditions, S_0. LIDs in which the recovery of mycorrhizae has been studied include almost every ecosystem, and almost every disturbance type, from volcanic activity to glacial retreat. In general, for our purposes, a key is to determine if nudation causes nearly complete removal of the topsoil (or burial to where none of the existing organisms can access the surface, in the case of a pyroclastic flow), resulting in primary succession, or only partial destruction, usually destroying some of the organisms, and often the topsoil, but allowing some to persist and create a new community, or secondary succession.

Residual Survival

The surviving pool or organisms range from none, in primary succession, to a very large fraction, in secondary succession. Survivors dictate the initial state, S_1, of the ecosystem. Following the severe fires on the Yellowstone Plateau in 1988, and although the aboveground biomass was nearly completely burned, EM fungi and roots survived, protected from the heat (506). The EM roots survived for a year, with the roots providing C to support the EM fungi until new seedlings emerged the following year, forming EM at S_1. Many EM fungi tolerate severe forest fire, such as *Rhizopogon olivaceotinctus* and species of *Tuber* (e.g., (143; 286)), and are ready to germinate as pine seeds immigrate and germinate. On a strip mine in Wyoming, AM fungal inoculum, spores, and root fragments survived soil removal and respreading, to initiate new AM infections (10), but that inoculum declined rapidly after three or more growing seasons in the absence of host plants (e.g., (502)).

Migration

Using the generalized successional model (Figure 8.1), both fungal and plant partners can be examined both independently and interactively. It is important to note that, in most cases, fungus and plant travel separately, but both by animals and wind. Animals often intentionally ingest spores by feeding on hypogeous sporocarps (and sometimes epigeous ones) designed to be nutritious by many fungi, and by fruits for many plants (5). In many cases, seeds themselves may be transported and buried, and often forgotten, initiating succession (708). It is likely that the same process can occur with sporocarps, although this phenomenon has never been reported. In a few cases, animals such as North American elk (*Cervus canadensis*) will pull plants up from their roots in the process of grazing, and transport both reproductive plant parts and mycorrhizal fungi simultaneously in their feces several kilometers to successional sites (33).

Wind dispersal is a complex challenge to measure. Wind usually disperses seeds and spores from stalks or sporocarps that emerge from the soil surface. But how high do they have to be in order for spores or seeds to be entrained? Wind is comprised of eddies that both entrain and deposit materials, regulated by turbulence. In a forested ecosystem, epigeous mushrooms (sporocarps) grow at least a few centimeters above the soil surface, so that a turbulent eddy can entrain the spores. But only minimal transport often occurs. When we measured wind and spore transport in a forest, only windspeeds that exceeded 2 m s^{-1} could generate surface wind speeds of greater than 0.3 m s^{-1}, necessary to actually move spores beyond 20 cm from the basidiocarp (46). At the other extreme, in a desert where horizontal windspeeds commonly exceeded 10 m s^{-1}, the 1 m s^{-1} of the turbulent eddies under a shrub were enough to entrain (56) and disperse AM fungal spores for more than one kilometer downwind (56) (Figure 8.2).

Another factor to consider is spore size, which constrains the ability of spores to migrate by wind (46). Windspeeds rarely exceed 10 m s^{-1}, a horizontal velocity that can result in a vertical soil surface windspeed of 100 cm s^{-1}, sufficient to entrain spores greater than 150 μm. The terminal velocity, the windspeed necessary to entrain particles including AM fungal spores greater than 200 μm, is limited to fungi in the Glomeraceae. This may be a limitation in AM fungal recolonization, especially in more mesic environments where larger-spored AM fungal taxa predominate.

In some cases, birds can carry spores of AM fungi on their feet or feathers. This process appears to be important, especially for establishing mycorrhizae on new islands at really long distances such as across oceans (180).

Together, survival and immigration determine a group of both seeds and fungal inoculum that comprise those organisms in the S$_1$ pool. But it is also important to remember that the migration process is continuous throughout the life of the stand. New taxa of both plants and fungi continually invade, attempting to establish both in occupied and open patches (such as by animal dispersal). As edaphic and vegetation structure conditions change during succession, there are always new invasions and new disturbance patches (Plate 5).

Ecesis

Ecesis is the establishment phase of succession. In this phase, the surviving and immigrating pool of plants and mycorrhizal fungi is reduced to those that can actually survive the environmental conditions of the newly

Figure 8.2 Measuring entrainment and deposition of spores. Shown are the equipment for a sagebrush steppe in southwestern Wyoming, USA, and a pine forest in northeastern Utah, USA. At the sagebrush-steppe, we measured windspeeds 10 m s^{-1} horizontally with 1 m s^{-1} vertical eddies, using a 2D sonic anemometer. The terminal velocity of an AM fungal spore 100 μm in diameter is 0.1 m s^{-1}. Alternatively, windspeeds in a forest are generally 0.01 m s^{-1} (10 cm s^{-1}), with surface windspeeds of 1 cm s^{-1}. With a turbulent eddy (or a fan), windspeeds at the forest floor can reach 2.0 m s^{-1}. The terminal velocity of a 12 μm EM fungal spore is 1.2 cm s^{-1}. It is likely that only when turbulent eddies reach the forest floor, in a gap or a storm, will measurable spore dispersal occur (27; 46; 56).

created environment, and may even reduce taxon richness from S_1 to S_2 (Figure 8.1). This step is often underappreciated, but critical. Egler's (219) old-field model of relay versus initial floristics succession was historically taught as the paradigm for succession. However, this model was based largely on old-field succession from abandoned agriculture, where the old-field soils already contain organic matter and mycorrhizal fungi. Here, initial floristics occurs as the propagules of many later seral plants arrive as soon as the field is abandoned, but whatever mycorrhizal fungi were present in the crop soils would likely remain. Alternatively, relay floristics occurs as new plant species successively invade from outside, and into a pre-existing inoculum. Most secondary successions are likely a combination of initial and relay floristics, with some late seral species colonizing during the initial stages of succession.

The role of ecesis has rarely been differentiated from the selecting effects of biotic interactions, but they are distinct processes. Propagules of many plants and fungi invade a new habitat but many do not survive. This differential survival is related to the new conditions following the disturbance event. Exposed soils without hosts preclude the establishment from spore rain by wind or animals for both EM and AM fungi. For basidiomycetous and ascomycetous fungi, long-distance spore dispersal may depend upon formation of a dikaryon, as only a few haploid spores are generally entrained into the atmosphere for long-distance transport (see Figure 8.2). How challenging dikaryon formation is, and what the differential role of homo- versus heterokaryon remains a controversial topic (see Chapter 5). But given the billions of spores released, many will land in a site where conditions preclude survival. An example is hyaline spores, sensitive to UV radiation, being entrained in high sunlight areas (mountains), and deposited on the soil surface, again with high radiation. Also, of the potential sources of AM fungi, normally only spores of smaller Glomales can be entrained and move long distance by wind (Figure 8.2). The first big question is where and how do they survive? Or, do many spores remain dormant, or live saprotrophically until a mycorrhiza can form? Probably both. In mined soils of Wyoming sagebrush steppe, both spore and hyphal inoculum (either independently or within dead roots) persisted for somewhere between 3 and 12 years (502).

Many plant species, when planted or dispersed (by wind, water, or animals) onto a newly disturbed site, fail to establish as they are intolerant of the temperatures, soil moisture, radiation regime, or lack of mycorrhizal inoculum! The reduced ability of pine seeds to germinate and grow

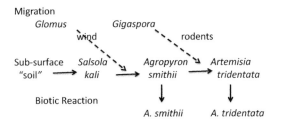

Figure 8.3 Mycorrhizae and succession on a severe disturbance (a restored coal strip mine) in a Wyoming, USA, shrub steppe. This pattern is documented in a suite of papers, summarized in text.

in the middle of pastures, versus in stone fence safe zones, was documented in the mid-nineteenth century (708). Many seeds remain dormant, and seedlings remain as seedlings sometimes for many years. Survival is a function of each population and the tolerance of the propagules to the environment in which they land.

But how do fungi and plants find each other? Occasionally, they co-disperse such as by elk grazing (33), but this appears to be rare. The dynamics of mycorrhizae may regulate the rate and pattern of relay floristics but is absolutely critical for initial floristics. Our Wyoming (USA) studies represent an example case of relay succession, although with few steps (Figure 8.3). Here, on the bare sub-surface mine "soil," species of *Glomus*, mostly immigrated by wind, and *Gigaspora*, by animals. The *Glomus* spores were collected in turbulent eddies, often captured by *Salsola kali*. However, *Glomus* was parasitic on *S. kali*, and simultaneously mutualistic on *Agropyron smithii*, enabling *A. smithii* to outcompete *S. kali* (11; 13; 43; 47). Over a longer time, *Gigaspora* immigrated, probably dispersed by rodents as spores are too heavy for wind dispersal (46). Both taxa of AM fungi immigrated, especially concentrating under *A. tridentata*. This shrub breaks up the eddies, extracting spores and organic matter, and provides cover for animals moving spores (56; 764). Both *Glomus* and *Gigaspora* improved growth of *A. tridentata*, but especially *Gigaspora* (606). *A. smithii* and *A. tridentata* both co-exist in the shrub steppe, but *S. kali* dropped out due to parasitism from *Glomus* and competition with *A. smithii*.

An example from our work in the Yucatán Peninsula, Mexico, demonstrates implications for both fungi and plant. Here, we inoculated several tree species with early- and late-seral AM fungi, and planted them back into a burned area with early and late-seral trees. Of the fungi reintroduced, the late-seral fungi Gigasporaceae, initially declined, but recovered after a year (15). Interestingly, when we repeated the experiment a year later, there was an incomplete burning due to high soil and plant moisture, retaining the soil organic layer. Here, the Gigasporaceae both survived and thrived (48). The plant species from both studies all initially established, providing a pool of both fungi and plants at S_2.

Initial floristics generally occurs on very harsh sites, which only a limited suite of organisms can tolerate, and both plant and fungus must somehow survive until both colonize at the same point and time. This type of succession is generally limited to primary succession. For example, *Alnus* and *Picea* seedlings could be found across glacier outwash (344), but it often took many years of seedling germination for a mycorrhizal fungus to find the seedling, allowing it to survive beyond that growing season and initiate plant community persistence. Often dating of sites is done from aging trees. However, how many years did it take for the fungus and tree to establish, after the glacial till was deposited?

Biotic Interactions

Biotic interactions are the final basis of understanding the role of mycorrhizae in succession. Mycorrhizae may be the single most critical type of symbiotic biotic interaction affecting succession. Both plant and mycorrhizal fungus need to invade and establish for obligately mycotrophic plant taxa to survive and create the S_3 pool.

Mycorrhizal fungi are obligately dependent upon a host plant, thus limiting their arrival at S_2. However, persistent living roots might provide a life-line between S_0 and S_2 (506). EM fungi such as *Rhizopogon olivaceotinctus* have a strategy of surviving fire and persisting through dormancy until a new host root grows (143). Deacon and Fleming (196) carefully describe the process of EM fungal species establishment, with early-stage EM fungi initiating EM upon the invasion of a tree, and late-stage EM fungi establishing as trees age. Importantly, the process of transition occurs both at the individual tree scale, as the initial tree develops a canopy and spreads outward, and at the whole stand-scale as the canopy closes and there are few "open" patches where the soil properties resemble those of the early seral soil at S_2. Although all EM are dependent upon finding a host, early-stage fungi appear to do well in early seral soils (low organic matter, available nutrients) and late-stage fungi thrive in late seral soils (high organic matter, organically bound nutrients).

In an interesting twist, models suggest that even a simple shift in the order of reinvasion by different fungi can change the growth responses of a host (75). William Swenson developed a stoichiometric model of different taxa of mycorrhizal fungi. By simply shifting the sequence of addition of fungi with differential traits, incorporating differential N access, immigration, and establishment, the outcome of plant productivity shifted. That complexity was exemplified in an experiment where transplants were grown in different seral stages on the retreating Exit Glacier, in southern Alaska (344). Here, the leaf N concentration varied with no pattern between successive EM fungal taxon richness.

AM fungi appear to exhibit some of the same patterns, but we understand less about successional mechanisms. For example, AM fungi persist as a parasite on the nonmycotrophic *Salsola kali* (44; 47), AM fungi appear to utilize the C to help persist until the fungus can establish on mycotrophic plants. There are preferential interactions among AM fungi and plants both across successional boundaries and plant taxa within a sere. Wilson and Tommerup (784) describe a number of characteristics and scales whereby the different AM fungi might compete for hosts, and whereby hosts might select among AM fungi. They described the importance of infectivity and aggressiveness as key characteristics to understand how the fungi might compete for host position. We know that different fungi differentially establish. For example, some taxa have few germ tubes emerging, others many, repeating across several days. There is a range in tolerance of temperatures and drought. AM fungal spores in Wyoming cold deserts have a dormancy period of several months, whereas those in the tropical seasonal forest germinate immediately. Different taxa can have very different effects on the growth of different hosts. But, at this point, we do not know how or for what AM fungi compete; the hypotheses of Wilson and Tommerup need far more experimental testing.

Plant responses to mycorrhizae vary during successional transitions in that different taxa range from nonmycotrophic to obligately mycotrophic. Nonmycotrophic plants, especially in the Amaranthaceae and Brassicaceae, can preferentially establish into newly disturbed soils, but do not compete well with mycotrophic species as succession proceeds. While these species may still be dominant at S_2, they rarely persist to S_3 (606). Facultative mycotrophs (as per (377)) have an advantage in that they can establish without the fungi, but become less competitive later as the fungi and more obligately mycotrophic species establish.

AM fungi require hosts but also appear to preferentially (although not exclusively) prefer early- or late-seral conditions. In the first Yucatán experiment, all plants preferentially grew with early seral fungi immediately after an intense burn (15), whereas in the second experiment, with a partial burn, no preference or even a late seral preference was exhibited (48); the fungal responses were more related to the soil conditions than to the specific host–fungus interaction. But then as the region recovered from the large fire in which the experiment was embedded, grazers, including deer, came back onto the plots. One of our planted species, the late-seral plant *Brosimum alicastrum*, a late-seral tree, was found to be especially sensitive to deer herbivory, and nearly disappeared from the plantings. In the second experiment, another species, *Cochlospermum vitifolium*, grew especially well with both late and early-seral AM fungi, but then was attacked and eliminated from the site by a *Fusarium* infection in the root system (48). Thus, the complexity across the ecosystem affected the detailed mycorrhizal responses and ultimate community composition on the small plots studied.

The establishment of a mycorrhizal fungal inoculum that can survive across seral stages is becoming a critical element of understanding both plant communities and succession. Dave Perry and colleagues (561) found EM fungi that survived fire in Douglas fir stands. These fungi formed arbutoid mycorrhizae with resprouting *Arbutus* plants. The same species of mycorrhizal fungi formed EM with the conifer seedlings, providing a continuing inoculum across the nudation, S0, into S1, and S2, and affecting the transition vegetation from arbutoid mycorrhizal shrubland to EM forest. They termed this "bootstrapping," where communities "pull themselves back up by their bootstraps," and argued that disturbed ecosystems rarely reestablish from scratch (Plate 8). This process evolved into the view that mycorrhizal fungi connect plants across broad networks, the Common Mycorrhizal Networks (Chapter 6).

More broadly, the mycorrhizal microbiome is a new take on a continuing research topic area that underlies processes of mycorrhizae in succession. In the 1960s and 1970s, Elroy L. Rice and colleagues studied N dynamics during succession. Blum and Rice (127) found that N_2 fixation was inhibited in late succession, and Rice and Pancholy (546) found that nitrification decreased, and ammonification remained high in late successional compared with early successional plots. This pattern agreed with the ideas of Odum (539) where the ecosystem shifted from an open system to a closed one. Interestingly, Rice and Pancholy (546) also noted that a mechanism was an increase in soil proteins, and soil proteins, including glomalin and other mycorrhizal-induced (or produced) proteins are at least partially responsible for the increase (582).

As we better understand the composition and the role of the mycorrhizal microbiome including the potential for horizontal transfer of critical genetic material, the interactions across organisms may well allow a far more detailed understanding of successional processes and mycorrhizae.

Mycorrhizae and Succession on Mount St. Helens

Here, I use the example of Mount St. Helens in the state of Washington, USA, where we studied mycorrhizae and succession since the 1980 eruption. The advantage of focusing on this LID is three-fold. There is a long research history spanning pre-eruption conditions, allowing us to get beyond space–for–time substitutions and evaluate changing conditions directly over time. There is a large area encompassing multiple initial conditions and multiple perturbations (Figure 8.4). And, importantly, every mycorrhizal type is represented – EM, AM, ericoid, and orchid, as well as also dark-septate fungi and mosses with fine endophyte mucoraceous fungi. My current study has been ongoing for 40 years of research, with each patch being studied independently across approximately 500 square kilometers (Plate 9).

Mount St. Helens erupted in May 1980, in a well-studied coniferous forest ecosystem, with many trees greater than 200 years old, and a meter or more in diameter. But embedded within that forest were natural meadows, along streams or at the edge of timberline, as well as clear-cuts, patches up to many hectares in size, that were in various stages of succession or reforestation at the time of eruption. The eruption created multiple initial conditions overlying each of the different pre-eruption conditions. The 1980 eruption initiated from small earthquakes commencing March 15, 1980 and continuing for over two months. A 5.1 earthquake initiated the collapse of the north flank of the volcano early on May 18, 1980, and resulted in a 2.5 km^3 debris flow, radically changing the geography of many square kilometers (693). After the collapse of the flank, a large explosion of hot gases (up to 300°C) blasted out, exceeding the speed of sound (>700 km h^{-1}), across the landscape and blew down forest trees over about 570 km^2, the Blast Zone. Mount St. Helens is described as a Plinean eruption, described by Pliny the Younger for its similarities to the Vesuvian eruption of 79 A.D. The column rose and fell during the day of May 18 of pumice material plus dense rock called a pyroclastic flow, depositing up to 40 m of sterile pumice across 0.3 km^2. The surface consisted of pumice pebbles about the size of golf balls, with scattered basalt larger rocks. This Pumice Plain

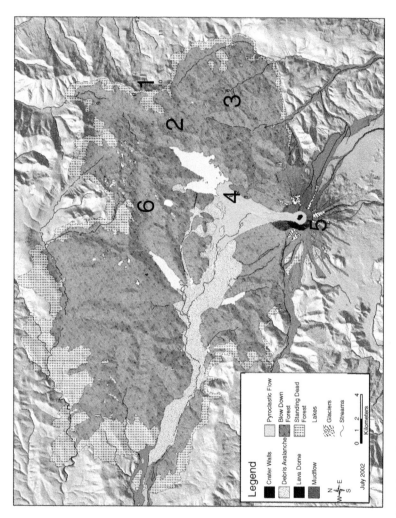

Figure 8.4 Map of the Mount St. Helens eruption zone disturbances. Shown are the sites described in the text, (1) Bear Meadow, (2) Meta Lake, (3) Smith/Bean Creek, (4) Pumice Plain, (5) Butte Camp, and (6) Johnson Ridge, using a 2014 map from (69).

is a true primary succession, with no surviving organisms. Surrounding the pyroclastic flow, from the blast area outward, deposited pumice and ash ranging from golf-ball sized pumice to fine, silt-sized particles, radiating outward and covered an area of 16 km². The fine particles extended outward in lesser amounts for thousands of kilometers. The morning after the eruption, I scraped ash off my car windshield in Lincoln, Nebraska. A decade later, while I was collecting in the eastern portion of the State of Washington, a distinct layer of ashfall could still be discerned.

What this means is that there were multiple areas with very different initial conditions or legacies. The pre-eruption initial conditions in the vicinity of the mountain had buried tephra from the eruption of 1800 covered by an organic layered forest floor. Even below the 1800 tephra layer, additional layers could be identified corresponding to previous eruptions, meadows, and forests. Across the post-1800 eruption, within the forests, clear-cut areas and natural meadows could be found. It is also very important to note that many of these forests were old growth, with trees a few hundred years old.

This volcanic eruption provides a comprehensive case study of succession, because it includes both primary and secondary succession, initial and relay floristics, and every known type of mycorrhiza. Many detailed papers have been published, and citations can be found in our summaries (69). I will focus on four very different sites across a ~ 10 km transect to the northeast from the volcano. These are the primary successional Pumice Plain; Smith Creek, a blowdown zone, with about 1 m of pumice deposited on top of an old clearcut meadow − with characteristics of both primary and secondary succession; Meta Lake, within the blowdown zone in an old-growth forest and with a meadow surrounding the lake − secondary succession; and the old-growth forest northeast of Bear Meadow, an old-growth forest inundated with ash deposited on the trees − secondary succession (Figure 8.4 and Plate 9).

The Bear Meadow and Adjacent Forest

This was the least affected site of this group. Ash deposited on all the needles of the conifers and covered the forest floor and meadow from 5 to 20 cm, burying the organic layer and all herbs. In the forest, these included achlorophyllous orchids (*Corallorhiza maculata*) and ericoid achlorophyllous plants (*Monotropa hypopitys*), and a diverse array of EM fungi. Predictions were made that the sunlight would penetrate the leaves, but heat would be reflected back inward, overheating and killing the leaves, and that the trees would not survive without their needles. The first part happened as predicted (652), needles all died and dropped onto the forest floor. However, needles normally ranged up to four years old in these trees. When the leaves died and dropped, the tree regrew new leaves. As new leaves are photosynthetically more active than old ones, tree productivity was higher than expected. Roots remained active under the new ash, with added nutrients from the eruption.

Examination of the soil showed that only long roots grew through the ash layer. This pattern was found from the 1980 as well as the 1880 eruption, and even in previous eruptions. Between each eruption layer, a new forest floor spread out of decomposing organic matter with primary and secondary roots and EM. From the leaf litter added from the eruption, a new forest floor emerged, with new EM short roots. Over 35 years of study, both *Corallorhiza* and *Monotropa* emerged in the new litter layer, with molecular sequences showing fungi that form orchid mycorrhizae and monotropoid EM. This site looks today very much like it did before the eruption, and virtually every short root is EM with a diverse suite of EM fungal morphotypes. S_3 (prior to the eruption) $= S_3$ (35 years post eruption); the entire secondary succession process took less than 35 years.

Bear Meadow, a clear-cut forest patch, underwent a more nuanced shift. There were no large trees remaining above the ashfall. However, because the eruption occurred in mid-May, when there was still a deep snow cover, fossorial organisms, including pocket gophers (*Thomomys talpoides*) and ants simply re-emerged. They mixed the buried old soil with roots and AM fungi (along with most of the soil community, including mites, nematodes, fungi, and bacteria) back to the surface, mixing with the newly deposited tephra. This new "soil" was nutrient rich and well aerated (473). As with the forest, while a layer of tephra still remains, plants rapidly returned, and a successional forest of EM alder and conifers has emerged (69).

Meta Lake

This has proven to be an interesting site. This was a blow-down site, where trees were toppled and hot gases rolled across the soil surface. The site was protected somewhat by the northern ridge, deep snow, and wet soil. While large trees were blown over, small trees with their EM and ericoid mycorrhizae were already bent under the weight of the snow and were simply covered over by the ash. Many of these trees re-emerged upon snowmelt and continued growing. By 1990, many EM sporocarps had appeared, including *Endogone pisiformis*, *Laccaria laccata*, and several species of *Boletus* (53; 156). AM, EM, and ericoid mycorrhizae were established within the first decade. Again, as with Bear Meadow, many of the organisms survived the eruption with soils remaining intact, indicating that S_3 largely equaled S_3 prior to the eruption.

Smith Creek

Smith Creek proved to be quite different. The site was mostly a sloping meadow that resulted from a prior clear-cut. This site experienced the blow-down winds and gases head on, as well as a heavy tephra fall. As the snow melted on the steep hillsides, tephra and ash eroded, creating deep ash layers (>1 m) of sterile material. Spores of Glomeromycotina were observed in the rills and gullies that emerged over the first few years. Gophers survived in patches where the ash eroded to depths of 10 cm or less. Over the first decade, gophers mixed ash and old soil, providing plant propagules and mycorrhizal inoculum. Plants, including *Lupinus latifolius*, *Chamerion angustifolium*, and *Anaphalis margaritaceae*, all formed AM. The *L. latifolius* was also nodulated, with red nodules, indicating active leghemoglobin, the hemoprotein that carries O_2 from the N_2 fixing region of the nodule. Interestingly, *Hieracium albiflorum*, a non-native AM plant, also became predominant by 2000, suggesting that some reinvasion of AM had also occurred. However, there remains little EM tree invasion. It appears that the bioturbation activity of the growing gopher population is maintaining the herbaceous plant community.

In the neighboring watershed of Bean Creek, a number of trees were planted. *Pisolithus tinctorius* emerged across the stand of seedlings, suggesting that some were inoculated in the nursery, and a *Melanogaster* sp., a basidiomycete truffle, suggesting that some native EM fungi also colonized the planted seedlings. Gophers and other rodents were common at this site.

The Pumice Plain

This is probably the most dramatic feature created from the eruption, certainly from satellite imagery. This feature resulted from the deep pyroclastic flow creating a sterile, primary landscape. No plants, fungi, or animals survived within this landscape. This landscape consisted of coarse tephra with no organic matter, virtually no water-holding capacity, and very low nutrient levels. But, within months after the eruption (1980), elk had wandered across the Pumice Plain, and birds had flown over. By the end of the growing season of 1981, a few individual seedlings were scattered across the landscape. By the second year after eruption (1982), a few scattered small patches of *Lupinus lepidus* were found. Each had a single large plant, with up to a dozen seedlings surrounding the large plant. All plants were nodulated, with red nodules, but were nonmycorrhizal, with no trace of mycorrhizal fungi. At the edges of the Pumice Plain, where erosion exposed pre-eruption patches of topsoil, herbs and willows began re-growing and small mammals were found. Presumably they had survived in small microsites around the edge and outward facing from the Pumice Plain.

In 1982, we helicoptered individual gophers trapped in the high-ash zone, and placed them within 1 m^2 enclosures surrounding individual *L. lepidus* and *L. latifolius* plants, to test the role of animals in succession (63). They remained in the enclosure for 24 hours, then were removed and returned to their original site. We examined feces of the gophers, and all scat examined had spores of *Glomus macrocarpum* and *Sclerocystis* sp. In 1983, we surveyed roots from these plants, the surrounding seedlings, and plants from other patches. The two inoculated plants were the only individuals found across the Pumice Plain that were mycorrhizal and had the same fungi that we found in the scat. Animals were the vectors for AM fungi. In 1983 and 1984, we found no other AM plants or AM fungi in the surrounding area, indicating initially slow dispersal of AM fungi.

Throughout the remainder of the decade, rodents continued to reinvade via streambanks across the Pumice Plain, bringing inoculum to many new patches, and increasing the diversity of AM fungi. By 2000, spores of species of *Glomus*, *Scutellospora*, *Acaulospora*, and *Sclerocystis* were found at the focal study plot. By 2003, Titus et al. (713) reported that many sites across the Pumice Plain had a diverse group of AM fungi. Patches of willow and alder began emerging as streams began to emerge. By 1987, we trapped air-borne spores of *Thelephora terrestris* as well as many unidentifiable spores. Individual seedlings of conifers began to persist, and all were EM (33). Newly forming streams began in the early 1980s, and by the mid-1980s, both willows (*Salix* sp) and alder (*Alnus* sp.) colonized the streams, and all were EM. EM fungi including *Laccaria bicolor* were found and all seedlings of both species examined were AM.

During the 1990s, the most noticeable characteristic of the Pumice Plain was the breakdown of the large pebbles of pumice, into smaller fragments and even into silt-sized particles. The silt began holding moisture, and in the plant patches, even organic matter. Halvorson and colleagues (e.g., (312; 313)) showed that patches of soil with mycorrhizae began accumulating glomalin as well as many other proteins and increasing C and N (see references in (69)). Newer molecular analyses undertaken by Mia Maltz and Michaela Phillips and colleagues are beginning to show that complex microbial communities have established well across the entire landscape.

Synthesis Points

Succession on Mount St. Helens demonstrates the complexity of disturbance, landscapes, succession mechanisms, and mutualisms. In going back to the Treseder model (Figure 1.1), we can see that disturbance types exist along a gradient in the x-axis, from the Pumice Plain, with a very low nutrient value, up to a relatively high value at the Bear Meadow site. The primary succession site, the Pumice Plain, initially had no mycorrhizae, until introduced by immigrating animals, and was occupied by facultative AM mycotrophic plants. As mycorrhizal fungi immigrated, obligately EM plant seedlings began to persist. This tells us that both relay and initial succession interact in mixed patches. Both mycorrhizal plants and animals, including gophers, continue to expand their range back up the volcano. Secondary successional sites continue to gain diversity in both species richness and in mycorrhizal types. I expect that these trends will continue, patch by patch.

But an interesting contradictory pattern has also emerged. In some patches, succession appears to have stalled out. Where patches of trees began to establish, around the Pumice Plain plot for example, the elk descended upon those patches, grazing virtually everything except for lupines out. This included both AM herbs and EM trees. Today, many of the lupine patches are again monotypic lupine patches. Where gophers reestablished, their feeding on roots, bulbs, and corms, and their bioturbation also appears to be creating nearly permanent early-seral patches. I find these patches scattered around the volcanic area, in such disparate places as Butte Camp on the south side of the volcano, and down Johnson Ridge.

As I began to understand these patches, I recognized these same early-seral patches remaining in other volcanoes, including Mount Vesuvius with its many well-documented eruptions, back to 79 A.D., the 2011 Puyehue-Cordon Caulle eruption in Chile with ash that drifted across to

Argentina, Mt. Lassen from 1930s, and even Crater Lake, in Oregon, that erupted 7,700 years ago. In reflecting back, while a postdoc at the University of Nebraska, I sampled mycorrhizae on the 9-mile prairie plots. These plots were established by John Weaver in the early twentieth century. Weaver was a student of Frederic Clements, who developed the theory of succession upon which our model (Figure 8.1) is based. In wandering around the adjacent prairie, one can still find patches of all successional seres, maintained by small-scale disturbances, and with the same mycorrhizal relationships described in Chapters 6 and 7 and the LIDs in this chapter. Patches, succession, and mycorrhizae are intimately interactive at all scales and remain in transition for millennia.

Summary

- Mycorrhizae are tightly coupled to succession. Following severe disturbance, both plant and fungi are reduced or eliminated, and species composition is altered. As both plants and fungi reestablish (ecesis), mycorrhiza-environment feedbacks result in changing ecosystem processes by regulating C throughput and tightening nutrient cycles.
- Migration by fungi to initiate the compositional change is largely by wind in arid ecosystems, and animals in mesic and wet ecosystems, although not exclusively. Indirect transport through ingestion of fungal propagules is also crucial to succession.
- The successional patterns of the Mount St. Helens Volcano in southwestern Washington State, USA, between 1980 and 2020, serve as an example of a complex array of mycorrhizal fungal–plant actors turning a landscape with regions of a pumice-covered sterile barren lands into a complex forest–meadow vegetation mosaic through both primary and secondary succession.
- Successional mechanisms and mycorrhizae are increasingly being used to reclaim or even restore both stands for fiber production and for wildlands for recovering biodiversity.

9 · *Global Change*

Mycorrhizal symbioses have existed since at least the Ordovician period, between 400 and 500 MYa. Associations, and even some plant–fungus species pairs, have existed in morphologically recognized combinations through every climate and substrate condition throughout that time period. Across that vast time, virtually every potential condition likely would have been encountered. Over the past eight chapters, I have covered nearly the entire range of conditions that mycorrhizal relationships have encountered: some ancient, some novel. For example, in studying Eocene Epoch environments, atmospheric CO_2 levels were high, making N limiting. By Pleistocene Epoch environments, atmospheric CO_2 declined, making C more limiting relative to N. But today, landscapes are experiencing both high and increasing atmospheric CO_2 and increasing N deposition; which direction is C:N stoichiometry going? Are we seeing something new occurring, as humans increasingly alter not just the patch on which individuals exist, but over large scales across the globe, where humans previously may never have stepped, such as Antarctica? And, will those changes alter the very nature of the mycorrhizal symbiosis that has persisted for nearly half-a-billion years? Here, we explore the characteristics of our rapidly changing globe, how we are altering the planet, and what potential shifting mycorrhizal relationship(s) might be occurring.

Lewis and Maslin (452) noted that the earth may be embarking on a new human-dominated geologic epic, the Anthropocene Epoch, where the rapid change in human population and exploitation of the earth's resources are dramatically altering basic evolutionary patterns and ecosystem functioning across the globe. Importantly, two time periods are identified, the Common Era (C.E.) of 1610 and 1964, as having markers of geological era-scale shift; periods dramatically altered by humans, *Homo sapiens*. These two time periods have special meaning for me and for this volume. I began my career torn between biology and archeology. My fascination with archeology blended my interest in wildlands, agriculture, and conservation, with interests in history. In the course of my undergraduate studies, I traveled on a class field trip for 6 weeks across Mexico, observing the impacts of exchanges and encounters between Europe and North America that fundamentally altered both. These ranged from crops such as maize, dry beans, peppers, and potatoes from the Americas, to wheat and rice from Eurasia, sugarcane and millet from Africa to the Americas, to animal husbandry from Eurasia and Africa to the Americas. They also included reciprocal translocations of diseases that locally decimated human populations, especially in the Americas. As a result of those exchanges, indigenous human populations initially crashed in the Americas, especially between 1570 and 1620, resulting in a massive forest regrowth as fields were abandoned, resulting in a 7 ppm to 10 ppm reduction in atmospheric CO_2 globally, and a shift in the atmospheric $\delta^{13}C$ isotopic signatures of CO_2 and CH_4. One postulated outcome was the deepening of the Little Ice Age with the lowered atmospheric CO_2. But then, use of the newly exchanged crops and animals began increasing; after 1620 human populations expanded to a billion by 1800 and 1.5 billion in 1900. Today, with forest and native grasslands converted to pasture and field, and fossil fuel (including the C sequestered by mycorrhizae in the Carboniferous Period, Chapter 3), the global human population is rapidly approaching 8 billion. Because of these impacts, it has been exceedingly challenging to study and understand how "natural" ecosystems work because, in essence, there are no natural systems; all are human affected (527).

The second date, 1964, resulted from the partial nuclear bomb test ban treaty, when a ^{14}C peak in the atmosphere culminated, facilitating short-term measurements of everything from climate markers (ice and sediment cores) to mycorrhizal fungal C ages (723; 745). But also, the 1950s represented a massive upswing in atmospheric CO_2 levels, largely from anthropogenic sources, which dominates the airwaves today. A large fraction of that change has occurred within my lifetime, and especially since I began studying mycorrhizae and CO_2 in graduate school in 1974. In 1952, atmospheric CO_2 was 312 ppm. In 1975, I used 330 ppm as my atmospheric baseline in mycorrhizae and photosynthesis studies (74). By 1985, when I purchased a new infrared gas analyzer to measure photosynthesis, I was initially surprised that the analyzer would not recalibrate at the 330-ppm level! This was because atmospheric CO_2 had subsequently increased to 345 ppm. As I write this chapter, peak atmospheric CO_2 broke 419, a 34 percent increase since my birth, and a 27 percent increase since I began my studies. Atmospheric CO_2 is projected to reach

450 ppm by 2030 to 2040. The last time the globe reached 450, there were AM *Sequoia* in Yellowstone National Park and EM *Nothofagus* forests in Antarctica. We are in a state of dramatic change, the Anthropocene Epoch.

How are mycorrhizae changing in the Anthropocene Epoch, how are mycorrhizae affecting the shifting conditions of the Anthropocene Epoch, and what are issues that we need to focus on as a research community? Here, I focus on the four relevant issues of global change that are directly tied to mycorrhizal functioning: atmospheric CO$_2$, warming temperatures, N deposition, and biodiversity changes including both loss of biodiversity and invasive species and their impacts on community and ecosystem dynamics.

Direct Atmospheric CO$_2$ Effects

Atmospheric CO$_2$ directly affects the rate of photosynthesis by increasing the partial pressure of CO$_2$, facilitating uptake into stomata. By stimulating photosynthesis, leaf C is increased. Stoichiometrically (Chapter 4), C:N and C:P then increase, inducing roots and mycorrhizae to increase nutrient uptake. Mycorrhizal plants respond by increasing percent infection, or by maintaining the same mycorrhizal frequencies but increasing root growth; in either case, *total* mycorrhizal activity is increased to maintain N and P concentrations. While total plant growth can increase (the CO$_2$–fertilization effect), either plant nutrient concentration falls or mycorrhizal activity increases. Numerous elevated CO$_2$ experiments have documented increased allocation of C to mycorrhizae, whether percent infection, total numbers of infection, or extramatrical hyphal production (see (703; 704) for a meta-analysis overview). Furthermore, mycorrhizae appear to be a major pipeline from atmospheric CO$_2$ to soil. Godbold et al. (289), using a combination of isotopic composition and ingrowth bags, found that mycorrhizal fungal hyphae produced 62 percent of the atmospheric C entering the soil organic matter, higher than fine root and leaf inputs.

However, it is important to remember that many subtle interactions may also intersect with the larger CO$_2$ response. Temporal and adaptive patterns may well emerge. For example, Klironomos et al. (413) found that when atmospheric CO$_2$ was increased immediately in an experiment, there was a dramatic increase in percent infection. However, if CO$_2$ levels were incrementally increased 57 percent over 21 generations (10 ppm per generation), no significant difference was observed.

Mycorrhizal adjustment to the rate of CO_2 *increase is a crucial research question that needs addressing!* Experiments raising CO_2 abruptly to 550 ppm, such as Free Air CO_2 Enrichment (FACE) studies, raise the atmosphere levels immediately upon startup. Remember that atmospheric CO_2 has increased 31 percent over the past seven decades and will increase 44 percent by the mid-2030s. This is well within a single generation for most trees and shrubs, although extends nearly 70 generations for annuals. We are left with a new question: *What constitutes a control, or baseline concentration?* Should we be contrasting between CO_2 levels of 190 ppm (Pleistocene Epoch minimum), 250 ppm (pre-industrial), 300 ppm (1950s), or 350 ppm (1990s)? Then, what comprises a generation?, 2,000 years for a *Sequoiadendron giganteum* or 5,000 years for *Pinus longaeva*; which takes us back to pre-industrial CO_2 levels, a nearly doubling within a generation!

Conversely, how might mycorrhizae affect atmospheric CO_2? First, mycorrhizal fungi comprise a sink for the fixed C, just as roots do. We do not know the size of this sink, and it changes with each species and each environment. However, we can develop some predictions that could be useful. In my dissertation research, I studied the effects of AM on photosynthesis (74). In my studies, I focused on a C_4 (warm season) grass, *Bouteloua gracilis*, back when atmospheric CO_2 was 330 ppm. Photosynthesis increased 68 percent, whereas dry mass of the plant only increased by 21 percent. This means that 47 percent of the added C fixed went to the fungus or to changes in the host respiration. The increase in C fixation was due to both a 51 percent reduction in gas-phase resistance and a 77 percent reduction in liquid-phase resistance. The liquid-phase resistance was due to increased P and N uptake, as this would be associated with more photophosphorylation activity – more energy (ATP) and more enzyme. This would be confirmed as AM plants had a higher organic P-to-inorganic P ratio in the leaves, and greater overall P uptake (73). The exact same study was done with a C_3 grass, *Agropyron smithii* (50). This plant, less responsive to AM, nevertheless showed a 27 percent increase in dry mass and a 47 percent increase in photosynthesis, thus 20 percent of the plant's fixation went to the fungus (or plant physiological changes). In the field, we found that mycorrhizal improvements in water flux in the field were similar to increases in the lab (12). Based on these responses, during the mid-1980s, AM would most likely account for around 20 percent of the fixed C, which could be either direct allocation to and increased physiological activity of the fungus, or increased activity and respiration of the host. To my knowledge, there is no data at the ecosystem scale for C_4 plants and AM, although we know that C_4 grasslands are highly dependent upon the N, P, and H_2O provided by AM fungi. For C_3 ecosystems, 20 percent allocation of C is within the range of the allocation to fungi, such as 20 percent for an EM *Pinus sylvestris* forest (366), an EM *Pinus edulis* stand (49), and 27 percent for a California mixed oak-conifer stand (59). Assuming that these figures represent C allocation for the 1970s to the early 2000s, how do these relate to the past environments, such as the pre-industrial world where atmospheric CO_2 was 250 ppm, and projections into the future, where atmospheric CO_2 will reach at least 450 ppm, with some projections reaching 550 ppm or even 750 ppm? How do we make projections about the role of mycorrhizae at these levels? The range for C allocation to mycorrhizal fungi ranges from 5 to 40 percent of the NPP, depending upon the CO_2 concentration, techniques used in the field or experimental approach used in the lab. We really do not know what the response is in ericoid, dark-septate, or orchid mycorrhizae. Experiments currently underway might provide illumination (although as my high-school physics teacher said many decades ago, we only poorly understand light, implying the illumination may not equal understanding).

My group began studying responses to elevated CO_2 in field chambers and in lab chambers. We found a number of responses, led by John Klironomos (e.g., (413)), Kathleen Treseder (e.g., (717; 718)), and Matthias Rillig (e.g., (608; 611)). All showed that increasing CO_2 alone triggered a major response by mycorrhizae. One way to summarize is to focus on our field studies (61; 612; 721). We studied a chaparral ecosystem emerging from a fire, where dominant shrubs were both AM and EM. Because of the prevalence of fire, chaparral shrubs have large, belowground C storage structures allowing them to re-grow rapidly. Our research was conducted using field chambers with increasing CO_2, from 250 ppm (simulating pre-industrial conditions) at 100 ppm intervals to 750 ppm. Here NEE changed linearly. At 250 ppm CO_2, the patch within the chamber lost over 1 kg C m^{-2} y^{-1} to the atmosphere in that less CO_2 was fixed by the plant than was respired by the soil. At 750 ppm CO_2, there was a net uptake of 300 g C m^{-2} y^{-1} into plant and soil (721). The standing crop of AM fungal hyphae did not appear to significantly increase with elevated CO_2 (61). The active fungi increased (comprised of both EM and saprotrophic fungi, as we could not morphologically disentangle them). But the interesting feature that emerged was that mite activity increased with increasing CO_2 concentrations, and above 550 ppm, and they grazed on so much fungal mass that the overall fungal mass declined. Turnover increased, not standing crop. Looking at fungal standing crop alone was an inadequate measure of the mycorrhizal contribution. If standing crop remained steady, but hyphal production and mortality increased, then there is a potential for greater C storage as glomalin and soil aggregation increase (e.g., (612; 717; 718; 749)).

More direct physiological data may also be useful. From our 2006 study (61), the data indicate that even arid ecosystems are sequestering CO_2 from the atmosphere, and that mycorrhizae may be contributing a great deal to this response. We can begin to tease this apart even more. For example, water–use efficiency appears not to change dramatically with AM (393; 585; 586), although some instances can show a significant response (587). The big response, as in the greenhouse studies, is a decrease in $\delta^{18}O$ ratio with AM, indicating a greater throughput of water (586). Assuming that most plants increase photosynthesis by 40 to 70 percent, and half of that is due to gas–phase change, we would expect a great deal more C fixed by and sequestered in soils with mycorrhizal fungi, as atmospheric CO_2 continues to increase. Allen and Rincón (71) noted that under 330 ppm, 1 molecule of CO_2 is fixed per 318 molecules of water transpired. That rate could fall to 1:197 by the end of the next century. As mycorrhizal plants access water under drier conditions (Chapter 4), we would expect a 50 percent or greater sequestration of C by mycorrhizae. Currently, current anthropogenic C inputs to the atmosphere average 50 g m^{-2} y^{-1} (on a global surface area basis). NEE ranges from 20 to over 100 g m^{-2} y^{-1} for mycorrhizal terrestrial ecosystems. Increasing C accumulation with the improved WUE over the long–term could have dramatic impacts on the overall global C budget, modulated by nutrient sufficiency.

Our question is, then, might there be enough of a CO_2 fertilization response to make a significant difference in mycorrhizal dynamics that might influence overall global C? Remember that mycorrhizae have persisted through far greater CO_2 fluctuations than the current projections provide (37). One mechanism is that mycorrhizae increase uptake of water and nutrients facilitating greater photosynthesis and overall C gain. As CO_2 increases WUE, nitrogen–use–efficiency also increased (585), sustaining photosynthesis. In our chaparral elevated CO_2 study (61), $\delta^{15}N$ of the leaf tissue decreased as *Adenostoma fasciculatum* mycorrhizae increased, as did the signal in *Ceanothus greggii*, an actinorhizal AM plant. These data indicate that both higher N uptake and increasing fraction, and higher N_2 fixation occurred, to keep up with the increasing CO_2 fixation. As the change in NEE amounted to 803 g C m^{-2} y^{-1}, and assuming mycorrhizal fungi accounted for 20 percent of that C, nearly 160 g C m^{-2} y^{-1}, the water-soluble aggregate fraction, which includes glomalin, is comprised of 65 g C m^{-2}. A similar hypothesis is the overflow CO$_2$ tap (342), where extra C is used to produce recalcitrant fungal compounds, such as glomalin (in AM fungi) and melanin in many EM fungi, like *Cenococcum geophilum*. These mechanisms are not mutually exclusive but demonstrate the many roles of mycorrhizae in sequestering atmospheric CO$_2$.

New Zealand forests currently have several tree species also found in the Miocene Epoch forests, when atmospheric CO$_2$ was approaching 550 ppm. Reichgelt and colleagues (603) used gas exchange measurements and modeling of these species, simulating the plant. WUE increased, facilitating the CO$_2$–fertilization effect. This implies that CO$_2$ fixation does increase with elevated atmospheric CO$_2$. Two conclusions emerge from this. First, the plant must take up the necessary N and P needed to keep the fixation rates up. Second, presumably mycorrhizal activity also increases, along with the sequestered C associated with the symbiosis. Mycorrhizae provide the mechanisms for increasing N and P (see Chapter 7). Globally, Shevliakova et al. (655) estimated that as global CO$_2$ increased, the enhanced plant photosynthesis has already reduced atmospheric CO$_2$ by as much as 85 ppm. Assuming that 20 percent of that 85 ppm is allocated to mycorrhizal fungi, then the fungi might directly account for a reduction of 17 ppm globally. This is my "back of the envelope" calculation but indicates a modeling approach focusing directly on atmospheric CO$_2$ and C allocation is needed. Approaches such as the development of the versions of the FUN model (Chapter 7) help us a great deal and show that mycorrhizae contribute greatly to overall NPP and C capture. Shi et al. (656), using the FUN model (Chapter 7), calculated that the total global uptake of N is 1.0 Pg (10^{15} g) y^{-1} requiring 2.4 Pg C y^{-1} to access that N, and that mycorrhizal root uptake is the largest C cost to acquire that N. As atmospheric CO$_2$ continues to increase, N demand will only continue to increase, making mycorrhizae more, not less important.

Interestingly, Shi et al. (656) also found that tropical ecosystems, largely AM, appear to be more P limited than N. This concept led to the revision of the FUN model to version 3.0, which explicitly added P acquisition costs as a function of N costs (23). Here they predicted that approximately 2 percent of the NPP was needed to produce the phosphatase necessary to acquire sufficient P for N$_2$ fixation. Given our estimates for the La Selva tropical forest that 10 percent of the total

NPP goes directly to AM fungi (Chapter 7), our measured estimates are probably not too far from modeled estimates. While more work is needed to project into the enriched CO_2 world of the future, the partnering of future measurements and models hold promise in teasing out the responses of mycorrhizae and impacts of mycorrhizae in our future greenhouse world.

Changing Temperatures and Precipitation Accompanies and Confounds the Direct Effects of CO_2

Global warming is often postulated to be the greatest single environmental threat to existing human activities. Elevated CO_2 appears to be tightly coupled to global temperature and precipitation spatial patterns as well as overall global rates of change. Temperature and precipitation regulate the distribution and boundaries of biomes and mycorrhizal types (Chapter 7). Although more information is needed, the ecosystem responses may well depend upon the degree of change, and the type of mycorrhiza. At high latitudes or high altitudes, CO_2 has a more dramatic effect on temperatures than at low latitudes and altitudes. This is because low latitudes and altitudes have greater atmospheric moisture, and the water vapor effect overrides the CO_2 greenhouse effect. Since some mycorrhizal types are tied to biomes, the responses may vary greatly. For example, increasing global temperatures (current increase has been approximately 6°C in some arctic regions over the past century) dry out and warm up boreal forest ecosystems, which is comprised of one of the largest pools of sequestered C in ericoid mycorrhizal heathlands.

Open-topped experimental chambers show that, with initial warming temperatures of 9°C and elevated CO_2 of 500 ppm above ambient, shrub roots and ectomycorrhizal fungal rhizomorphs increased, and the below-ground active growing season, with active soil respiration, increased by 62 days (201; 761). This led to a rapid loss of total ecosystem C. R_s, R_r, and fine root (EM) turnover in *Picea mariana* all showed a highly significant temperature response.

How are C stocks globally likely to be affected? We undertook a modeling effort at the James Reserve, in the mixed deciduous–coniferous forest of southern California. First, we examined the projections for climate change scenarios exhibited by multiple climate models (60). Elevated CO_2 increases reflected long-wave radiation, currently estimated at a rate of 3 W m^{-2}. During a mid-summer day, with radiation inputs up to 1,400 W m^{-2}, this represents a small increment for a site like ours. However, the addition of 3 W m^{-2} occurs all day and night, worldwide; a loss of this much energy input would put us into an ice age, so these small overall changes have dramatic impacts. According to all models tested, temperatures will rise steadily by at least 3°C over the next few decades. So then we simulated a number of ecosystem parameters using the DayCent model, and varied precipitation by ±20 percent. Three responses stand out for our purposes. First, heterotrophic respiration decreased by 12 percent with reduced spring precipitation, indicating a drop in microbial activity. (Approximately 50 percent of soil respiration has a $\delta^{14}C$ signal for heterotrophic C.) Fine root NPP, 40 percent of which is EM fungal mass (60), decreased by 11 percent, and overall NEP declined by 10 percent. We then began working on modeling change in mycorrhizal hyphal length. Because we measured daily hyphal lengths using our SEO (58), and these measurements were coupled to sensors for T and θ, I modeled hyphal responses to climate change, using *in situ* measurements. At the James Reserve (Chapter 7) where we projected increasing soil temperature, at soil moisture content of 10 percent, a temperature increase from 15° to 20° resulted in a drop from 9.8 to 8.7 m cm^{-3} (based on 806 observations). If we increased moisture from 10 to 20 percent, at 15°, hyphal length rose from 9.8 to 10.6 m cm^{-3}. Simultaneously, if we increased the soil temperature of 15° to 20° and reduced moisture from 20 to 10 percent, the 10.6 m cm^{-3} hyphal length dropped to 8.8 m cm^{-3}. In southern California, we are already seeing higher temperatures at higher elevations and extended drought. Maybe a more realistic scenario would be to go from 20 percent soil moisture to 5 percent moisture, thereby dropping the hyphae to 8.0 m cm^{-3}, or an 11 percent drop in the extramatrical hyphae standing crop, which would be coupled to a 10 percent drop in fine root NPP and NEP. This pattern was also observed in a semiarid shrubland using molecular approaches (277; 446).

In a tropical rainforest, Clark et al. (168) estimated that there would be little CO_2 fertilization effect, because the postulated small increment would be overwhelmed by the greater minimum temperature, and drought. I again evaluated root and hyphal change. As this is a wet, tropical ecosystem, CO_2 impacts on temperature are less apparent. But temperature has a dramatic impact on hyphal production. Because the clay soils are saturated most of the year, increasing the temperature by 3° had little impact on root elongation but that same temperature change increased hyphal growth by 22 percent. One projection is that Central America will have a more variable, seasonal, El Niño climate. In the

2016 El Niño, soil moisture dropped to 25 percent (88). Based on our modeling (Chapter 7), root length dropped by half, but hyphal length increased by 70 percent with the reduced soil saturation. How overall the projected change will affect soil respiration (CO_2 loss to the atmosphere) versus C sequestration through glomalin and other complex compounds is a topic in need of research.

One projection that is emerging is an increased annual variability in temperatures and variability in precipitation patterns, coupled with increasing storm size and/or intensity (535). As oceans warm, one mechanism for heat transfer back to the upper atmosphere is storms such as hurricanes and typhoons. Drought causes increasing fire frequency and intensity, and can increase albedo in a subsequent year, thereby increasing cloud formation and precipitation. So, a changing global climate is likely to have many unpredicted effects on mycorrhizae and ecosystem functioning. Annual variation, especially in precipitation, has significant impacts on mycorrhizal activity and subsequent ecosystem responses. The El Niño, La Niña, Southern Oscillation (ENSO) phenomena dramatically affect precipitation patterns, that feed a downstream temperature and albedo effect, all of which affect mycorrhizae and plant responses. In California and Wyoming, USA, El Niño conditions generally increase precipitation and mycorrhizal production, whereas La Niña generally produces drought, reducing root production and mycorrhizae (11; 590; 766). Alternatively, in the Pacific Northwest of the United States, and in the eastern lowlands of Central America, the opposite is true. If conditions shift, as predicted by some global models, then long-term tracking of mycorrhizae could have significant implications to global NPP and biome structure.

There are few direct studies of mycorrhizae and severe storms, but three impacts are known that alter mycorrhizal relationships, all of which alter gap or successional patterns. The first outcome is a likely increase in gap formation (21; 744). Severe storms such as hurricanes cause a drop in barometric pressure and de-gassing of soil CO_2. As we previously discussed (Chapter 8), these small treefalls serve to provide new patches for colonization, but overall maintain a mature forest species composition.

However, other storms can hammer sites more broadly, with interesting implications. These large storms, projected to increase with continued global warming, fit the definition of LEDs (Chapter 8) and, with increasing frequency, will have broad impacts on global biomes. We began working in Mexico's Yucatán Peninsula in 1995, with the outbreak of hurricanes in the La Niña years of 1995–6. These hurricanes produced extensive dead timber across the seasonal forests, resulting in numerous fires, especially in 1999, a strong La Niña year, across the region, including the El Eden Research Station, 40 km inland from Cancún, Quintana Roo, Mexico. These fires resulted in a range of forest ages, from newly burned patches to some old-growth patches (within the wetlands), with trees up to several hundred years in age, that we began studying (290).

We installed one of our early sensor networks to study mycorrhizae and forest structure in the seasonal tropics of Mexico's Yucatán Peninsula, in August 2005 (Plate 10). Hurricane Wilma hit the El Eden Ecological Reserve in October 2005. It was the strongest (to date) hurricane in the Atlantic Basin, with a barometric pressure drop from 1,010 to 970 mbar, sustained winds between 190 and 210 km h^{-1}, and nearly a meter of rainfall. The barometric pressure drop and resulting wind was so extreme that the majority of leaves were stripped off their branches, creating a dense dead litter layer, and virtually no shade within the canopy. Over 25 percent of the trees in the mature forest were destroyed. The response of this ecosystem to the severe storm was very interesting. When we were able to return to the site in mid-December, we recovered much of the sensor data, and were able to reset the equipment (738). At that point, we began to focus on mycorrhizal dynamics of a recovering ecosystem.

Of special interest was the response of soils to this disturbance. Without a canopy, the soil surface warmed and dried, releasing large amounts of soil organic matter as CO_2 to the atmosphere, especially in early seral forest ecosystems. In mature forest patches, the dead leaf layer created a temporary thick organic layer. We studied the source of C to mycorrhizal fine roots using $\delta^{14}C$, from the changing atmospheric $\delta^{14}C$ (the "bomb" carbon (723; 745)). Trees recovered in two ways. First, in early seral ecosystems with small trees, leaf regeneration was rapid, and thereafter, new mycorrhizal fine roots regenerated from newly fixed CO_2. Alternatively, in the mature forest, trees allocated old, stored C to new mycorrhizal roots, and then with newly absorbed N and P, dramatically regrew the forest canopy (745).

The N dynamics were particularly interesting. In the early-seral patches, soil N was always low, and much of the plant regrowth consisted of legumes. By mid-seral conditions, pre-hurricane soil N was high, but rapidly declined. The $\delta^{15}N$ signature of the roots was less enriched than in the early seral forest, suggesting a gain of newly fixed N. In both systems, $\delta^{15}N$ of roots was ^{15}N-depleted compared with the soil, suggesting continued access toNH_4^+, increased N_2 fixation, and N fractionation by AM during transfer (see Chapter 7). But, in late seral ecosystems, there was a depletion of ^{15}N in the roots, but less so following the hurricane. In the mature forest, with the large leaf litter input, the soil $\delta^{13}C$ initially decreased with the new leaf C, but rapidly increased, indicating a rapid turnover (333). Interestingly, in the mature forest, in particular, a high incidence of EM fungi emerged, particularly Boletaceae. A strong fractionation of ^{15}N between fungal sporocarps and tree leaves, and a $\delta^{13}C$ signature of autotrophic respiration in the fungi, suggested a high EM response (330). We had previously observed occasional EM fungi in small patches (71), but the dramatic increase following Hurricane Wilma was far greater than observed earlier.

In a more subtle shift, we observed that 30 to 60 percent of the mycorrhizal canopy epiphytes, whether Bromeliaceae, with AM or a dark-septate endophyte, or orchids like *Vanilla insignis* and *Rhyncholaelia digbyana*, with their specialized mycorrhizae, and even epiphytic cacti including AM *Epiphyllum strictum*, were lost with the winds and the subsequent solar exposure of the tree limbs where these plants established (293). Recovery was very slow, as the exposure of seeds and seedlings to the high vapor-pressure deficits of unshaded forest inhibited establishment (294). These drying impacts would also have inhibited the establishment of mycorrhizal fungi of epiphytes.

Even passing glances by storms can alter the forest structure and mycorrhizal activity. Mycorrhizae and forest dynamics are tightly intertwined such that any event, from a "hot moment" such as a large snowfall, freezing soil, or monsoonal rainstorm, or a prolonged change, such as a drought, can dramatically alter the soil physical structure, gas exchange, and mycorrhizae. In arid regions, monsoonal precipitation can result in improved soil moisture with a rapid increase in AM (12) and EM fungal diversity (695). Following snowfall or heavy rain, a water layer across the soil surface builds up, preventing diffusion of CO_2 from the soil to the atmosphere, increasing CO_2 concentrations for roots and mycorrhizal fungal hyphae, until air gaps can form renewing O_2 diffusion (232; 373). Alternatively, drought can initially stimulate AM activity and facilitate plant growth, but if prolonged, result in plant mortality and reduction in mycorrhizae (e.g., (12; 42)).

EM and AM systems are governed by very different variables, which will have a dramatic impact on vegetation responses at the global level. With global climatological change, some biomes are likely to see relatively larger shifts in temperature, and others precipitation. To test this hypothesis, 50 FLUXNET sites were studied for 236 site-years for variations on gross primary production (GPP) and ecosystem respiration (R_{eco}). (FLUXNET is part of a global eddy-covariance system network to study climate and vegetation). To evaluate temperature versus precipitation, structural equation modeling was undertaken to evaluate a suite of potential predictor variables. Briefly summarizing, sites limited by temperature tend to be dominated by EM, whereas those controlled by precipitation are dominated by AM. What this suggests is that the way any particular biome responds to global change is likely to be strongly influenced by its mycorrhizae (743).

What this points out is a need for high-resolution spatial modeling of climate change. Coupling these observations with limiting resource shifts suggests that entire ecosystems can transition from water and P-limitations, to N following a perturbation such as a hurricane. As events continue to increase, how does that play on global change? Integrating these results with the outputs coming from FUN modeling will become great fun!

Nitrogen Deposition

Nitrogen deposition has a large and direct effect on mycorrhizal functioning. The role that mycorrhizae play in increasing N uptake for plants goes back to the original studies defining what constitutes a mycorrhiza (Chapter 1). By the 1970s, the impacts of N deposition were beginning to emerge as a global problem (see review in (751)). As legislation aimed at reducing acid rain began to take effect, following the Clean Air Act (1970), and a reduction in sulfuric acid began to take hold, researchers began to see that increasing nitric acid (HNO_3) inputs were, in some cases, keeping soil and water acidity levels high downwind of factory inputs, and increasing in anthropogenic watersheds (e.g., (688)). Efforts such as the NADP (National Atmospheric Deposition Program 2019), which began in 1978, included measuring and modeling N inputs, both NO_3^- and NH_4^+. A weakness was that the initial measurements were based on wet deposition, and largely missed dry deposition. Consequently, arid regions, such as Los Angeles, CA, USA, were classified as having low deposition as late as the mid-1990s. But research in both mesic and arid lands has identified N deposition as a major factor changing both mycorrhizae and ecosystems. In part because N is directly related to photosynthesis and plant production, and because there is a linear stoichiometry between C and N, whatever affects N affects

C. Overall, the two primary impacts are forms of N and the total amount of N, both of which change with deposition. Researchers in the field have determined critical loads, those deposition values wherein significant changes in ecosystems occur. Pardo et al. (547) evaluated critical loads for mycorrhizal responses for North America. The values range from a low of 5–7 kg ha^{-1} y^{-1} in tundra and boreal forest, to 5–10 kg ha^{-1} y^{-1} in temperate coniferous forests, and up to 12 kg ha^{-1} y^{-1} in eastern deciduous forests and Great Plains grasslands. Values above these critical loads caused a decline in mycorrhizal infection and/or species richness. As a note, one early critique leveled: fertilization experiments using an addition of 100 kg N ha^{-1} y^{-1} are well above most measured critical loads. However, this level is actually what has been measured downwind of urban and agricultural regions in areas as different as Europe, China, Mexico City, and even in the mountains above Los Angeles, CA, USA. Across the globe, in many terrestrial ecosystems, N deposition exceeds the local critical load (526; 547). However, in many of the urban areas, efforts are underway to reduce the NO$_x$ pollution, thereby reducing N deposition. (NO$_x$ is produced by internal combustion from transportation and industry systems. NO$_x$ commonly as NO$_2$ (the reddish-brown haze), is a pollutant, causing respiration issues for humans, among others. Some is deposited and is often leached into the groundwater, where it is converted to NO$_3{}^-$, a fertilizer, and NO$_2{}^-$. NO$_2$ is also split in sunlight, releasing an O that attaches to O$_2$, creating O$_3{}^+$, a toxin when breathed.) Important to mycorrhizae, NO$_3{}^-$ deposited in soil reduces the dependency of plants on mycorrhizae, being directly taken up by the host plants (Chapter 8). However, as industrial-scale agriculture increases globally to keep up with human populations, there is an increasing input of NH$_4{}^+$ to lands surrounding agricultural and livestock feeding operations (311; 526). NH$_3$ deposition largely occurs locally, mostly within 2–5 km from a livestock source (249). This deposition can have a major impact on EM/AM relationships between trees and herbs (20). Interestingly, both AM and some EM are important for NH$_4{}^+$ uptake as this molecule is not soluble and binds to soil particles. So global N deposition, either as NO$_3{}^-$ or as NH$_4{}^+$, alters N cycling with impacts on the ecosystem. How have these changes reverberated across ecosystems?

The study of mycorrhizae began as research expanded on understanding N cycling and plant uptake (Chapter 1). Today, we understand much about the critical role of mycorrhizae in N uptake (Chapter 7). However, the role of N deposition, especially when coupled to atmospheric CO$_2$ enrichment, makes understanding N deposition of ever-greater importance. During the late twentieth century, first Arnolds (87) documented a drop in EM sporocarps across a N deposition gradient, and then Egerton-Warburton and Allen (215) documented a reduction in AM fungal spore taxon richness along an increasing N deposition gradient. Subsequently, literally hundreds of papers were published showing N impacts on mycorrhizae, and ecosystem processes. Next, I distill the various complex reactions, but I recommend Lilleskov et al. (460) for a comprehensive overview.

Fungal Response

In many cases, especially where N is the dominant limiting resource, diversity of mycorrhizal fungi declines, whether AM or EM (e.g., (216; 458; 459)). Interestingly, in some studies EM diversity measured by root tip sequencing may not decline, even as sporocarp diversity does (e.g., (400)). Interestingly, some clades, especially *Rhizopogon*, *Cortinarius*, and *Russula* decline in the face of N deposition, whereas others, particularly *Cenococcum*, appear unaffected (65; 661). In some ecosystems, it might be different functional groups that respond to N (716; 733), especially mat-forming and long-distance exploration guilds of EM fungi. In some cases, these overlap and have similar responses to N, in others, not. Using archived soils from 1937 to 1999 that had been subject to N deposition in the mountains outside Los Angeles, Egerton-Warburton et al. (216) found that with a tripling of atmospheric NO_x loading between 1937 and 1975, Acaulosporaceae and Gigasporaceae disappeared, but Glomeraceae increased. After the Clean Air Act of 1970, N deposition declined and spore numbers began to recover, but the Gigasporaceae remained missing. Are these particular fungi more sensitive to the increasing drought stress (766)?

In many cases, the different mycorrhizal fungi with similar traits can be organized into guilds based on morphologies or enzymatic characteristics (460; 766) that are generally related phylogenetically. Are there particular guilds that are missing in our analyses of N deposition?

Plant Response

Limitations to ecosystem NPP are first temperature and water, then N. Increasing N, through deposition or fertilization, increases plant production. However, pushing N beyond stoichiometric optimal levels can cause a number of environmental problems, from high deposition of

NO_x groundwater pollution to a reduction in mycorrhizal associations – along with the other benefits the symbiosis provides, including P and water uptake that shifts the competitive relationships among plants. One result of these changes is the creation of conditions optimal for invasive species. I will deal with competition and invasive species in the last section of this chapter; here I focus on direct N input effects.

Fertilization of plants in low-deposition areas provides the best comparative study opportunities for field studies. Using a comparative cross-biome approach, Pregitzer et al. (576) studied fine root structure, following experimental application of 10 g m^{-2}. Mycorrhizal infection and diversity characteristics can be found in Lansing (443). Interestingly, fertilization only altered root N in two taxa. In the AM *Liriodendron tulipifera*, fertilization reduced root N concentration, but this had no significant effect on percent AM of roots. Likely, there was simply a reduced allocation of C to roots where N was less limiting. In *Pinus edulis*, N fertilization increased root tip N. Added N decreased numbers of EM root tips, but increased N in the leaf tissue (49). In all other EM and AM species, from Alaska to Georgia, added N simply increased overall production of roots and shoots without changing the percent mycorrhizal infection levels. The overall response in forests suggested that, in North America, highly productive forests either are affected by other factors such as ozone, or they remain N-limited even with high fertilization levels (229).

Feedback Responses and Novel Ecosystems?

The most basic stoichiometric relationship between plants and fungi is the C:N ratio, largely because of the overwhelming abundance of the enzyme RuBisCo and its role in fixation of CO_2 in photosynthesis (Chapter 4). In considering the Earth's biotic history, these two elements have largely been diametrically limiting – as CO_2 goes up, N becomes limiting, and as CO_2 goes down, N becomes less limiting. In today's global change environment, as CO_2 increases simultaneously with an increase in N deposition, humans have inadvertently created a unique experiment, where neither CO_2 nor N availability are limiting. Increasing CO_2 results in increasing mycorrhizae due to nutrient (especially N) limitations, but that N deposition increments reduce mycorrhizae by providing N and lessening the host plant's dependency upon N. How do both increasing CO_2 and N deposition simultaneously alter mycorrhizal dynamics? Overall, we don't know. However, there are opportunities for additional studies. CO_2 has been increasing, but there were signals such as isotope ratios from the 1950s and 1960s, before the big jumps currently occurring. N deposition, while global, varies along suburban, industrial, and agricultural gradients.

To my knowledge, there are a limited number of larger field studies for mycorrhizal dynamics from simultaneous incremental CO_2 and N. Field studies that have manipulated one element have rarely controlled for the other. As atmospheric CO_2 increased from 311 ppm in 1937 to 325 in 1970 to 410 today, above the LA Basin, at the San Dimas Experimental Station (California), average daily NO_x went from 0.036 ppm, to a peak of 0.095 ppm in 1970, and then declined to around 0.045 by 1998 (216). The International Biome Project began, in 1964, to initiate global ecosystem research, which included N fertilization studies, where CO_2 was already increasing. FACE studies, initiated in 1989, were already in place as N deposition was higher than 1937 levels. So, for every CO_2 or N-deposition study, the other parameter was changing. For some species, such as annual herbs, these changes were occurring multigenerational, for trees, within a generation. These inadvertent dynamics makes understanding global C: N stoichiometry processes challenging.

Pritchard et al. (580) showed that there was a marked increase in mycorrhizal tip production in only the first year of N fertilization within an elevated CO_2 study. Thereafter, temperature appeared to be a critical variable. They did note that the fungal community composition could have shifted. In another study, both mycorrhizal mass and composition were highly dynamic (e.g., (550)). In laboratory and small-scale studies, CO_2 by N-addition treatments show major interactions. For example, Klironomos et al. (415) studied a shrub, *Artemisia tridentata*, and found that with the addition of CO_2, root mass and AM arbuscules increased; with added nutrients, bacteria increased (along with increasing bacterial-feeding nematodes); and with the addition of both CO_2 and nutrients, bacterial and fungal mass increased, but arbuscules declined. Both nematodes and microbivorous microarthropods increased, indicating greater flow through soil energy channels. However when measured across generations, with an incremental CO_2 addition, no change in mycorrhizae was detected (413). Jing Zhang et al. (797) measured glomalin with CO_2 by N addition. Both additions increased glomalin, especially when both treatments (CO_2 to 700 ppm plus N addition of 10 g m^{-2}) were increased. Across gradients, Averill et al. (93) postulate that EM trees are more sensitive to N deposition than AM trees, resulting in an overall drop in soil C. But the vast majority of conifers in the database are EM, and a high fraction of the angiosperms are AM. Despite the more common gymnosperms being EM on a per area basis (Pinaceae), Cupressaceae form AM, including *Juniperus* and *Calocedrus*. When fertilized with N in New Mexico, EM *Pinus edulis* died, with needles resembling a "burning" response as in lawns with over-fertilization by N. Alternatively, the interspersed AM *Juniperus monospermum*, when fertilized, simply increased production without any observable drought symptoms (49). In southern California, during the early 2000s drought, in high N depositional locations, when Pinaceae suffered widespread mortality, we observed no "drought-induced" mortality of AM *Calocedrus*. van der Linde et al. (461) noted that a 5 to 8 kg ha^{-1} N deposition threshold of EM communities occurred across European communities, but we don't know how that might be offset by the increasing CO_2.

In another interesting aspect, Smith and Wan (665) studied EM-dominated ecosystems with high soil C:N ratios. They applied the Tilman Resource-ratio theory (R$\star$, Chapter 6) to evaluate the Gadgil effect. Their model indicates that EM N uptake increased C:N ratios, slowing decomposition. This is important for three reasons. First, they found no evidence for the Gadgil effect in temperate or tropical broadleaf forests, even when EM. Second, EM coniferous ecosystems appear to be especially sensitive to N deposition, and these are the systems that appear to be especially affected by NO_x deposition (93). Finally, one impact of O_3 on coniferous forests is to shorten needle lifespans, thereby increasing litter accumulation. But NO_x deposition alone increases decomposition (662). What is likely needed is a comprehensive experiment and modeling effort to integrate mycorrhizae, C accumulation and turnover, N form and dynamics, and even O_3. The ecosystem grows ever more complex!

Lastly, limiting resources shift in time and space (Chapter 7), but they also shift with changing C and N. Recalling the Redfield ratio (Chapter 1) of C:N:P of 106:16:1, simply as C increases in the partial pressure of elevated atmospheric CO_2, the increase in available N, either as NO_3^- or NH_4^+, then P (or K, Ca, Fe, Mg) becomes limiting. When P becomes the limiting factor, many early studies and most models propose that AM become of greater importance (see (23; 222)). As CO_2 increases, again, N becomes limiting. But as N also increases, the most likely need is P. Both AM and EM increase P uptake (Chapter 4), both in organic and inorganic forms. It is likely that as CO_2 and N are enriched, we predictably should see patchy shifts such as in taiga forest bogs with ericoid mycorrhizal systems to EM, in taiga forests from EM to AM, temperate forests from EM to AM, and a potential for increasing AM activity in AM tropical forests. The challenging aspect is that this is likely to be patchy, and also spill over to as yet unknown limits. For example, in much of southern California, the geologically young soils can be P rich, ranging from 10 to 30 mg kg^{-1} of available P and 800 to 950 mg kg^{-1} total P (55). Upon disturbance, the available P can then increase to levels that support plant growth without mycorrhizae. In arid ecosystems, we can observe patchy areas without mycorrhizae (Chapter 3).

Biome Shifts

Biomes are large vegetation units with measurable ecosystem dynamics (see Chapter 7). Because global climate is always changing, the boundaries are naturally fuzzy, but major changes have a potential to react with feedbacks that can further enhance or reduce the overall change. For instance, the globe is experiencing a rise in temperature, in response to rising atmospheric CO_2. Most projections suggest that global temperatures will rise by 1°C to 5°C, depending upon the models. These models are relatively consistent in direction, even if varying in amount (171). As temperatures rise, overall precipitation increases, but the complexity in circulation patterns means that temperature change will not be uniform, and precipitation will be highly variable in time and space. But some directional predictions are interesting and useful. One is that regions where atmospheric moisture is low, high latitudes and altitudes, and desert regions, the effects of elevated CO_2 are stronger. This is because atmospheric moisture in itself is a strong greenhouse factor, although highly temporally and spatially variable. This provides us with some predictability in both change and mechanisms.

In evaluating change, there are two approaches that are useful. The first is to use past vegetation distributions as an indicator of potential directional change. One analogy is to look at the last time when the Earth was at 450 ppm CO_2, the late Eocene–early Miocene Epochs. Today, Ramshorn Peak in the Gallatin Range in Montana, and the Lamar Valley of Yellowstone National Park in Montana and Wyoming, USA, are cold, semiarid shrubland–conifer woodlands. However, in the Miocene Epoch, a volcanic eruption buried a forest floor intact, and showed that this area was dominated by *Sequoia sempervirens*, indicating a warm, wetter climate, approaching a temperate rainforest. Since the nineteenth century, the Kenai Fjords glaciers extended well into the fjords of the Gulf of Alaska, altering the distribution of plant communities. Between 1977 and 2006, dominant plant species moved about 65 m up the slope of the Santa Rosa Mountains of southern California. Each of these shifts mean not only changes in plant composition, but also in mycorrhizal types, species, and guilds indicating dramatic shifts in ecosystem functioning.

Using Figure 7.2 as a baseline (today's biome distribution), biomes are coupled not only to vegetation type, but also to mycorrhizal types (600). Probably the most noticeable change is occurring at the boundary of the boreal forest and the arctic tundra. With warming at high latitudes, the tundra–tiaga interface across northern Europe, Siberia, northern Alaska and Canada, and Greenland appears to be shifting northward. In the arctic tundra, dominated by *Sphagnum* and other mosses and lichens, and Ericaceae, where ericoid mycorrhizae, the Mucoromycotina AM, and infrequent Glomeromycotina AM are found (237), photosynthetic rates exceed decomposition rates and peat formation is common. With warming of only a few degrees, EM shrubs and trees (*Salix*, *Betula*) expand (201; 758). Decomposition and N turnover increase. In some cases, given the high surface moisture, methane is respired, again increasing the positive global warming feedback. Black spruce, *P. mariana*, is common in bogs across boreal forests (627). I examined roots of several species in the Brooks Range foothills in Alaska. Grasses and herbs formed AM, whereas shrubs, largely *Betula* and *Salix*, mostly formed EM. Again, C stored in peats are being mineralized, whereas EM and AM, while sequestering C differently, release more C than the tussock tundra and northern European heath communities can absorb (734).

It is important to note, as we discussed (Chapter 2), ericoid mycorrhizae and Mucoromycotina AM are beginning to appear in Antarctica. How the warming earth is likely to alter Antarctica is unknown. However, during the Cretaceous Period and Eocene Epoch, Antarctica had extensive forests (Chapter 3). In the future, Antarctica could become a C sink, not a source.

At temperate and tropical high elevations, the impacts of warming on alpine tundra are similar to that of the arctic tundra biome. Probably the largest difference is that alpine ecosystems are especially sensitive to drought because the extremely low air pressure (at 4,000 m, barometric pressure is 0.56 atm) creates a high diffusion coefficient, and rapid drying. Warming accentuates the drying impacts on plants. There is a potential for shifts from fens forming peat with *Sphagnum* and *Pedicularis* with no mycorrhizae, and ericaceous mycorrhizal *Vaccinium*, to EM *Betula* and *Salix*, then to a mix of AM dry meadows with patches of EM conifers (174; 329; 340).

Arid ecosystems are responding dramatically to climate change, although the processes are different from mesic and wet ecosystems at local scales. CO_2 absorbs long-wave radiation, reducing heat energy re-radiation and leading to global warming. But water vapor is an even stronger greenhouse gas. In tropical rainforests, the local effect of increased CO_2 is low compared with water vapor. However, deserts have low atmospheric water vapor content, making CO_2 a larger greenhouse gas compared with water vapor. Desert plants mostly grow slowly, except during wet events. Many phreatic plants with deep roots and deep AM are sequestering C both as organic C and by inorganic C, by respiring CO_2, much of which becomes fixed as $CaCO_3$, or caliche, before diffusing back to the atmosphere. As temperatures continue to rise, even desert trees may not tolerate the high temperatures and transpiration demands, leading to annuals with short growth cycles and nonmycotrophic annuals that do not depend on AM adapted to these extreme conditions (Plate 2). The sequestration by woody plants in deserts, continuous at least since the Pleistocene Epoch, may well largely shut down. Are mycorrhizal fungi directly affected by temperature changes of only a few degrees, or only indirectly affected through host dynamics?

Invasive Species

The genus *Homo sapiens*, as hunter-gatherers, had occupied nearly all potential suitable habitat globally by the early Holocene Epoch, over 8,000 years ago, when atmospheric CO_2 was somewhere around 200 ppm. The carrying capacity was somewhere around 9 million individuals. At this point, there was nowhere to go. The southern tip of South America was occupied, and humans were island hopping eastward across the Pacific. Malthusian population dynamics would have constrained human populations at probably less than 10 million. As the goal of every population is to increase, in order to do so, humans needed to undertake a fundamental shift in resource acquisition. One crucial step was the undertaking of horticulture, then agriculture (see (165) for a detailed description). Broadly, we tilled ever-increasing amounts of land, first in shrub-grasslands, then forests and wetlands. In the process, we created ever-larger opportunities for groups of plants that rejected mycorrhizae, such as the Amaranthaceae and Cruciferae, as early as the Miocene

Epoch. In the process of tilling soils, we inadvertently chopped up mycorrhizal fungal hyphal networks, creating ever-growing opportunities for these nonmycotrophic and facultatively mycotrophic species (Chapter 3). Interestingly, Amaranthaceae have more C_4 species than any other plant family and C_4 is a photosynthetic system designed to function nearly as well at low atmospheric CO_2 as high. Through agriculture and forestry, humans opened up patches in which new mycorrhizal fungi could establish, at least those with adequate dispersal mechanisms. What this meant was not only a beginning in the growth of atmospheric CO_2 beyond a background natural variation responsive to Milankovich glaciation cycles and plate tectonics, but also new resources for invasive species, both plants and fungi, globally.

Humans not only created the opportunity for establishment of invasive species, we also carried many plants and fungi as we dispersed to new regions. Agricultural peoples from the Levant and the Caspian steppes brought cereals and domesticated animals to Europe some 5,000 to 6,000 years ago. Forested ecosystems were converted to crop- and grazing lands across Europe. As Europeans colonized North and South America, and Australia, they brought both crops and animals with them. Amer-Indians developed their own crops, especially maize, potatoes, peppers, tomatoes, and squash that have become staples globally and become new basic ingredients in "ethnic" foods well beyond their origins. (Who could imagine Italian pizza without tomatoes, or Szechwan chicken without peppers?) Today, there are more hectares of wheat in the Central Plains of the USA and Canada, and western Australia than in Europe. Many crops are often highly bred to respond to super-fertilizers, and can be less responsive to mycorrhizae, such as wheat and maize (86; 347; 397). Simultaneously, high N fertilizers reduce mycorrhizal response and C allocation, and tillage break up the hyphal matrix, reducing and changing soil organic matter and soil aggregation. How these factors play into global C and N models remains to be elucidated.

Along with intentional introductions, numerous species hitched rides. California is known for its annual grasslands. These are comprised of Mediterranean grasses, including species of *Avena* and *Bromus*. These facultative AM plants respond to high NO_x deposition resulting in a reduced diversity of AM fungi, trending towards a rhizophillic guild of AM fungi. In Europe, *Solidago canadensis*, a facultative AM species, is invasive and altering native grassland ecosystems. It is estimated that there are approximately 4,300 invasive plant species in the USA, and a brief scan shows that the vast majority are facultative AM or occasionally nonmycotrophic. In nearly every hike I have taken across the USA (49 of the 50 states), I have observed invasive plants, again almost exclusively NM or AM. This is likely due to the lack of specificity between AM fungi and plants, despite the observed preferences (Chapter 5).

EM symbioses are very different. Although specificity is not 100 percent, there appears to be some very specific symbioses that are indicative of transport, whether intentional or not, designed to enhance woody plant establishment. In the 1850s, during the 1850s California gold rush, native trees were decimated for fuel and fiber. Australian *Eucalyptus* spp. were introduced, because they rapidly grew in arid regions, and planted statewide. In a eucalyptus stand in the Santa Margarita Experimental Station (San Diego State University), we found a number of sporocarps of exotic mycorrhizal fungi (with the help of Orson Miller), including a *Pisolithus tinctorius* with a different RFLP pattern from those found in native Fagaceae, *Hydnangium separabile* from Tasmania, and *Laccaria amethystina* from Australia. None of these fungi were found in symbiosis with native plants (54). *Suillus luteus* has been transported from North America across the southern hemisphere in plantings of *Pinus*, especially *P. radiata* (see Chapter 5).

In other cases, interesting host shifts have been identified. Wolfe and Pringle (786) found that *Amanita phalloides* appears to have shifted from European oaks (its native symbiont) to pines in eastern USA and to *Quercus agrifolia*, an endemic live oak, in California. Pringle et al. (577) and Dickie et al. (205) have excellent overviews of fungal and plant invasions.

Animal introductions can also be invasive and shift ecosystems, in part through their actions on mycorrhizae. Often these are referred to as ecosystem engineers as, once they are introduced, they become invasive and spread their impacts well beyond the initial introduction point. These two species provide useful examples and point to the need for more work on other potential engineers: earthworms and beavers. Charles Darwin (192) described the importance of earthworms in converting litter to soil. In coniferous forests of North America, native earthworms are small, and do not necessarily turnover large amounts of litter. In the Canadian Rockies, invasive earthworms rapidly converted needle litter to organic soil. As a result, AM plants were able to take advantage of the increased N availability, whereas without the introduced earthworms, EM short-circuiting of the N cycle was far more important (128). Simply the invasion of the non-native earthworms shifted the role of mycorrhizae and subsequently altered plant competition. A number of studies scattered in many ecosystems are documenting this widespread impact.

Beavers present another very interesting example. Beavers (*Castor canadensis* in North America, *C. fiber* in Eurasia) were driven nearly to extinction during the fur trade days of the late eighteenth and early nineteenth century. They occurred across northern Asia and Europe, and from New England across the Great Plains to the Pacific and Arctic Oceans in North America, continually damming streams and creating wet meadows dominated by EM poplar, willow and aspen. Once removed, hillside incision was initiated, lowering water tables, and allowing for the invasion of AM grass- and shrublands (99). In 1946, Canadian beaver were introduced to the Patagonian Forests of Argentina and Chile, this time eating through forests and forming lakes and ponds over patches of old-growth *Nothofagus* trees, adapted to more upland EM conditions. AM and NM grasses, shrubs, and forbs became dominant.

Summary

- Mycorrhizae have been a component of the global environment since land invasion in the Paleozoic Era to human perturbation of the atmosphere in today's Anthropocene Epoch, sequestering C from the atmosphere into the soil.
- Using stoichiometric relationships, mycorrhizae become especially important with high atmospheric CO_2 of times such as the Paleozoic and Mesozoic Eras, and the Eocene Epoch, as soil nutrients, especially N and P, become more limiting. During low atmospheric CO_2, such as the Pleistocene Epoch, nutrients are relatively less limiting, and mycorrhizae appear of lesser importance and C sequestration declines. The Anthropocene Epoch has resulted in increasing atmospheric CO_2 globally, but also increased N deposition and P availability in local patches. How these factors interact at projected scales needs additional research.
- As temperature and precipitation confound but simultaneously accompany elevated CO_2 warming, storm intensities and frequencies are likely changing. How different mycorrhizal fungi and types, especially EM, AM, NM, and ericoid mycorrhizae respond both directly and indirectly through hosts remains a suite of open research questions.
- How managers utilize, or fail to integrate mycorrhizae, is increasingly critical for managing both vegetative and atmospheric C at local to global scales.

10 · *Conservation, Restoration, and Re-wilding*

Mycorrhizae as a Cornerstone

Wildlands continue to decline globally at a rapid rate. As hunter-gatherers, the human population probably approached a global K value, estimated at 8.6 million, approximately 10,000 years ago, and continues to grow, from 2.5 billion when I was born, to rapidly approaching 8 billion as I write (65). Deforestation rates are in the range of 10 million hectares annually. Atmospheric CO_2 is rapidly increasing and growing-season days are projected to move as much as $10°$ poleward higher in latitude over the next three decades (407). In the environmental media, the idea of re-wilding (reintroducing absent or declining species into relatively intact habitat) has been gaining strength as the likelihood of mass extinction materializes. While one extreme view is the reincarnation of extinct large animals, *another is that we should simply design wildland ecosystems for biodiversity, climate mitigation, and ecosystem sustainability, within the constraints of our changing global environment.* We now understand that mycorrhizal fungi are highly diverse, with each population containing unique genetic variation. Each time we bulldoze a patch of wildland, we probably eliminate at least a unique individual, if not a unique population or even an entire taxon that provides distinctive benefits to the ecosystem (65). For example, the potential impacts of habitat destruction or range shifts in *Tuber magnatum* or *T. melanosporum* could have dramatic economic impacts (148) as well as on competitive relationships between host and non-host plants (523; 571).

Although theoretically, negative-feedback (parasitic) relationships confer stability within a community, mycorrhizae, through multiple positive-feedback (mutualistic) interactions, may also enhance stability across a community (Chapter 6). This perspective represents another way to examine the role of mycorrhizae in designing land management strategies for overcoming detrimental impacts of human resource development. Social scientists have noted that complex interactions in themselves dramatically alter system behavior. David Christian (165), in his perspective on *Big History*, noted that in transitions across levels of complexity, positive feedbacks transform systems well beyond any simple additive effect. For example, Horticulture, with the domestication of wheat followed by goats and peas nearly 11,000 years ago, (315) increased food production, and therefore bumped human population growth well beyond that sustained by hunting–gathering approaches. Populations jumped again as agriculture dramatically increased production beyond horticultural technologies. The development of new pesticides, fertilizers, and crops from the "Green Revolution" of the 1960s facilitated a tripling of the human population during my lifetime and a new stabilization where famine is more of a political distribution problem than food production problem. Each of these steps was translational, and dependent upon new complexities in technology and understandings of agroecosystem characteristics.

But the land use has also changed, resulting in a loss of functionality of ecosystems. From a matrix of wildlands with scattered human settlements, most regions transitioned to a matrix of human-dominated landscapes with scattered wildland matrices. Any flight or satellite search over mesic forest biomes will show largely patchy mixed-species forests within a production fiber or agricultural matrix across most of the globe. Light maps of the earth show urban areas globally except in a few extreme deserts, polar, or mountainous regions. Conservation biologists are discovering that even large national parks such as the Serengeti (Tanzania), Yellowstone (USA), or Wood Buffalo (Canada) are too small such that large game populations are subject to detrimental boundary impacts. This pattern has critical implications for protecting biodiversity and the ecosystem services that nature provides.

Understanding and management of mycorrhizal symbioses, positive feedback relationships between plants and fungi, in my view, hold a key to overcoming problems with degrading environment, through coupling new and old, familiar management strategies for conservation, restoration, and agriculture. The evolution of each new mycorrhiza created not within-community stability, but rather the ability to flourish in new habitats. Stebbins (680) postulated that evolution of plants occurred especially at the margins of their niche. Mycorrhizae predominate where nutrient or water stress levels are at the edge of a plant taxon's tolerance. Interestingly, Treseder et al. (722) noted that older phyla of fungal communities preferred lower and warmer latitudes than did younger phyla, perhaps reflecting Cretaceous Period or even Eocene Epoch conditions? Therefore, I assert that it is not stability *per se* that we should focus on, but how mycorrhizae as a key state variable will shift the system

Box 10.1 *Mycorrhizae and Designer Ecosystems*

Designer ecosystems are managed ecosystems that are managed for specific human goals. In essence, croplands are designer ecosystems for food production. But these can also have specific conservation goals (470), examples are managed wetlands such as the Bosque del Apache for winter habitat for sandhill cranes in southern New Mexico, USA, in which plants that support birdlife are planted at low water levels during the summer, then surface water levels are raised and highly managed to support winter crane (and other species) habitat. While restoration for habitat is often challenging, creating specific conditions wherein succession is directed toward specific goals is sometimes achievable (17). Mycorrhizae play a critical role in such efforts. For example, in a tropical seasonal forest, recurring fires within a dense, fern-dominated stand results in arrested succession (22). The result is species extirpation and loss of C sequestration capacity. However, restoring mycorrhizal fungi and structuring architecture can redesign the ecosystem such that both functions are recovered (15; 737). Incorporating mycorrhizae into designer ecosystems has the potential to add to many of the ecosystem functions that are important to global health.

from resource depleting to resource-sustaining. This system will flourish under careful management when conservation, restoration, and management are integrated within a *designer* ecosystem concept. With this concept in mind, and understanding the role of mycorrhizae in ecosystems (Chapter 7), I believe we can utilize mycorrhizae to manage habitats to provide useful biotic resources, restore disturbed habitats, improve efficiency in agriculture, facilitate carbon sequestration, and ultimately contribute to biodiversity (Box 10.1).

Conservation of Biological Resources

Mycorrhizae are a crucial element in the conservation of endangered species and biological diversity as part of the infrastructure that maintains critical ecosystems (Chapter 7). As an example, at the La Selva Biological Station in Costa Rica, the late successional tall trees, up to 30 meters in height, are all obligately AM, or nearly so (378). Without AM, especially for P, but also N (for direct uptake of both nutrients, and for P, Fe, and Mg needed for N_2 fixation) and other nutrients, these forests would nowhere obtain the height needed for endangered species such as Geoffrey's Spider Monkey (*Ateles geoffroyi*). The canopy height is needed for predator protection (there are both jaguar and cougars on the ground), and the different tree species are food sources that fruit at different times of the year (Plate 11).

But what is the biodiversity of mycorrhizal fungi and how is it being impacted by habitat loss? One impact on mycorrhizal fungi is the transition from wildland old-growth to managed forests. As the matrix has shifted from an old-growth forest to managed lands, what biodiversity might have been lost? An old-growth managed forest, suitable for harvest, is often 50 to 100 years in age. It is rare for European or eastern North American temperate forests to exceed 100-year stands. European old growth is often considered 100 years, and the last stand of old-growth *Pinus strobus* was destroyed by a hurricane in 1939 in New Hampshire, USA. Varenius et al. (736) found that the EM fungal composition differed between "natural" versus 50 year old stands of *Pinus sylvestris*, and Rosinger et al. (623) suggested that the peak richness of the 100-year-old stands was due to a peak diversity when early- and late-stage fungi overlapped. This transition is potentially critical and points to the need for conserving truly old growth (>200-year stands). Glassman et al. (287) reported only 96 OTUs across forests that are less than 50 years old and in the process of moving upward in elevation with global climate change. What is the richness of true old-growth stands that have been lost? I estimated that there could be 500 taxa of EM fungi for a large stand of ancient oaks (>200 years) based on island biogeography (Chapter 6). Trappe (715) estimated 5,000 species of EM fungi for *Pseudotsuga menziesii* (Douglas fir), and Jaroszewicz et al. (382) noted measurements of 1,380 species of fungi for the ancient *Białowieża forests*, with *Quercus rober*, *Carpinus betulus*, *Tilia cordata*, and *Picea abies*, where some trees are known to exceed 400 years in age. They estimated that there could be more than 5,000 species of fungi, many of which are EM. So, it is challenging to estimate what has been lost, and how we might conserve biodiversity. Sky "islands" in many areas of the world would make an excellent research effort.

In a tropical rainforest, there are nearly 5,000 species of plants, including more than 700 tree species, where the vast majority are AM. In a very small survey, Hannah Schulman, Emma Aronson, and I have thus far found 9,506 fungal OTUs including 226 Glomeromycota, only four of which could be assigned a species name (*Acaulospora mellea, A. lacunosa, A. laevis,* and *Gigaspora margarita*). There are also scattered likely EM fungi in the Thelephorales, Boletales, and Cortinarius, potentially associated with Myrtaceae trees, and rare species such as *Scleroderma sp.* with the primitive gymnosperm liana, *Gnetum leyboldii* (Plate 2). We have a long way to go to even describe the basic biodiversity of even a small patch of tropical forest.

Surprisingly, even after more than a century and a half of applied studies, mycorrhizal fungi represent a resource that is still underappreciated, and broadly poorly understood. Mycorrhizal fungi are far more diverse than we recognized in the past. In every sequence analysis I have undertaken, no individual taxon that I sequenced matched exactly that of the type specimen or the other deposited sequences for the fungus identified. The matches are generally (if lucky) in the 85 to 98 percent range. What this means is that very few fungal taxa in different locations would be considered the same species in higher species (such as primates), if we expect a 98 percent similarity. New species are emerging from even well-known taxa. For example, in surveying *Pinus ponderosa* in the San Bernardino Mountains of southern California (661), five distinct taxa of Amanitaceae (both present as sporocarps and EM root tips) were found. The BLAST search for each of the five taxa ranged from a 99 percent (*A. pantherina*), one 96 percent (unculturable Amanitaceae), two 95 percent (*A. gemmata* and *A. rubescens*), and a 90percent (*A. citrina*). Of the 145 taxa distinguished as EM fungi following sequencing, we could only identify 29 clades above the family, and another 24 to genus but not species.

The continued loss of biodiversity is potentially enormous. In tropical EM forests, most patches have simply not been surveyed at all despite extensive and continuing deforestation. In every survey, unknown taxa are revealed. As I hike the different trails across New England reserves, protected by conservation organizations from towns to the state forests, every hike brings at least one new mushroom taxon to learn that I hadn't seen in prior days. Each new condition, from the high pH limestone outcrop in Bartholomew's Cobble, to the neutral pH metamorphic phyllite of Mount Greylock, to the low pH granite around Berry Pond, has unique taxa.

Every region has distinctive taxa. However, in few cases is there any recognition of the need for protection. There are few reserves, such as the National Nature Monument Luční in the Czech Republic, designed to protect mushrooms, such as *Boletus rhodopurpureus*, a rare mycorrhizal species with *Quercus*, found in central Europe (Plate 12). This reserve is "designed" like wildlife preserves for waterfowl for mushroom protection using managed ponds and oak plantings on the dikes. Other taxa are not considered rare. For example, *Boletus* (=*Baorangia*) *bicolor*, found in the northeastern USA and in northeastern China, is considered an Asa Gray (grayan) disjunction, and even *Boletus edulis* is considered the "same" species ranging from Europe (steinpilz) across North America (also called cep or porcini). Given the low degree of molecular similarity among isolates of the "same" species, it is apparent that both rare and common species have distinct ecotypes or subspecies, and likely every reserve has rare, novel taxa. This level of diversity needs more study, both within taxa and across their ranges.

Mycorrhizal fungi, including many that are unknown, offer the potential to provide useful products. Many mycorrhizal fungi produce unique antibiotics as they compete for root tips or occupy a patch of ground. Others produce polysaccharides that are useful for treating diabetes, obesity, and cancer (e.g., (261)), and we have only surveyed a few taxa for useful attributes. It is also important to remember that biochemical products are, at least as of now, all compounds that have evolved in nature; new biotechnological approaches start with the naturally produced compounds and develop more accurate or more rapid, or even more efficient mechanisms to produce them. Nature's labs have already developed most of what we need, and mycorrhizal fungi are amongst those laboratories. Finally, mycorrhizal fungi produce organic acids and enzymes (Chapter 7) that catalyze a number of elemental transformations crucial for food and fiber production. These are intimately integrated in soil C cycling (Chapters 7 and 9) and C sequestration.

Carbon Sequestration

Mycorrhizae moderate the levels of atmospheric CO_2. The long drawdown occurring from the Ordovician through the Carboniferous Periods, and the more recent one from the Eocene to the Miocene Epochs (Chapter 3) is coupled to mycorrhizal activity, whether indirectly by supporting plants that fixed CO_2, or by direct sequestration from recalcitrant C compounds buried in soil (Chapter 9). What is the extent of that contribution, and how does it play into C sequestration strategies? One example is to understand mycorrhizae and C in a tropical ecosystem, with a view toward widespread conservation efforts in all ecosystems. Recent estimates suggest that globally, structurally intact tropical forest ecosystems sequester nearly 15 percent of the anthropogenic-derived CO_2 emissions, with 17 to 37 percent going belowground (369). They further estimated that Amazonia amounts to a 0.24 Mg C ha^{-1} y^{-1} sink. At the La Selva Biological Station, a research station situated on the edge of the Braulio Carrillo National Park, within the Área de Conservación Central, in Costa Rica, David Janos (377; 378) demonstrated that disturbance reduces AM activity, and that successional and mature forests are highly mycorrhizal. In the mature forest, 3.2 percent of the total soil C in the top 10 cm, or 1.45 Mg C ha^{-1} was glomalin (467). Here, the annual NEE was estimated at 4.5 Mg ha^{-1} (464). These data clearly show, through multiple approaches and multiple measurements, that a protected mature rainforest is sequestering both above and belowground C, and in large part due to AM.

Globally, anthropogenic C inputs to the atmosphere are estimated at 0.5 Mg ha^{-1} y^{-1}. In semiarid-to-arid regions of California alone, C sinks range from 0.2 to 0.5 Mg ha^{-1} y^{-1}, whereas in boreal forests, the sink could be as much as 0.7 Mg ha^{-1} y^{-1}, and in other tropical rain forests, 1.0 Mg ha^{-1} y^{-1}. Conservation lands therefore have a role beyond just protecting biodiversity, but also in sequestering and reducing the detrimental levels of CO_2 that humans are contributing towards global warming. In wildland soils, Rillig et al. (612) and Treseder et al. (721) found that C accumulation into glomalin and soil aggregates increased with elevated CO_2; mycorrhizae become of even greater importance in an elevated atmospheric CO_2 environment. Even as C is released (as CH_4 and CO_2) in boreal fens, mycorrhizal fungal hyphae increase, which could mitigate at least some of the global-warming potential (201).

Studies of C sequestration focus on two aspects of the biosphere: the aboveground canopy structure and biomass, and belowground organic C to between 20 cm (a normal plow depth) and 1 m, approximately the maximum obtained by remote sensing. However, wildland ecosystems rarely limit C accumulation to these depths, nor do mycorrhizae. Fine roots and mycorrhizal fungi grow tens of meters into the soil, through cracks in bedrock, and even into weathered granite (Chapter 2). Even in the tropical seasonal forest, roots and mycorrhizae extend deep into the soil. In other ecosystems, the rate of CO_2 fixation is not nearly so high, but neither is ecosystem respiration. Especially in seasonal environments where drought limits production, roots often penetrate deep into soil and even bedrock to extract water. Deep C in roots and mycorrhizal fungi only returns to the atmosphere very slowly, often at geological time scales. Ecosystems with deep-rooted plants may well represent large CO_2 sinks when calculated globally. With my colleagues, I have studied two ecosystems that provide some insights into the topic of deeply stored C, an area in far greater need of study: Mediterranean-climate shrub-forests, and extreme hot deserts. These are not end points for research but represent initial forays into global C exchange needs. Here, I describe two systems on which I have worked, that will illustrate distinctly different roles of mycorrhizae that contribute to the sequestration of C.

Southern California Natural Reserves

I have already introduced our work in the James Reserve in Chapter 7, where we studied mycorrhizae and C fluxes. But the work lends itself to a broader development of a strategy to manage for accumulation of C. In southern California, earthquakes are a part of life, like all active regions of the earth's crust. Even smaller earthquakes crack the "bedrock", creating a space for roots and hyphae to grow several meters deep. Within the conifer–oak vegetation zone, mycorrhizal fungi accounted for 27 percent of the NPP in 2011 (60). But this value was derived from the depth to bedrock. Depth to bedrock was highly variable (based on ground-penetrating radar output), ranging from 10 to 70 cm.

What happens down in the bedrock? Southern California has a "Mediterranean" climate, that is, a cool, wet winter and a hot, dry summer. Here, temperatures are too cold for high production when moisture (much as snow) is available in the winter, and the soils too dry to support trees during the warm summer. Using the rock fractures, mycorrhizal roots explore deep for water. Indeed, approximately 25 percent of the water transpired during the year is groundwater, several meters deep (408), and acquired during the dry summer. Sixty-one percent of the respired CO_2 is autotrophic, that is mycorrhizal and root. That means enough C is allocated deep into the bedrock through roots and mycorrhizal fungi to support the forest during maximal production. For oaks, in particular, up to 3.9 g C kg^{-1} is deposited in bedrock fractures and 0.3 g C kg^{-1} within the bedrock matrix, as decomposing fine roots and mycorrhizal hyphae (131). The total depth is unknown from our sites, but roots of oaks are known to explore up to 45 meters deep for water and nutrients (375).

Across the mountains from the Mediterranean-type forests and into the rain shadow desert, even a larger fraction of the plant's C is allocated to deep mycorrhizal roots to acquire water (750). In analyzing desert phreatophyte trees in the desert, we found that *Parkinsonia florida* acquired 70 percent of its water from deep groundwater (Allen, unpublished data).

Soil respiration provides information as to the C sequestration capability. For example, in the conifer ecosystem, EM hyphal mass was 14.3 g C m^{-2}. Assuming that 10 percent of the hyphae are active (516), then these hyphae respire 0.45 μmol m^{-2} s^{-1}, compared with a root respiration of 6 μmol m^{-2} s^{-1}, or 5 to 8 percent. A large fraction of the difference potentially becomes C stored as organic matter. Experimental CO_2 enrichment studies further indicate that as atmospheric CO_2 increases, the amount sequestered by mycorrhizae also increases (724).

A second factor then becomes important to C sequestration. If Ca^{2+} is present, the respired CO_2 forms $CaCO_3$, also known as caliche (Chapter 7). Under oak canopies, a $CaCO_3$ signature can be found up to 4 meters deep in granite bedrock, with mycorrhizae (131). In the southern California deserts, the $\delta^{13}C$ signal of $CaCO_3$ is a strong rhizo-sphere microbial fractionation signal (57; 638) indicating that much of the CO_2 was respired by the myco-rhizosphere system and fixed by the Ca. In the California deserts, as much as 8.3 kg m^{-2} of C occurs as $CaCO_3$, and globally, $CaCO_3$ comprises 500 GT, or an amount equal to all of the aboveground plant C. The southern California desert is not the only ecosystem in which the myco-rhizosphere inorganic C becomes crucial. The oasis agricultural ecosystems of western China are found along the silk roads in the interface of the Gobi Desert and the Himalayan and Tianshan mountains. Here, there is an accumulation of C deep in the groundwater beneath the Tarim Basin that has been on-going for at least 2,000 years, below irrigated agriculture. The C leaching has sequestered as much as 1,000 Pg of C, potentially the third largest active pool on land (455). Here, the ^{14}C of the dissolved inorganic C, the C sink, was from respiration of the mycorhizosphere, not the soil organic C. Hence, mycorrhizae play a large role in the overall C global C sequestration even in ecosystems with minimal aboveground C accumulation.

What these types of data tell us is that the wildland biosphere and some agroecosystems largely sequester C, albeit slowly. In current biomes, more fixation than respiration occurs. The conditions that drive CO_2 release and global warming are largely those processes not in equilibrium such as fossil fuel burning, rapid respiration with large-scale deforestation and land turnover for mechanized agriculture, human land development and urbanization, and de-greening of large landscapes. The solutions exist within known land management strategies coupled with new efficiencies in forestry and agriculture. I will address these in the con-cluding two sections.

Restoration of Disturbed Lands

Bradshaw and Chadwick (135) called restoration of disturbed lands the acid test of our understanding of, and ability to recover functioning ecosystems (see (34)). Prior to the creation of the field of "Restoration," afforestation technologies demonstrated that mycorrhizae are essential players in recovery of disturbed lands, throughout terrestrial ecosystems. Afforestation efforts for EM pine plantations across the southern hemisphere, to eucalyptus plantations in wide-ranging lands from California to Argentina to Ukrainian shelterbelts (654), and even pines in the Nebraska National Forest in the middle of the American Great Plains (296) have repeatedly demonstrated the essential role of EM fungal inoculation in establishing novel communities. Inoculation of seedlings then became an important step in the restoration of forests after logging, mining, or other Large Infrequent Disturbance (Chapter 8, see (306; 484; 645; 715) for summaries). AM work lagged behind largely because of the challenges of pure culture isolation and mass culture production. AM were being studied for their role in agriculture production, but there was little effort to integrate AM fungal inoculation into early restoration programs. In part, this was because most AM plants used in restoration are facultatively mycorrhizal, that is, plants that do not require AM for initial establishment.

That perspective changed during the late 1970s and into the 1980s. In his pioneering work, Tom Nicolson (532) put AM into a successional context; that is, AM are not present in the initial plants colonizing newly formed dunes, but become an indispensable bridge for mid- and late-seral plant establishment (see also (45; 533)). Restoration, as opposed to reclamation or afforestation, became a research field (9; 135). As large-scale restoration of mined lands, especially in AM shrublands, grasslands, and forests, became legally mandated in the 1970s, work on AM became a priority. Aldon (6) and Williams and Aldon (781) showed that to establish perennial Amaranthaceae (*Atriplex canescens*) successfully in New Mexico mined lands, AM inoculation, and out-planting of seedlings was a critical step. Daft and Hacskaylo (188) improved plant establishment on mine spoils with inoculation, and Kormanik and colleagues (427) demonstrated similar roles for restoration and reforestation in eastern (US) deciduous forests. In the late-1970s, I found that across the Hanna and Shirley Basins of Wyoming, USA, where coal and uranium mining exposed large heaps of subsurface overburden material with no AM fungal inoculum, attempts to replant grasses and shrubs repeatedly failed. These sites were colonized by nonmycotrophic annuals, mostly in the Amaranthaceae, especially *Salsola tragus* (Russian thistle). In the Powder River Basin of Wyoming, we (10) then found that if topsoil were immediately replaced overtop of this overburden material, mycotrophic grasses would readily re-establish forming productive new stands. R. Michael Miller and colleagues showed that short-term storage and replacement of topsoil could also provide effective inoculum when re-spread back over the overburden (502). AM especially helped the facultative mycotrophic grasses in competing with the nonmycotrophic annuals (11; 13; 43; 45), and even further facilitated the establishment of more responsive perennial shrubs, particularly *Artemisia* and *Atriplex*, and forbs including legumes such as *Hedysarum boreale* (155; 606). For successful revegetation, it may be crucial to distinguish amongst different species of nonmycotrophic plants that interact with AM fungi in different ways. While AM can survive and even reproduce with annual Amaranthaceae (44), annual Brassicaceae appear to be inhibitory to AM fungi (187), potentially due to the glycosides (Chapter 3). Additional simultaneous work by colleagues in the eastern deciduous forest with stands of AM trees including *Liriodendron tulipifera*, *Liquidambar styraciflua*, and *Fraxinus pennsylvanica* showed that inoculation effectively enhanced tree establishment (394; 427).

Early efforts were made to inoculate seedlings prior to out-planting. These ranged from rolling seeds in forest duff (such as acorns in oak leaves prior to out-planting in Ukrainian shelterbelts (654)), to planting seedlings inoculated (from background contamination) in nurseries prior to out-planting, then to inoculation of tree seedlings with successful EM fungal taxa, such as *Pisolithus tinctorius* isolate Marx 270, by Marx and colleagues and *Laccaria laccata* by Molina and colleagues (508). Inoculation of nursery seedlings became common for EM trees, where many of the EM fungal taxa can be grown in pure culture, and scaled-up for mass production (see (510)).

AM fungi have proven less tractable for mass production. Even today, AM fungi cannot be grown without a host plant, and a limited suite of taxa appear suitable for root organ culture (425). Nevertheless, several new companies produce inoculum for specific, especially limited-scale, restoration efforts in many areas of the world (for example, run a computer search for mycorrhizal fungi for sale). These are typically produced by inoculating an annual host plant with spores of a common AM fungus such as *G. intraradices*, using pesticides to which AMF are resistant to reduce potential pathogen load, and harvesting the root-rhizosphere soil mix to sell as inoculum. While there are numerous sources, each will have preferential regions or soils, or plant taxa for which their product works best. Corkidi et al. (176) found a wide variation in effectiveness of commercial AM inocula, even amongst good companies with good products. Clearly, with any inoculum product, tests for specific applications are needed to find the optimal inoculum based on plant and fungal taxon, soil, and climate condition.

A second argument then emerged: should inoculation be undertaken, especially in restoration of native wildlands? Peitsa Mikola (500) found that inoculation with an exotic EM species could inhibit the replacement by local EM taxa, especially under severe conditions. Alternatively, Don Marx (485) argued that inoculation with *P. tinctorius* facilitated tree establishment, and that the fungus would be rapidly replaced as the stand recovered. Even today, many seedling out-plantings are not pre-inoculated with EM fungi. For example, across many large nurseries, artificial inoculation is not common, but seedlings rapidly form EM with local fungi upon out-planting. Alternatively, in some places, such as mine tailings with high concentrations of heavy metals in Portman, Spain, naturally colonizing *Pisolithus* sp. dominated as the trees gradually established. *Pisolithus* sp. was found in a planted pine stand on Mount St. Helens, but has not been found in native forests surrounding the area (Chapter 8). How to proceed? Do we transplant mycorrhizal fungi to new regions, given what we know about invasive species (Chapter 9), or do we focus on local inocula, in the same manner as we do plants?

AM inoculation has proven to be functionally a different suite of issues. AM associations show little specificity, although preferential responses are common (Chapter 6). A key question emerges, then, does inoculum survive and is it effective? This issue could also emerge as differentially crucial in whether succession on a site is dominated by initial or relay floristics (Chapter 6). Using an introduced AM fungus, *Gigaspora margarita*, Friese and Allen (265) found that the exotic inoculum survived for two years following transplanting. Over three growing seasons, Weinbaum et al. (767) found that native AM fungal populations survive better with their site plus host or origin. Although these fungi persisted in new habitats and with new host populations, the AM fungi declined when transplanted with the new population and into a new site. The AM fungal data suggest that transplanting may not be optimal, especially as these fungi may well be adapted to the local community, soils, and environment.

The challenges of producing large-scale inoculum (788) led to a unique approach to recovering AM during restoration. Again, the key point is that the AM plants are less specialized than other symbioses (including EM), such that late-seral plants and early seral plants both can use the same fungus (Chapter 8). Based on the reestablishment potential for simply moving soil from the adjacent area to the new disturbance (43), and assuming that the vegetation and soils were nearly identical, we initiated restoring AM simply by adding a thin layer of soil from the adjacent site. AM rapidly re-established on host plants. In a related effort, Kormanik et al. (427) inoculated tree seedlings and planted along with a cover crop of sorghum, which rapidly built up the inoculated AM fungus. In a tropical seasonal forest, we inoculated early- and late-seral trees with early- and late-seral AM fungi, following burning that removed vegetation and soil organic matter. Here, the early seral fungi rapidly re-established where seedlings were inoculated (15). They also recolonized slowly on control seedlings. The result was that, after 3 months, all trees were mycorrhizal, but those inoculated had a head start and improved growth. Interestingly, while undertaking the same study a year later, under wetter conditions (48) along with a less effective burn where soil litter layer remained too wet to burn, late-seral AM fungi were just as effective in promoting reestablishment as early seral inoculum. Today, many restoration projects consider mycorrhizae, whether direct inoculation of seedlings with mycorrhizae, or some form of large-scale recovery, management, and redistribution of topsoil coupled with mycorrhizal assessment. In one study, native drought-adapted AM fungi (eight *Glomus* taxa) proved to be more effective than an exotic *Glomus claroideum* (586). A broader metanalysis showed that use of local inoculum or topsoil gave the best results, and exotic inoculum had no beneficial effect (477).

The success of AM inoculation and carry over between seral stages also has application to reforestation or restoration in EM ecosystems. David Perry and colleagues (561) introduced the "bootstrapping" concept, whereby plants that recover after fire support the pre-fire EM fungal inoculum, and those fungi then shift to later-seral species facilitating forest recovery. Fire rarely destroys soil inocula, either mycorrhizal or saprotrophs, except in a few surface millimeters. The soil insolates soil organisms. Even though dependent upon living roots, both roots and mycorrhizal fungi survive through fire and can even last until the following growing season (286; 288; 417; 506). As new arbutoid or EM plant seeds are dispersed or survive at depth, they germinate, and the fungi re-establish rapidly. With time, they then switch to the alternate host (Figure 8.4). What this means is that following disturbance of an existing forest or wildland, there is rarely a need to replant inoculated seedlings. Wildland succession (Chapter 8) is well positioned to recover mycorrhizae adapted to the site and species, and inoculation probably only serves to introduce exotic species of fungi (in addition to novel genotypes of host plants) assuming that source populations are available.

When is inoculation useful and important? Likely only in cases where the soils are dramatically different from the pre-disturbance event, such as massive changes in chemistry like heavy metals or extreme pH conditions may require novel mycorrhizal fungi (see Chapter 4), or in urban ecosystems where the vegetation restoration is no longer spatially connected to the pre-disturbance ecosystem. An example is closed landfills where a limited inoculum reinvasion constrains mycorrhizal reestablishment (551).

An interesting new application of mycorrhizal inoculation is in urban settings. Urban environments present unique ecological conditions, severe and long-term disturbance, high CO_2 and nitrogen deposition, heavy metals, heat, and altered moisture regimes due to both enhanced precipitation and irrigation coupled with constrained rooting space. The plants often chosen are AM, possibly from evolving under high CO_2 conditions, such as *Ginkgo biloba*, *Liquidambar styraciflua*, and species of *Platanus* that are among the most common trees planted in urban environments. Urban environments are also home to taxa including *Quercus*, *Populus*, and *Tilia* that can be AM as seedlings (and on edges of the canopy) and EM when mature. Here, mycorrhizae were often eliminated, and have few sources for invasion by existing mycorrhizal forests. Gardens and lawns are often over-fertilized and over-irrigated, reducing mycorrhizae. These environments represent unique opportunities to incorporate mycorrhizae towards re-establishing plant patches to enhance biodiversity, moderate local climate anomalies, and enhance carbon sequestration.

Forestry and Agriculture: Integrating Mycorrhizae and Management Strategies for the Future

The importance of mycorrhizae in food and fiber production cannot be overestimated (625). Here, I provide the reader an entrance into how mycorrhizae regulate community and ecosystem functioning of managed ecosystems. There are many other recent books, such as Robson et al. (618), Siddiqui and Futai (657), Solaiman and Mickan (671), Goss et al.

(295), and hundreds of papers that address the role of mycorrhizae in agriculture and forestry production in detail, and I refer the reader to those, and many subsequent publications. But here I will address a view that understanding and managing mycorrhizae can play a critical role, not only in increasing food production, but also in managing for sustainable agriculture and forestry.

In part, those insights reveal a need to understand thresholds whereby perturbations of soils, climate, and edaphic factors become irreversible. And, how management for mycorrhizae can be optimized in new "designer ecosystems" that simultaneously produce important products, sequester C, and sustain biodiversity across landscapes, as a prelude to large-scale solutions for much of our environmental crisis. Utilizing the dynamics of mycorrhizae illuminate a new way of addressing our collective future. From the discussion early in this chapter, mycorrhizae promote new levels of complexity using positive feedback. Based on this understanding, we can incorporate mycorrhizae in the design of wildland, agricultural, and forest ecosystems using innovative new technologies of plant and fungal genetic manipulation, but also incorporating ancient horticultural technologies. David Christian (165) examines what he terms "life-ways" at the interface of the hunter–gatherer to the horticultural, and the horticultural to agricultural revolutions. Although these shifts occur at different times on different continents, the impacts represent the creation of new, human-managed biomes, represented by new artificial vegetation combinations, a major loss of wildland, and an increase in human population. The "Green Revolution" with its dramatic increase in chemical inputs beginning in the 1960s, represents a distinct new biome characterized by minimal internal nutrient and energy cycling, reduced soil health, and dramatically increasing crop yields but while rapidly increasing human populations. Technology may be delaying the Malthusian crash, but is it simply pushing the limiting biosphere to a higher human population level?

Many of the old technologies remain viable in places like the New Guinea Highlands and the Yucatán Peninsula. Beach et al. (109) described the "Mayacene" in the Yucatán Peninsula, when the culture and population of the Maya rose to astonishing heights. The population of the region during the Classical Period rose to a level that was nearly as high as today. Turner and colleagues (729) estimated that the carrying capacity in the late Classic period was up to 100 people per kilometer squared. Today's population is 22 people per kilometer squared. The basic agricultural technology of today is not dramatically different from that described by Fray Diego de Landa in approximately 1566 (440). Some of the tools have become iron instead of wood, fires can be easily ignited using matches instead of using rotational energy between sticks, and new varieties of critical crops, including maize, squash, and dry beans are used. But the procedures are still swidden agriculture, a rotational system of one-hectare patches (newly cleared and burned forest per campesino [farmer] per year), maintained in production for 2–3 years, followed by 7–15 or more years of fallow. These procedures were and are at the forefront of minimum tillage and organic agricultural production methods. Importantly, nearly 40 percent of the forest may have remained intact old growth, even in Tikal, Guatemala, one of the most densely inhabited areas in the American continent.

Although the Classic Maya period collapsed about a thousand years ago, in part due to climate change, storms, and likely leadership limitations, the Maya remain the most populous Amer-Indian group in North America. The population declined and 90 percent of the forest returned and fluctuated around a stable point of one to three people per kilometer squared, until recent decades. Are there lessons we can learn? In particular, can we integrate ecological and agricultural sciences to form a sustainable landscape, and what role do mycorrhizae play?

The primary reason for loss of mycorrhizal fungi in ecosystems is disturbance, and the Mayan forests were primarily disturbed by agriculture, fire, and hurricanes. My colleagues Arturo Gomez-Pompa, Juan Jimenez-Osornio, Emmanuel Rincon, Edith Allen and I initiated work on understanding successional dynamics and agricultural practices within seasonal tropical forest, in the Yucatán Peninsula, from research sites in the western lowlands of the State of Yucatán, Mexico across to the El Eden Ecological Reserve, Quintana Roo, Mexico. At the time our efforts commenced (1993), the forests surrounding the reserve had been damaged by a series of hurricanes (following Gilbert in 1988) resulting in extensive downed timber, and widespread fires. The vegetation that recovered was dense, early seral, and subject to recurring fires and limited forest recovery. Initially, we studied two features of succession. The first was the role of early- versus late-seral AM fungi by early- versus late-seral trees. In the second, we undertook thinning of a dense early-seral stand to create a more open understory stand, while closing the canopy. During the thinning process, one goal was to retain known, useful plants. The overall goal was to hasten succession, including of mycorrhizal fungi, to a more stable community. Both studies showed that mycorrhizae played key roles in the re-establishment of a more mature forest stand (15; 22; 48; 741).

But these studies also showed a linkage to the traditional agricultural practices of the Maya campesinos (384). The traditional Maya milpa, or maize-frijole-calabaza, is interspersed with patches of other crops such as pimienta or tomate, following swidden agriculture; a pattern that has been undertaken for over three millennia. Here, a forest patch of approximately one hectare and at least 15–20 years old would be burned, using the same manner as our experimental studies (22). These sites would be planted using a digging stick (today a metal rod), for up to three years, each year preparing an additional new hectare. After three years, the hectare would be abandoned, to allow succession to proceed, and regrow as forest until the patch was needed again. Importantly, as the spatial rotation of crop–successional forest–older forest occurred, many useful plants for food, medicines, and fuel are provided by identifiable successional seres (290). Protein and shelter in the form of animals hunted for food or skins, and wood were acquired old-growth forests.

Mycorrhizal relationships are a critical resource in sustaining agriculture and agro-forestry in the Maya region. I know of no nonmycotrophic plant taxa cultivated in traditional fields of the Maya. Within the milpa ecosystem, mutualisms, especially mycorrhizal fungi and N_2-fixing mutualisms, can play a critical role in sustainable food production (25). Mycorrhizae reduce leaching (90; 177; 226) of both NO_3^- and P and have/facilitate a more efficient utilization of both N and P (73; 585; 589). What we have found is that the burned patch maintains mycorrhizal fungi – they survive in cracks deeper in the rock matrix or deeper patches of soil. Using a digging stick is simply a less mechanized version of minimum-tillage agriculture. We know from numerous studies that the AM fungal hyphal network retains much of its physical structure with minimal tillage agriculture (e.g., (396)). By using the three-crop milpa, and coupled to mycorrhizae, this system provides exchangeable resources. Maize provides the greatest amount of C fixed for the AM fungus as well as itself, supporting the fungal network. The frijoles (beans) fix N_2 and transport excess N to the AM fungus and to the interconnected maize, as well as calabaza (squash). The calabaza has a prostrate growth habit that provides a canopy below the crops and competes with invading weeds. That system gradually runs down as the N_2 fixed is not enough for continued high levels of maize production, and weeds and microbial and insect pests continue to invade. Upon abandonment, many useful trees including legumes reinvade, rebuilding soil N and restoring late-seral AM fungi. Soil organic C also continues to build (479). These forests sustain a wide range of animals and plants supporting a diverse biodiversity, ranging from spider monkeys (*Ateles geoffroyi*) to jaguars (*Panthera onca*).

The Yucatán forests also contribute to the stability of these ecosystems. Coincident with the high human populations at the end of the Late PreClassic period, both drought and hurricane activity appeared to intensify (642). As Hurricane Wilma hit the El Eden Ecological Reserve there was a large deposition of leaf litter, coupled with dying roots from the severed trees (Chapter 8). Without mycorrhizae, there would have been a large loss of nutrients through decomposition and leaching. However, AM activity increased, and there was a flush of EM fungi (333). EM could take up and transfer a large fraction of the N, thereby reducing N loss from these ecosystems.

Allen et al. (48) described the usefulness of the Yucatán forests. In restoring a less-flammable designer forest ecosystem, a longer-term C sequestration, a more biodiverse forest for both humans and nature could emerge. Further, Jiménez-Osornio et al. (383) designed a net-worked milpa–agroforestry–conservation forest structure that interconnected individual villages with a larger network, the MesoAmerica Biological Corridor. The goal here is to conceptualize a managed designer ecosystem that fulfills goals of sustainable food and fiber production, and biodiversity, from microbial diversity to large jaguars that roam hundreds of miles. Mycorrhizal fungi, from the mycorrhizal vanilla orchid *Tulasnella*, to EM fungi in the Boletaceae in wetlands and emerging after hurricane-soaked storms, to AM fungi dominating across the forests and agricultural lands, are prime organisms regulating the sustainability of these ecosystems.

11 · *Conclusion and Summary*

Now travel back to your nearest nature reserve, but simultaneously look at the adjacent agriculture field or human urban neighborhood. Instead of just looking at what you can readily see, the diverse array of plants and animals; the complex architecture and the smells of the flowers and the diverse array of shapes and colors, double that, or maybe even increase it 10-fold by imagining that it is connected to what you cannot see below-ground! Diverse plant and fungal compositions and structures are inter-connected through multiple networks transferring resources and chemical signals. They may persist for only a day or even a few minutes, but alternatively maybe for decades, or even centuries. Even observing that part of the ecosystem, whether through minirhizotrons and ground-penetrating radar, or sampling and microscopy, you will only visualize a tiny fragment of that world. In the world of mycorrhizae, imagination may be the single most useful tool!

Studying mycorrhizae also provides continuity with the history of the world. The mycorrhizal symbiosis began more than 400 million years ago, as plants and fungi together first explored the incredible, but open resource that is the terrestrial environment. As the global ecosystems reduced CO_2, and increased O_2, in part through mycorrhizal relation-ships, the terrestrial world became increasingly occupied and more com-plex. New organisms and relationships, parasitic, competitive, and mutualistic evolved and spread. The opportunity to study soil biology and mycorrhizae, during the exploration of the heavens, through sputnik (I still remember as a child seeing this man-made object, like a planet, traversing the sky overhead), and now seeing the images and data collected and sent back, describing Mars through the robotic ecologists Sojourner, Spirit and Opportunity, and Curiosity and Perseverance, searching for traces of the life that might well resemble our own from 1,000 million years ago, creates an incredible opportunity for our science. While our findings to date represent an exciting and interesting body of knowledge, as David Read pointed out at the 1990 North American

Conference on Mycorrhizae, we have really only done the easy parts. Continually toggling back and forth between simplified theory and complexified natural history, we are gaining an ever-more sophisticated understanding of both mycorrhizae and our surrounding ecosystems. New tools, from isotope signatures, to molecular tools for distinguishing organisms and physiological processes, to observation systems from microscopic scale minirhizotrons to satellite imagery, provide new ways to look at the structure and functioning of mycorrhizae. Just as importantly, all can be brought to bear simultaneously to derive new theory, from elemental transfers across membranes to global atmospheric–biospheric exchanges.

But studying mycorrhizae is more than just science. From replanting forests, to managing agricultural fertilizers and pesticides, to regulating atmospheric CO_2 and even methyl halides, understanding mycorrhizae plays key practical management roles. Thus, the study of both the theory and natural history of mycorrhizae become essential tools in recovering biodiversity, providing the human population with food and fiber, and managing even our atmosphere and climate.

I wish the next generations as much collegiality and fun as the past century and a half of mycorrhizal research has provided to those of us privileged to have participated.

Bibliography

1. Abbott L, Johnson NC. 2017. Introduction: Perspectives on mycorrhizas and soil fertility, pp. 93–105. In NC Johnson, C Gehring, J Jansa, eds. *Mycorrhizal Mediation of Soil: Fertility, Structure, and Carbon Storage*. Amsterdam: Elsevier Press.
2. Abbott L, Robson A. 1982. The role of vesicular arbuscular mycorrhizal fungi in agriculture and the selection of fungi for inoculation. *Australian Journal of Agricultural Research* 33:389–408.
3. Agerer R. 2006. Fungal relationships and structural identity of their ectomycorrhizae. *Mycological Progress* 5:67–107.
4. Ågren GI, Bosatta E. 1998. *Theoretical Ecosystem Ecology: Understanding Element Cycles*. Cambridge: Cambridge University Press.
5. Aguirre F, Nouhra E, Urcelay C. 2021. Native and non-native mammals disperse exotic ectomycorrhizal fungi at long distances from pine plantations. *Fungal Ecology* 49:101012.
6. Aldon EF. 1975. Endomycorrhizae enhance survival and growth of fourwing saltbush on coal mine spoils. USDA Forest Service Research Note R.M. 294.
7. Alexander I. 1989. Mycorrhizas in tropical forests. *Special publications of the British Ecological Society*.
8. Alexopoulos CJ, Mims CW, Blackwell M. 1979. *Introductory Mycology*, 3rd ed. New York: John Wiley and Sons.
9. Allen EB. 1988. *The Reconstruction of Disturbed Arid Ecosystems*. Boulder, CO: Westview Press.
10. Allen EB, Allen MF. 1980. Natural re-establishment of Vesicular Arbuscular Mycorrhizae following stripmine reclamation in Wyoming. *Journal of Applied Ecology* 17:139–47.
11. Allen EB, Allen MF. 1984. Competition between plants of different successional stages: Mycorrhizae as regulators. *Canadian Journal of Botany* 62:2625–9.
12. Allen EB, Allen MF. 1986. Water relations of xeric grasses in the field: Interactions of mycorrhizas and competition. *New Phytologist* 104:559–71.
13. Allen EB, Allen MF. 1988. Facilitation of succession by the nonmycotrophic colonizer Salsola-kali (Chenopodiaceae) on a harsh site: Effects of mycorrhizal fungi *American Journal of Botany* 75:257–66.
14. Allen EB, Allen MF. 1990. The mediation of competition by mycorrhizae in successional and patchy environments, pp. 367–89. In JB Grace, GD Tilman, eds. *Perspectives on Plant Competition*. San Diego: Academic Press.

15. Allen EB, Allen MF, Egerton-Warburton L, Corkidi L, Gómez-Pompa A. 2003. Impacts of early-and late-seral mycorrhizae during restoration in seasonal tropical forest, Mexico. *Ecological Applications* 13:1701–17.

16. Allen EB, Allen MF, Helm DJ, Trappe JM, Molina R, Rincon E. 1995. Patterns and regulation of mycorrhizal plant and fungal diversity *Plant and Soil* 170:47–62.

17. Allen EB, Brown JS, Allen MF. 2001. Restoration of animal, plant and microbial diversity. *Encyclopedia of Biodiversity* 5:185–202.

18. Allen EB, Chambers JC, Connor KF, Allen MF, Brown RW. 1987. Natural reestablishment of mycorrhizae in disturbed alpine ecosystems. *Arctic and Alpine Research* 19:11–20.

19. Allen EB, Cunningham GL. 1983. Effects of vesicular–arbuscular mycorrhizae on Distichlis spicata under three salinity levels. *New Phytologist* 93:227–36.

20. Allen EB, Egerton-Warburton LM, Hilbig BE, Valliere JM. 2016. Interactions of arbuscular mycorrhizal fungi, critical loads of nitrogen deposition, and shifts from native to invasive species in a southern California shrubland. *Botany* 94:425–33.

21. Allen EB, Rincon E, Allen MF, Perez-Jimenez A, Huante P. 1998. Disturbance and seasonal dynamics of mycorrhizae in a tropical deciduous forest in Mexico. *Biotropica* 30:261–74.

22. Allen EB, Violi HA, Allen MF, Gomez-Pompa A. 2003. Restoration of tropical seasonal forest in Quintana Roo, pp. 587–98. In A Gomez-Pompa, MF Allen, S Fedick, JJ Jimenez-Osornio, eds. *Lowland Maya Area: Three Millennia at the Human–Wildland Interface*. New York: Haworth Press.

23. Allen K, Fisher JB, Phillips RP, Powers JS, Brzostek ER. 2020. Modeling the carbon cost of plant nitrogen and phosphorus uptake across temperate and tropical forests. *Frontiers in Forests and Global Change* 3:43.

24. Allen M. 1992. *Mycorrhizal Functioning: An Integrative Plant–Fungal Process*. New York: Chapman & Hall.

25. Allen M, Allen E. 2017. Mycorrhizal mediation of soil fertility amidst nitrogen eutrophication and climate change, pp. 213–31. In NC Johnson, C Gehring, J Jansa, eds. *Mycorrhizal Mediation of Soil: Fertility, Structure, and Carbon Storage*. Amsterdam: Elsevier Press.

26. Allen M, Clouse S, Weinbaum B, Jeakins S, Friese C, Allen E. 1992. Mycorrhizae and the integration of scales: From molecules to ecosystems, pp. 488–515. In MF Allen, ed. *Mycorrhizal Functioning*. New York: Chapman & Hall.

27. Allen M, Hipps L. 1985. Long-distance dispersal of mycorrhizal fungi: A comparison of vectors of VAM and ectomycorrhizal fungi from predictable versus unpredictable habitats. *Proceedings of the 17th Conference on Agricultural and Forest Meteorology and 7th Conference on Biometeorology and Aerobiology*, Scottsdale, AZ, May 21–24.

28. Allen M, Klironomos J, Harney S. 1997. The epidemiology of mycorrhizal fungal infection during succession, pp. 169–83. In G Carroll, P Tudzynski, eds. *The Mycota*, vol VB. Berlin: Springer.

29. Allen M, St John T. 1982. Dual culture of endomycorrhizae, pp. 85–89. In NC Schenck, ed. *Methods and Principles of Mycorrhizal Research*. St. Paul, MN: American Phytopathological Society.

30. Allen MF. 1980. Physiological alterations associated with vesicular-arbuscular mycorrhizal infection in Bouteloua gracilis. PhD thesis. University of Wyoming.

31. Allen MF. 1982. Influence of vesicular arbuscular mycorrhizae on water-movement through Bouteloua gracilis. *New Phytologist* 91:191–6.

32. Allen MF. 1983. Formation of vesicular–arbuscular mycorrhizae in Atriplex gardneri (Chenopodiaceae): Seasonal response in a cold desert. *Mycologia* 75:773–6.

33. Allen MF. 1987. Reestablishment of mycorrhizae on Mount St. Helens: Migration vectors. *Transactions of the British Mycological Society* 88:413–17.

34. Allen MF. 1988. Belowground structure: A key to reconstructing a productive arid ecosystem, pp. 113–35. In EB Allen, ed. *The Reconstruction of Disturbed Arid Ecosystems.* Boulder, CO: Westview Press.

35. Allen MF. 1988. Re-establishment of VA-mycorrhizas following severe disturbance: Comparative patch dynamics of a shrub desert and a subalpine volcano. *Proceedings of the Royal Society of Edinburgh Section B–Biological Sciences* 94:63–71.

36. Allen MF. 1991. *The Ecology of Mycorrhizae.* Cambridge: Cambridge University Press.

37. Allen MF. 1996. The ecology of arbuscular mycorrhizas: A look back into the 20th century and a peek into the 21st. *Mycological Research* 100:769–82.

38. Allen MF. 2001. Modeling arbuscular mycorrhizal infection: Is % infection an appropriate variable? *Mycorrhiza* 10:255–8.

39. Allen MF. 2006. Water dynamics of mycorrhizas in arid soils, pp. 74–97. In GM Gadd, ed. *Fungi in Biogeochemical Cycles.* Cambridge: Cambridge University Press.

40. Allen MF. 2007. Mycorrhizal fungi: Highways for water and nutrients in arid soils. *Vadose Zone Journal* 6:291–7.

41. Allen MF. 2009. Bidirectional water flows through the soil–fungal–plant mycorrhizal continuum. *New Phytologist* 182:290–3.

42. Allen MF. 2011. Linking water and nutrients through the vadose zone: A fungal interface between the soil and plant systems. *Journal of Arid Land* 3:155–63.

43. Allen MF, Allen EB. 1986. Utah State Project Exemplifies Restoration Ecology Approach to Research. *Restoration and Management Notes* 4:64–7.

44. Allen MF, Allen EB. 1990. Carbon source of VA mycorrhizal fungi associated with Chenopodiaceae from a semiarid shrub-steppe. *Ecology* 71:2019–21.

45. Allen MF, Allen EB. 1992. Mycorrhizae and plant community development: Mechanisms and patterns, pp. 455–79. In. GC Carroll, DT Wicklow, eds. *The Fungal Community.* New York: Marcel Dekker.

46. Allen MF, Allen EB, Dahm C, Edwards F. 1993. Preservation of biological diversity in mycorrhizal fungi: Importance and human impacts, pp. 81–108. In G Sundnes, ed. *International Symposium on Human Impacts on Self-Recruiting Populations.* Trondheim, Norway: The Royal Norwegian Academy of Sciences.

47. Allen MF, Allen EB, Friese CF. 1989. Responses of the non-mycotrophic plant Salsola-Kali to invasion by vesicular arbuscular mycorrhizal-fungi *New Phytologist* 111:45–9.

48. Allen MF, Allen EB, Gomez-Pompa A. 2005. Effects of mycorrhizae and nontarget organisms on restoration of a seasonal tropical forest in Quintana Roo, Mexico: Factors limiting tree establishment. *Restoration Ecology* 13:325–33.

49. Allen MF, Allen EB, Lansing JL, Pregitzer KS, Hendrick RL, et al. 2010. Responses to chronic N fertilization of ectomycorrhizal pinon but not arbuscular mycorrhizal juniper in a pinon-juniper woodland. *Journal of Arid Environments* 74:1170–6.

50. Allen MF, Allen EB, Stahl PD. 1984. Differential niche response of Bouteloua-Gracilis and Pascopyrum-Smithii to VA mycorrhizae. *Bulletin of the Torrey Botanical Club* 111:361–5.

51. Allen MF, Allen EB, West NE. 1987. Influence of parasitic and mutualistic fungi on artemesia tridentata during high precipitation years. *Bulletin of the Torrey Botanical Club* 114:272–9.

52. Allen MF, Boosalis MG. 1983. Effects of two species of vesicular-arbuscular mycorrhizal fungi on drought tolerance of winter wheat *New Phytologist* 93:67–76.

53. Allen MF, Crisafulli C, Friese CF, Jeakins SL. 1992. Re-formation of mycorrhizal symbioses on Mount St. Helens, 1980–1990: Interactions of rodents and mycorrhizal fungi *Mycological Research* 96:447–53.

54. Allen MF, Egerton-Warburton L, Treseder K, Cario C, Lindahl A, Lansing J, Querejeta I, Karen O, Harney S, Zink T. 2005. Biodiversity and mycorrhizal fungi in southern California. In B Kus, JL Beyers, eds. *Planning for Biodiversity: Bringing Research and Management Together: Proceedings of a Symposium for the South Coast Ecoregion*, March 2000, Pomona, CA. USDA Forest Service Pacific Southwest Research Station General Technical Report PSW-GTR-195.

55. Allen MF, Figueroa C, Weinbaum BS, Barlow SB, Allen EB. 1996. Differential production of oxalate by mycorrhizal fungi in arid ecosystems. *Biology and Fertility of Soils* 22:287–92.

56. Allen MF, Hipps LE, Wooldridge GL. 1989. Wind dispersal and subsequent establishment of VA mycorrhizal fungi across a successional arid landscape. *Landscape Ecology* 2:165–71.

57. Allen MF, Jenerette GD, Santiago LS. 2013. *Carbon Balance in California Deserts: Impacts of Widespread Solar Power Generation: Final Project Report.* Publication number: CEC-500-2013-063. California Energy Commission.

58. Allen MF, Kitajima K. 2013. In situ high-frequency observations of mycorrhizas. *New Phytologist* 200:222–8.

59. Allen MF, Kitajima K. 2014. Net primary production of ectomycorrhizas in a California forest. *Fungal Ecology* 10:81–90.

60. Allen MF, Kitajima K, Hernandez RR. 2014. Mycorrhizae and global change, pp. 37–59. In M Tausz, N Grulke, eds. *Trees in a Changing Environment: Ecophysiology, Adaptation, and Future Survival.* Dordrecht: Springer.

61. Allen MF, Klironomos JN, Treseder KK, Oechel WC. 2005. Responses of soil biota to elevated CO_2 in a chaparral ecosystem. *Ecological Applications* 15:1701–11.

62. Allen MF, MacMahon JA. 1985. Impact of disturbance on cold desert fungi: Comparative microscale dispersion patterns. *Pedobiologia* 28:215–24.

63. Allen MF, MacMahon JA. 1988. Direct VA mycorrhizal inoculation of colon-izing plants by pocket gophers (*Thomomys talpoides*) on Mount St. Helens. *Mycologia* 80:754–6.

64. Allen MF, MacMahon JA, Andersen DC. 1984. Reestablisment of Endogonaceae on Mount St. Helens: Survival of residuals. *Mycologia* 76:1031–8.

65. Allen MF, Mishler, BD. In press. A phylogenetic approach to conservation: Biodiversity and ecosystem functioning for a changing globe. In B Swartz, BD Mishler, eds. *Speciesism and the Future of Humanity: Biology, Culture, and Sociopolitics.* Springer.

66. Allen MF, Moore TS, Christensen M. 1980. Phytohormone changes in *Bouteloua gracilis* infected by vesicular-arbuscular mycorrhizae. 1. Cytokinin increases in the host plant. *Canadian Journal of Botany* 58:371–4.

67. Allen MF, Moore TS, Christensen M. 1982. Phytohormone changes in *Bouteloua gracilis* infected by vesicular-arbuscular mycorrhizae. 2. Altered levels of gibberellin-like substances and abscisic-acid in the host plant. *Canadian Journal of Botany* 60:468–71.

68. Allen MF, Moore TS, Christensen M, Stanton N. 1979. Growth of vesicular-arbuscular mycorrhizal and non-mycorrhizal *Bouteloua gracilis* in a defined medium. *Mycologia* 71:666–9.

69. Allen MF, O'Neill MR, Crisafulli CM, MacMahon JA. 2018. Succession and mycorrhizae on Mount St. Helens, pp. 199–216. In VH Dale, CM Crisafulli, eds. *Ecological Responses at Mount St. Helens: Revisited 35 Years after the 1980 Eruption.* Berlin: Springer.

70. Allen MF, Richards JH, Busso CA. 1989. Influence of clipping and soil-water status on vesicular–arbuscular mycorrhizae of two semi-arid tussock grasses. *Biology and Fertility of Soils* 8:285–9.

71. Allen MF, Rincon E. 2003. The changing global environment and the lowland Maya: Past patterns and current dynamics, pp. 13–29. In A Gomez-Pompa, MF Allen, S Fedick, JJ Jimenez-Osornio, eds. *Lowland Maya Area: Three Millennia at the Human–Wildland Interface.* New York: Haworth Press.

72. Allen MF, Rincon E, Allen EB, Huante P, Dunn JJ. 1993. Observations of canopy bromeliad roots compared with plants rooted in soils of a seasonal tropical forest, Chamela, Jalisco, Mexico. *Mycorrhiza* 4:27–8.

73. Allen MF, Sexton JC, Moore TS Jr, Christensen M. 1981. Influence of phosphate source on vesicular-arbuscular mycorrhizae of Bouteloua gracilis. *New Phytologist* 87:687–94.

74. Allen MF, Smith WK, Moore TS, Christensen M. 1981. Comparative water relations and photosynthesis of mycorrhizal and non-mycorrhizal Bouteloua gracilis. *New Phytologist* 88:683–93.

75. Allen MF, Swenson W, Querejeta JI, Egerton-Warburton LM, Treseder KK. 2003. Ecology of mycorrhizae: A conceptual framework for complex inter-actions among plants and fungi. *Annual Review of Phytopathology* 41:271–303.

76. Allen MF, Vargas R, Graham EA, Swenson W, Hamilton M, et al. 2007. Soil sensor technology: Life within a pixel. *BioScience* 57:859–67.

77. Allen TF, Starr TB. 1982. *Hierarchy: Perspectives for Ecological Complexity.* Chicago, IL: University of Chicago Press.

78. Alvarez W. 2008. *T. Rex and the Crater of Doom*. Princeton, NJ: Princeton University Press.

79. Anacker BL, Klironomos JN, Maherali H, Reinhart KO, Strauss SY. 2014. Phylogenetic conservatism in plant-soil feedback and its implications for plant abundance. *Ecology Letters* 17:1613–21.

80. Anderson R, Homola RL, Davis RB, Jacobson GL Jr. 1984. Fossil remains of the mycorrhizal fungal Glomus fasciculatum complex in postglacial lake sediments from Maine. *Canadian Journal of Botany* 62:2325–8.

81. Anderson R, Liberta A, Dickman L. 1984. Interaction of vascular plants and vesicular–arbuscular mycorrhizal fungi across a soil moisture–nutrient gradient. *Oecologia* 64:111–17.

82. Anonymous. 1931. Establishing pines. Preliminary observations on the effects of soil inoculation. *Rhodesia Agricultural Journal* 28:185–7.

83. Antibus RK, Kroehler CJ, Linkins AE. 1986. The effects of external pH, temperature, and substrate concentration on acid phosphatase activity of ecto-mycorrhizal fungi. *Canadian Journal of Botany* 64:2383–7.

84. Antoninka AJ, Ritchie ME, Johnson NC. 2015. The hidden Serengeti: Mycorrhizal fungi respond to environmental gradients. *Pedobiologia* 58: 165–76.

85. Antunes P, Koyama A. 2017. Mycorrhizas as nutrient and energy pumps of soil food webs: Multitrophic interactions and feedbacks, pp. 149–73: In NC Johnson, C Gehring, J Jansa, eds. *Mycorrhizal Mediation of Soil: Fertility, Structure, and Carbon Storage*. Amsterdam: Elsevier Press.

86. Aquino SD, Scabora MH, Andrade JAD, da Costa SMG, Maltoni KL, Cassiolato AMR. 2015. Mycorrhizal colonization and diversity and corn genotype yield in soils of the Cerrado region, Brazil. *Semina-Ciencias Agrarias* 36:4107–17.

87. Arnolds E. 1991. Decline of ectomycorrhizal fungi in Europe. *Agriculture, Ecosystems & Environment* 35:209–44.

88. Aronson EL, Dierick D, Botthoff JK, Oberbauer S, Zelikova TJ, et al. 2019. ENSO-influenced drought drives methane flux dynamics in a tropical wet forest soil. *Journal of Geophysical Research-Biogeosciences* 124:2267–76.

89. Asai T. 1943. Die Bedeutung der Mykorrhiza für das Pfianzenleben. *Japanese Journal of Botany* 12:359–436.

90. Asghari HR, Cavagnaro TR. 2011. Arbuscular mycorrhizas enhance plant interception of leached nutrients. *Functional Plant Biology* 38:219–26.

91. Augé RM, Stodola AJW, Tims JE, Saxton A. 2001. Moisture retention properties of a mycorrhizal soil. *Plant and Soil* 202:87–97.

92. Augé RM, Toler HD, Saxton AM. 2015. Arbuscular mycorrhizal symbiosis alters stomatal conductance of host plants more under drought than under amply watered conditions: A meta-analysis. *Mycorrhiza* 25:13–24.

93. Averill C, Dietze MC, Bhatnagar JM. 2018. Continental-scale nitrogen pollution is shifting forest mycorrhizal associations and soil carbon stocks. *Global Change Biology* 24:4544–53.

94. Baar J, Horton TR, Kretzer A, Bruns TD. 1999. Mycorrhizal colonization of Pinus muricata from resistant propagules after a stand-replacing wildfire. *New Phytologist* 143:409–18.

95. Bae K-S, Barton LL. 1989. Alkaline phosphatase and other hydrolyases produced by Cenococcum graniforme, an ectomycorrhizal fungus. *Applied and Environmental Microbiology* 55:2511–16.

96. Bago B, Azcón-Aguilar C, Goulet A, Piché Y. 1998. Branched absorbing structures (BAS): A feature of the extraradical mycelium of symbiotic arbuscular mycorrhizal fungi. *New Phytologist* 139:375–88.

97. Bago B, Pfeffer PE, Zipfel W, Lammers P, Shachar-Hill Y. 2002. Tracking metabolism and imaging transport in arbuscular mycorrhizal fungi. Metabolism and transport in AM fungi. *Plant and Soil* 244:189–97.

98. Bahram M, Harend H, Tedersoo L. 2014. Network perspectives of ectomycorrhizal associations. *Fungal Ecology* 7:70–7.

99. Bainbridge D. 2007. *New Hope for Arid Lands: A Guide for Desert and Dryland Restoration*. Washington, DC: Island Press.

100. Baldocchi DD. 2020. How eddy covariance flux measurements have contributed to our understanding of global change biology. *Global Change Biology* 26:242–60.

101. Balogh-Brunstad Z, Keller CK, Shi Z, Wallander H, Stipp SL. 2017. Ectomycorrhizal fungi and mineral interactions in the rhizosphere of Scots and red pine seedlings. *Soils* 1:5.

102. Barabási A-L, Albert R. 1999. Emergence of scaling in random networks. *Science* 286:509–12.

103. Barbosa MV, Pereira EA, Cury JC, Carneiro MA. 2017. Occurrence of arbuscular mycorrhizal fungi on King George Island, South Shetland Islands, Antarctica. *Anais da Academia Brasileira de Ciências* 89:1737–43.

104. Barceló M, van Bodegom PM, Soudzilovskaia NA. 2019. Climate drives the spatial distribution of mycorrhizal host plants in terrestrial ecosystems. *Journal of Ecology* 107:2564–73.

105. Barea JM, Azcon R, Azcón-Aguilar C. 1992. 21 Vesicular–arbuscular mycorrhizal fungi in nitrogen-fixing systems. *Methods in Microbiology* 24:391–416.

106. Barrows CW, Murphy-Mariscal ML, Hernandez RR. 2016. At a crossroads: The nature of natural history in the twenty-first century. *BioScience* 66:592–9.

107. Bartnicki-García S. 2002. Hyphal tip growth: Outstanding questions, pp. 29–58. In HD Osiewacz, ed. *Molecular Biology of Fungal Development*. New York: Marcel Dekker.

108. Barton LL, Northup DE. 2011. *Microbial Ecology*. Hoboken, NJ: Wiley.

109. Beach T, Luzzadder-Beach S, Cook D, Dunning N, Kennett DJ, et al. 2015. Ancient Maya impacts on the Earth's surface: An early anthropocene analog? *Quaternary Science Reviews* 124:1–30.

110. Becquer A, Guerrero-Galan C, Eibensteiner JL, Houdinet G, Bucking H, et al. 2019. The ectomycorrhizal contribution to tree nutrition. *Molecular Physiology and Biotechnology of Trees* 89:77–126.

111. Beilby J, Kidby D. 1980. Sterol composition of ungerminated and germinated spores of the vesicular-arbuscular mycorrhizal fungus, *Glomus caledonius*. *Lipids* 15:375–8.

112. Beiler KJ, Durall DM, Simard SW, Maxwell SA, Kretzer AM. 2010. Architecture of the wood-wide web: *Rhizopogon* spp. genets link multiple Douglas-fir cohorts. *New Phytologist* 185:543–53.

113. Beiler KJ, Simard SW, Durall DM. 2015. Topology of tree-mycorrhizal fungus interaction networks in xeric and mesic Douglas-fir forests. *Journal of Ecology* 103:616–28.

114. Bellgard S, Whelan R, Muston R. 1994. The impact of wildfire on vesicular–arbuscular mycorrhizal fungi and their potential to influence the re-establishment of post-fire plant communities. *Mycorrhiza* 4:139–46.

115. Bellgard SE. 1991. Mycorrhizal associations of plant-species in Hawkesbury Sandstone vegetation *Australian Journal of Botany* 39:357–64.

116. Bennett JA, Maherali H, Reinhart KO, Lekberg Y, Hart MM, Klironomos J. 2017. Plant-soil feedbacks and mycorrhizal type influence temperate forest population dynamics. *Science* 355:181–4.

117. Berkeley M. 1846. *Observations, Botanical and Physiological, on the Potato Murrain.* Reprinted as *Phytopathological Classics 8.* Saint Paul, MN: APS Press.

118. Bethlenfalvay GJ, Brown MS, Pacovsky RS. 1982. Parasitic and mutualistic associations between a mycorrhizal fungus and soybean: Development of the host plant *Phytopathology* 72:889–93.

119. Bever JD. 2002. Negative feedback within a mutualism: Host-specific growth of mycorrhizal fungi reduces plant benefit. *Proceedings of the Royal Society B–Biological Sciences* 269:2595–601.

120. Bidartondo MI. 2005. The evolutionary ecology of myco-heterotrophy. *New Phytologist* 167:335–52.

121. Bitterlich M, Rouphael Y, Graefe J, Franken P. 2018. Arbuscular mycorrhizas: A promising component of plant production systems provided favorable conditions for their growth. *Frontiers in Plant Science* 9:154.

122. Björkman E. 1942. Über die bedingungen der Mykorrhizabildung bei Kiefer und Fichte. *Acta Universitatis Upsaliensis* 2:1–191.

123. Björkman E. 1960. *Monotropa hypopitys* L.: An epiparasite on tree roots. *Physiologia Plantarum* 13:308–27.

124. Bledsoe CS, Allen MF, Southworth D. 2014. Beyond mutualism: Complex mycorrhizal interactions. *Progress in Botany* 75:311–34.

125. Blom C, Voesenek L. 1996. Flooding: The survival strategies of plants. *Trends in Ecology & Evolution* 11:290–5.

126. Bloss H, Walker C. 1987. Some endogonaceous mycorrhizal fungi of the Santa Catalina mountains in Arizona. *Mycologia* 79:649–54.

127. Blum U, Rice EL. 1969. Inhibition of symbiotic nitrogen-fixation by gallic and tannic acid, and possible roles in old-field succession. *Bulletin of the Torrey Botanical Club* 96:531–44.

128. Bohlen PJ, Scheu S, Hale CM, McLean MA, Migge S, et al. 2004. Non-native invasive earthworms as agents of change in northern temperate forests. *Frontiers in Ecology and the Environment* 2:427–35.

129. Bonfante-Fasolo P, Scannerini S. 1977. Cytological observations on the mycorrhiza *Endogone flammicorona - Pinus strobus. Allionia* 22:23–34.

130. Bonfante-Fasolo P, Scannerini S. 1977. A cytological study of the vesicular arbuscular mycorrhiza in *Ornithogalum umbellatum. Allionia* 22:5–21.

131. Bornyasz MA, Graham RC, Allen MF. 2005. Ectomycorrhizae in a soil-weathered granitic bedrock regolith: Linking matrix resources to plants. *Geoderma* 126:141–60.

132. Bosatta E, Ågren GI. 1996. Theoretical analyses of carbon and nutrient dynamics in soil profiles. *Soil Biology and Biochemistry* 28:1523–31.

133. Boulet FM, Lambers H. 2005. Characterisation of arbuscular mycorrhizal fungi colonisation in cluster roots of shape Hakea verrucosa F. Muell (Proteaceae), and its effect on growth and nutrient acquisition in ultramafic soil. *Plant and Soil* 269:357–67.

134. Boyce MS. 1979. Seasonality and patterns of natural-selection for life histories. *American Naturalist* 114:569–83.

135. Bradshaw AD, Chadwick MJ. 1980. *The Restoration of Land: The Ecology and Reclamation of Derelict and Degraded Land.* Berkeley: University of California Press.

136. Branco S, Bi K, Liao HL, Gladieux P, Badouin H, et al. 2017. Continental-level population differentiation and environmental adaptation in the mushroom Suillus brevipes. *Molecular Ecology* 26:2063–76.

137. Branco S, Gladieux P, Ellison CE, Kuo A, LaButti K, et al. 2015. Genetic isolation between two recently diverged populations of a symbiotic fungus. *Molecular Ecology* 24:2747–58.

138. Bravo A, Brands M, Wewer V, Dormann P, Harrison MJ. 2017. Arbuscular mycorrhiza-specific enzymes FatM and RAM2 fine-tune lipid biosynthesis to promote development of arbuscular mycorrhiza. *New Phytologist* 214:1631–45.

139. Bray JR, Curtis JT. 1957. An ordination of the upland forest communities of southern Wisconsin. *Ecological Monographs* 27:325–49.

140. Briscoe CB. 1959. Early results of mycorrhizal inoculation of pine in Puerto Rico. *Carribean Forester* July–December:73–77.

141. Brosed M, Jabiol J, Gessner MO. 2017. Nutrient stoichiometry of aquatic hyphomycetes: Interstrain variation and ergosterol conversion factors. *Fungal Ecology* 29:96–102.

142. Bruns TD, Corradi N, Redecker D, Taylor JW, Öpik M. 2018. Glomeromycotina: What is a species and why should we care? *New Phytologist* 220:963–7.

143. Bruns TD, Hale ML, Nguyen NH. 2019. Rhizopogon olivaceotinctus increases its inoculum potential in heated soil independent of competitive release from other ectomycorrhizal fungi. *Mycologia* 111:936–41.

144. Brusatte S. 2018. *The Rise and Fall of the Dinosaurs: A New History of Their Lost World.* New York: Harper Collins.

145. Brzostek ER, Fisher JB, Phillips RP. 2014. Modeling the carbon cost of plant nitrogen acquisition: Mycorrhizal trade-offs and multipath resistance uptake improve predictions of retranslocation. *Journal of Geophysical Research: Biogeosciences* 119:1684–97.

146. Buchmann NB, Brooks JR, Ehleringer JR. 2002. Predicting daytime carbon isotope ratios of atmospheric CO_2 within forest canopies. *Functional Ecology* 16:49–57.

147. Buller A. 1909. *Researches on Fungi.* Vol I. London: Longmans, Green and Company.

148. Buntgen U, Lendorff H, Lendorff A, Leuchtmann A, Peter M, et al. 2019. Truffles on the move. *Frontiers in Ecology and the Environment* 17:200–1.

149. Burrola-Aguilar C, Garibay-Orijel R, Argüelles-Moyao A. 2013. *Abies religiosa* forests harbor the highest species density and sporocarp productivity of wild edible mushrooms among five different vegetation types in a neotropical temperate forest region. *Agroforestry Systems* 87:1101–15.

150. Burton AJ, Pregitzer KS, Ruess RW, Hendrik RL, Allen MF. 2002. Root respiration in North American forests: Effects of nitrogen concentration and temperature across biomes. *Oecologia* 131:559–68.

151. Cabello M, Gaspar L, Pollero R. 1994. *Glomus antarcticum* sp- nov, a vesicular–arbuscular mycorrrhizal fungus from Antarctica. *Mycotaxon* 51:123–8.

152. Caldwell MM, Eissenstat DM, Richards JH, Allen MF. 1985. Competition for phosphorus: Differential uptake from dual-isotope labelled soil interspaces between shrub and grass *Science* 229:384–6.

153. Cannon JP, Allen EB, Allen MF, Dudley LM, Jurinak JJ. 1995. The effects of oxalates produced by Salsola tragus on the phosphorus nutrition of Stipa pulchra. *Oecologia* 102:265–72.

154. Cario CH. 2005. Elevated atmospheric carbon dioxide and chronic atmospheric nitrogen deposition change nitrogen dynamics associated with two Mediterranean climate evergreen oaks. PhD dissertation. University of California, Davis.

155. Carpenter AT, Allen MF. 1988. Responses of *Hedysarum boreale* Nutt to mycorrhizas and Rhizobium: Plant and soil nutrient changes in a disturbed shrub-steppe. *New Phytologist* 109:125–32.

156. Carpenter SE, Trappe JM, Ammirati Jr J. 1987. Observations of fungal succession in the Mount St. Helens devastation zone, 1980–1983. *Canadian Journal of Botany* 65:716–28.

157. Caruso T, Hogg ID, Uffe NN, Bottos EN, Lee CK, et al. 2019. Nematodes in a polar desert reveal the relative role of biotic interactions in the coexistence of soil animals. *Communications Biology* 2:63.

158. Chagnon P-L, Bradley RL, Maherali H, Klironomos JN. 2013. A trait-based framework to understand life history of mycorrhizal fungi. *Trends in Plant Science* 18:484–91.

159. Chambers SM, Williams PG, Seppelt RD, Cairney JWG. 1999. Molecular identification of *Hymenoscyphus* sp. from rhizoids of the leafy liverwort *Cephaloziella* exiliflora in Australia and Antarctica. *Mycological Research* 103:286–8.

160. Chang Y, Desiro A, Na H, Sandor L, Lipzen A, et al. 2019. Phylogenomics of Endogonaceae and evolution of mycorrhizas within Mucoromycota. *New Phytologist* 222:511–25.

161. Chapela IH, Osher LJ, Horton TR, Henn MR. 2001. Ectomycorrhizal fungi introduced with exotic pine plantations induce soil carbon depletion. *Soil Biology and Biochemistry* 33:1733–40.

162. Chiarello N, Hickman JC, Mooney HA. 1982. Endomycorrhizal role for interspecific transfer of phosphorus in a community of annual plants. *Science* 217:941–3.

163. Chilvers GA, Lapeyrie FF, Horan DP. 1987. Ectomycorrhizal vs. endomycorrhizal fungi within the same root-system. *New Phytologist* 107:441.

164. Christensen M. 1969. Soil microfungi of dry to mesic conifer–hardwood forests in northern Wisconsin. *Ecology* 50:9–27.
165. Christian D. 2011. *Maps of Time. An Introduction to Big History.* Berkeley: University of California Press.
166. Claridge AW, Trappe JM, Mills DJ, Claridge DL. 2009. Diversity and habitat relationships of hypogeous fungi. III. Factors influencing the occurrence of fire-adapted species. *Mycological Research* 113:792–801.
167. Clark AL, St Clair SB. 2011. Mycorrhizas and secondary succession in aspen–conifer forests: Light limitation differentially affects a dominant early and late successional species. *Forest Ecology and Management* 262:203–7.
168. Clark DA, Clark DB, Oberbauer SF. 2013. Field-quantified responses of tropical rainforest aboveground productivity to increasing CO2 and climatic stress, 1997–2009. *Journal of Geophysical Research: Biogeosciences* 118:783–94.
169. Clausen J, Keck DD, Hiesey WM. 1948. *Experimental studies on the nature of species. III: Environmental responses of climatic races of Achillea.* Publication 581. Washington, DC: Carnegie Institution of Washington.
170. Clements F. 1916. *Plant Succession.* Publication 242. Washington, DC: Carnegie Institute of Washington.
171. Collins M, Knutti R, Arblaster J, Dufresne J-L, Fichefet T, et al. 2013. Long-term climate change: Projections, commitments and irreversibility, pp. 1029–136. In *Climate Change 2013: The Physical Science Basis: Contribution of Working Group I to the Fifth Assessment Report of the Intergovernmental Panel on Climate Change.* Cambridge: Cambridge University Press.
172. Colpaert JV, Vandenkoornhuyse P, Adriaensen K, Vangronsveld J. 2000. Genetic variation and heavy metal tolerance in the ectomycorrhizal basidiomycete *Suillus luteus. New Phytologist* 147:367–79.
173. Cooke JC, Gemma J, Koske R. 1987. Observations of nuclei in vesicular-arbuscular mycorrhizal fungi. *Mycologia* 79:331–3.
174. Cooper DJ, Sueltenfuss J, Oyague E, Yager K, Slayback D, et al. 2019. Drivers of peatland water table dynamics in the central Andes, Bolivia and Peru. *Hydrological Processes* 33:1913–25.
175. Cooper WS. 1923. The recent ecological history of Glacier Bay, Alaska: The present vegetation cycle. *Ecology* 4:223–46.
176. Corkidi L, Allen EB, Merhaut D, Allen MF, Downer J, et al. 2005. Effectiveness of commercial mycorrhizal inoculants on the growth of Liquidambar styraciflua in plant nursery conditions. *Journal of Environmental Horticulture* 23:72–76.
177. Corkidi L, Merhaut DJ, Allen EB, Downer J, Bohn J, Evans M. 2011. Effects of mycorrhizal colonization on nitrogen and phosphorus leaching from nursery containers. *HortScience* 46:1472–9.
178. Corradi N, Brachmann A. 2017. Fungal mating in the most widespread plant symbionts? *Trends in Plant Science* 22:175–83.
179. Corrales A, Henkel TW, Smith ME. 2018. Ectomycorrhizal associations in the tropics: Biogeography, diversity patterns and ecosystem roles. *New Phytologist* 220:1076–91.
180. Correia M, Heleno R, da Silva LP, Costa JM, Rodríguez-Echeverría S. 2019. First evidence for the joint dispersal of mycorrhizal fungi and plant diaspores by birds. *New Phytologist* 222:1054–60.

181. Cowan M, Lewis B, Thain J. 1972. Uptake of potassium by the developing sporangiophore of Phycomyces blakesleeanus. *Transactions of the British Mycological Society* 58:113–26.

182. Cowles HC. 1899. The ecological relations of the vegetation on the sand dunes of Lake Michigan. *Botanical Gazette* 27:95–391.

183. Cowling RM, Richardson DM. 1995. *Fynbos: South Africa's Unique Floral Kingdom*. Capetown: Fernwood Press.

184. Cox G, Tinker PB. 1976. Translocation and transfer of nutrients in vesicular–arbuscular mycorrhizas. 1. Arbuscule and phosphorus transfer: Quantitative ultrastructural study. *New Phytologist* 77:371.

185. Craine JM, Elmore AJ, Aidar MPM, Bustamante M, Dawson TE, et al. (2009) Global patterns of foliar nitrogen isotopes and their relationships with climate, mycorrhizal fungi, foliar nutrient concentrations, and nitrogen availability. *New Phytologist* 183:980–92.

186. Cromack Jr K, Sollins P, Graustein WC, Speidel K, Todd AW, et al. 1979. Calcium oxalate accumulation and soil weathering in mats of the hypogeous fungus *Hysterangium crassum*. *Soil Biology and Biochemistry* 11:463–8.

187. Crowell HF, Boerner RE. 1988. Influences of mycorrhizae and phosphorus on belowground competition between two old-field annuals. *Environmental and Experimental Botany* 28:381–92.

188. Daft M, Hacskaylo E. 1976. Arbuscular mycorrhizas in the anthracite and bituminous coal wastes of Pennsylvania. *Journal of Applied Ecology* 13:523–31.

189. Dangeard P. 1896. Une maladie du peuplier dans l'ouest de la France. *Le Botaniste* 5:38–43.

190. Dangeard P. 1900. Le Rhizophagus populinus. *Le Botaniste* 7:285–7.

191. Darwin C. 1859. The Origin of Species by Means of Natural Selection. Available from http://darwin-online.org.uk/content/frameset?itemID=F373&viewtype=text&pageseq=1. Last accessed November 16, 2021.

192. Darwin C. 1881. *The Formation of Vegetable Mould through the Action of Worms*. London: John Murray.

193. Darwin C, Wallace AR, Lyell SC, Hooker JD. 1858. On the tendency of species to form varieties and on the perpetuation of varieties and species by natural means of selection, *1858: Journal of the Proceedings of the Linnean Society of London*. Available from http://darwin-online.org.uk/content/frameset?itemID=F350&viewtype=text&pageseq=1Zoology 3:45–50. Last accessed November 16, 2021.

194. Davidson DE, Christensen M. 1977. Root-microfungal and mycorrhizal associations in a shortgrass prairie, pp. 279–87. In JK Marshall, ed. *The Belowground Ecosystem: A Synthesis of Plant-Associated Processes*. Fort Colloins: Colorado State University.

195. De Mazancourt C, Schwartz MW. 2010. A resource ratio theory of cooperation. *Ecology Letters* 13:349–59.

196. Deacon J, Fleming L. 1992. Interactions of ectomycorrhizal fungi, pp. 249–300. In MF Allen, ed. *Mycorrhizal Functioning*. New York: Chapman and Hall Press.

197. DeBary A. 1853. *Untersuchungen über die Brandpilze und die durch sie verursachten Krankheiten der Pflanzen mit Rücksicht auf das Getreide und andere Nutzpflanzen*. Berlin: GWF Müller.

198. DeBary A. 1879. *Die erscheinung der symbiose.* Strasbourg: Verlag von Karl J Trübner.

199. DeBary A. 1887. *Comparative Morphology and Biology of the Fungi, Mycetozoa and Bacteria* (English translation of the 1884 edition). Oxford: Clarendon Press.

200. Defrenne CE, Abs E, Longhi Cordeiro A, Dietterich L, Hough M, et al. 2021. The Ecology Underground coalition: Building a collaborative future of belowground ecology and ecologists. *New Phytologist* 229:3058–64.

201. Defrenne CE, Childs J, Fernandez CW, Taggart M, Nettles WR, et al. 2021. High-resolution minirhizotrons advance our understanding of root-fungal dynamics in an experimentally warmed peatland. *Plants, People, Planet* 3:640–52.

202. DeMars BG, Boerner RE. 1995. Mycorrhizal status of *Deschampsia antarctica* in the Palmer Station area, Antarctica. *Mycologia* 87:451–3.

203. Di Fossalunga AS, Lipuma J, Venice F, Dupont L, Bonfante P. 2017. The endobacterium of an arbuscular mycorrhizal fungus modulates the expression of its toxin–antitoxin systems during the life cycle of its host. *The ISME Journal* 11:2394–8.

204. Dickie IA, Alexander I, Lennon S, Opik M, Selosse MA, et al. 2015. Evolving insights to understanding mycorrhizas. *New Phytologist* 205:1369–74.

205. Dickie IA, Bufford JL, Cobb RC, Desprez-Loustau ML, Grelet G, et al. 2017. The emerging science of linked plant–fungal invasions. *New Phytologist* 215:1314–32.

206. Dickie IA, Xu B, Koide RT. 2002. Vertical niche differentiation of ectomy-corrhizal hyphae in soil as shown by T-RFLP analysis. *New Phytologist* 156:527–35.

207. Dickson S. 2004. The Arum–Paris continuum of mycorrhizal symbioses. *New Phytologist* 163:187–200.

208. Dominguez LS, Sérsic A. 2004. The southernmost myco-heterotrophic plant, Arachnitis uniflora: Root morphology and anatomy. *Mycologia* 96: 1143–51.

209. Dominik T. 1951. Badanie mykotrofizmu roślinności wydm nadmorskich i śródlądowych. *Acta Societatis Botanicorum Poloniae* 21:125–64.

210. Downing JL, Liu H, McCormick MK, Arce J, Alonso D, Lopez-Perez J. 2020. Generalized mycorrhizal interactions and fungal enemy release drive range expansion of orchids in southern Florida. *Ecosphere* 11:e03228.

211. Driver JD, Holben WE, Rillig MC. 2005. Characterization of glomalin as a hyphal wall component of arbuscular mycorrhizal fungi. *Soil Biology and Biochemistry* 37:101–6.

212. Droser ML, Gehling JG, Dzaugis ME, Kennedy MJ, Rice D, Allen MF. 2014. A new Ediacaran fossil with a novel sediment displacive life habit. *Journal of Paleontology* 88:145–51.

213. Duddridge J, Malibari A, Read D. 1980. Structure and function of mycorrhizal rhizomorphs with special reference to their role in water transport. *Nature* 287:834–6.

214. Egerton-Warburton L, Graham R, Hubbert K. 2003. Spatial variability in mycorrhizal hyphae and nutrient and water availability in a soil-weathered bedrock profile. *Plant and Soil* 249:331–42.

215. Egerton-Warburton LM, Allen EB. 2000. Shifts in arbuscular mycorrhizal communities along an anthropogenic nitrogen deposition gradient. *Ecological Applications* 10:484–96.

216. Egerton-Warburton LM, Graham RC, Allen EB, Allen MF. 2001. Reconstruction of the historical changes in mycorrhizal fungal communities under anthropogenic nitrogen deposition. *Proceedings of the Royal Society B–Biological Sciences* 268:2479–84.

217. Egerton-Warburton LM, Johnson NC, Allen EB. 2007. Mycorrhizal community dynamics following nitrogen fertilization: A cross-site test in five grasslands. *Ecological Monographs* 77:527–44.

218. Egerton-Warburton LM, Querejeta JI, Allen MF. 2008. Efflux of hydraulically lifted water from mycorrhizal fungal hyphae during imposed drought. *Plant Signaling & Behavior* 3:68–71.

219. Egler FE. 1954. Vegetation science concepts I. Initial floristic composition, a factor in old-field vegetation development with 2 figs. *Vegetatio* 4:412–17.

220. Ek H, Andersson S, Söderström B. 1997. Carbon and nitrogen flow in silver birch and Norway spruce connected by a common mycorrhizal mycelium. *Mycorrhiza* 6:465–7.

221. Ellenberg D, Mueller-Dombois D. 1974. *Aims and Methods of Vegetation Ecology*. New York: Wiley.

222. Elser JJ, Bracken ME, Cleland EE, Gruner DS, Harpole WS, et al. 2007. Global analysis of nitrogen and phosphorus limitation of primary producers in freshwater, marine and terrestrial ecosystems. *Ecology Letters* 10:1135–42.

223. Enloe HA, Graham RC, Sillett SC. 2006. Arboreal histosols in old-growth redwood forest canopies, northern California. *Soil Science Society of America Journal* 70:408–18.

224. Espeleta J, Clark DA. 2007. Multi-scale variation in fine-root biomass in a tropical rain forest: A seven-year study. *Ecological Monographs* 77:377–404.

225. Espiñeira J, Novo Uzal E, Gómez Ros L, Carrión J, Merino F, et al. 2011. Distribution of lignin monomers and the evolution of lignification among lower plants. *Plant Biology* 13:59–68.

226. Fang F, Wang C, Wu F, Tang M, Doughty R. 2020. Arbuscular mycorrhizal fungi mitigate nitrogen leaching under poplar seedlings. *Forests* 11:325.

227. Fasolo-Bonfante P, Brunel A. 1972. Caryological features in a mycorrhizal fungus: Tuber melanosporum Vitt. *Allionia* 18:5–11.

228. Favre-Godal Q, Gourguillon L, Lordel-Madeleine S, Gindro K, Choisy P. 2020. Orchids and their mycorrhizal fungi: An insufficiently explored relationship. *Mycorrhiza* 30:5–22.

229. Fenn ME, Poth MA, Johnson DW. 1996. Evidence for nitrogen saturation in the San Bernardino Mountains in southern California. *Forest Ecology and Management* 82:211–30.

230. Fernandez CW, Koide RT. 2012. The role of chitin in the decomposition of ectomycorrhizal fungal litter. *Ecology* 93:24–8.

231. Fernandez CW, Koide RT. 2014. Initial melanin and nitrogen concentrations control the decomposition of ectomycorrhizal fungal litter. *Soil Biology and Biochemistry* 77:150–7.

232. Fernandez-Bou AS, Dierick D, Allen MF, Harmon TC. 2020. Precipitation–drainage cycles lead to hot moments in soil carbon dioxide dynamics in a Neotropical wet forest. *Global Change Biology* 26:5303–19.

233. Fernandez-Bou AS, Dierick D, Swanson AC, Allen MF, Alvarado AGF, et al. 2019. The role of the ecosystem engineer, the Leaf-Cutter Ant Atta cephalotes, on soil CO_2 dynamics in a wet tropical rainforest. *Journal of Geophysical Research – Biogeosciences* 124:260–73.

234. Fetcher N, Oberbauer SF, Chazdon RL, McDade LA, Bawa KS, et al. 1994. Physiological ecology of plants, pp. 128–41. In LA McDade, KS Bawa, HA Hespenheide, GS Hartshorn, eds. *La Selva: Ecology and Natural History of a Neotropical Rain Forest.* Chicago: University of Chicago Press.

235. Field CB, Chapin III FS, Matson PA, Mooney HA. 1992. Responses of terrestrial ecosystems to the changing atmosphere: A resource-based approach. *Annual Review of Ecology and Systematics* 23:201–35.

236. Field KJ, Pressel S, Duckett JG, Rimington WR, Bidartondo MI. 2015. Symbiotic options for the conquest of land. *Trends in Ecology & Evolution* 30:477–86.

237. Field KJ, Rimington WR, Bidartondo MI, Allinson KE, Beerling DJ, et al. 2016. Functional analysis of liverworts in dual symbiosis with Glomeromycota and Mucoromycotina fungi under a simulated Palaeozoic CO2 decline. *ISME Journal* 10:1514–26.

238. Finlay R, Read D. 1986. The structure and function of the vegetative mycelium of ectomycorrhizal plants: I. Translocation of 14C-labelled carbon between plants interconnected by a common mycelium. *New Phytologist* 103:143–56.

239. Finlay R. Söderström B. 1992. Mycorrhiza and carbon flow to the soil, pp. 134–62. In MF Allen, ed. *Mycorrhizal Functioning.* New York: Chapman and Hall Press.

240. Fisher JB, Sitch S, Malhi Y, Fisher RA, Huntingford C, Tan SY. 2010. Carbon cost of plant nitrogen acquisition: A mechanistic, globally applicable model of plant nitrogen uptake, retranslocation, and fixation. *Global Biogeochemical Cycles* 24.

241. Fitter A. 1977. Influence of mycorrhizal infection on competition for phosphorus and potassium by two grasses. *New Phytologist* 79:119–25.

242. Fitter A. 1986. Effect of benomyl on leaf phosphorus concentration in alpine grasslands: A test of mycorrhizal benefit. *New Phytologist* 103:767–76.

243. Fitter A, Graves J, Watkins N, Robinson D, Scrimgeour C. 1998. Carbon transfer between plants and its control in networks of arbuscular mycorrhizas. *Functional Ecology* 12:406–12.

244. Fitter A, Heinemeyer A, Staddon P. 2000. The impact of elevated CO_2 and global climate change on arbuscular mycorrhizas: A mycocentric approach. *New Phytologist* 147:179–87.

245. Fitter A, Sanders I. 1992. Interactions with the soil fauna, pp. 333–56. In MF Allen, ed. *Mycorrhizal Functioning.* New York: Chapman and Hall Press.

246. Fitter AH. 2006. What is the link between carbon and phosphorus fluxes in arbuscular mycorrhizas? A null hypothesis for symbiotic function. *New Phytologist* 172:3–6.

247. Fogel R, Hunt G. 1979. Fungal and arboreal biomass in a western Oregon Douglas-fir ecosystem: Distribution patterns and turnover. *Canadian Journal of Forest Research* 9:245–56.

248. Fogel R, Hunt G. 1983. Contribution of mycorrhizae and soil fungi to nutrient cycling in a Douglas-fir ecosystem. *Canadian Journal of Forest Research* 13:219–32.

249. Fowler D, Pitcairn CER, Sutton MA, Flechard C, Loubet B, et al. 1998. The mass budget of atmospheric ammonia in woodland within 1 km of livestock buildings. *Environmental Pollution* 102:343–8.

250. Fox T, Comerford N. 1990. Low-molecular-weight organic acids in selected forest soils of the southeastern USA. *Soil Science Society of America Journal* 54:1139–44.

251. Fox T, Comerford N. 1992. Influence of oxalate loading on phosphorus and aluminum solubility in spodosols. *Soil Science Society of America Journal* 56:290–4.

252. Francis R, Finlay R, Read D. 1986. Vesicular arbuscular mycorrhiza in natural vegetation systems: IV. Transfer of nutrients in inter- and intra-specific combinations of host plants. *New Phytologist* 102:103–11.

253. Francis R, Read D. 1984. Direct transfer of carbon between plants connected by vesicular–arbuscular mycorrhizal mycelium. *Nature* 307:53–6.

254. Francl LJ, Dropkin VH. 1985. Glomus fasciculatum, a weak pathogen of Heterodera glycines *Journal of Nematology* 17:470–5.

255. Frank A. 1894. Die Bedeutung der Mykorrhizapilze für die gemeine Kiefer. *Forstwissenschaftliches Centralblatt* 16:185–90.

256. Frank B. 1885. Über die auf Wurzelsymbiose beruhende Ernährung gewisser Bäume durch unterirdische Pilze. *Berichte der Deutsche Botanische Gesellschaft* 3:128–45.

257. Frank B. 1887. Über neue Mycorhiza-Formen. *Berichte der Deutsche Botanische Gesellschaft* 5:395–422.

258. Frank B. 1888. Über die physiologische Bedeutung der Mycorhiza. *Berichte der Deutsche Botanische Gesellschaft* 6:248–68.

259. Frank B. 1891. Über die auf Verdauung von Pilzen abzielende Symbiose der mit endotrophen Mykorhizen begabten Pflanzen, sowie der Leguminosen und Erlen. *Berichte der Deutsche Botanische Gesellschaft* 9:244–53.

260. Frey SD. 2019. Mycorrhizal fungi as mediators of soil organic matter dynamics. *Annual Review of Ecology, Evolution, and Systematics* 50:237–59.

261. Friedman M. 2016. Mushroom polysaccharides: Chemistry and antiobesity, antidiabetes, anticancer, and antibiotic properties in cells, rodents, and humans. *Foods* 5:80.

262. Friedmann E, Ocampo-Friedmann R. 1984. The Antarctic cryptoendolithic ecosystem: Relevance to exobiology. *Origins of Life* 14:771–6.

263. Friese C, Allen M. 1993. The interaction of harvester ants and vesicular–arbuscular mycorrhizal fungi in a patchy semi-arid environment: The effects of mound structure on fungal dispersion and establishment. *Functional Ecology* 7:13–20.

264. Friese CF, Allen MF. 1991. The spread of VA mycorrhizal fungal hyphae in the soil: Inoculum types and external hyphal architecture. *Mycologia* 83:409–18.

265. Friese CF, Allen MF. 1991. Tracking the fates of exotic and local VA mycorrhizal fungi: Methods and patterns. *Agriculture, Ecosystems & Environment* 34:87–96.

266. Friese CF, Allen MF, Martin R, Vanalfen NK. 1992. Temperature and structural effects on transfer of double-stranded-RNA among isolates of the chestnut blight fungus (Cryphonectria–Parasitica). *Applied and Environmental Microbiology* 58:2066–70.

267. Frydman I. 1957. Mykotrofizm roślinności pokrywającej gruzy i ruiny domów Wrocławia [Mycotrophic properties of plants growing on ruins in Wrocław]. *Acta Societatis Botanicorum Poloniae* 26:45–60.

268. Gadd GM. 1993. Interactions of fungi with toxic metals *New Phytologist* 124:25–60.

269. Gadgil RL, Gadgil P. 1971. Mycorrhiza and litter decomposition. *Nature* 233:133.

270. Gadgil RL, Gadgil PD. 1975. Suppression of litter decomposition by mycorrhizal roots of Pinus radiata. *New Zealand Journal of Forest Science* 5:35–41.

271. Gadkar V, Rillig MC. 2006. The arbuscular mycorrhizal fungal protein glomalin is a putative homolog of heat shock protein 60. *FEMS Microbiology Letters* 263:93–101.

272. Gallaud I. 1905. Études sur les mycorrhizes endophytes. *Revue Genéral de Botanique* 17:5–500.

273. Gan T, Luo T, Pang K, Zhou C, Zhou G, et al. 2021. Cryptic terrestrial fungus-like fossils of the early Ediacaran Period. *Nature Communications* 12:1–12.

274. Garcia K, Delaux PM, Cope KR, Ane JM. 2015. Molecular signals required for the establishment and maintenance of ectomycorrhizal symbioses. *New Phytologist* 208:79–87.

275. Gardes M, Bruns TD. 1993. ITS primers with enhanced specificity for basidiomycetes: Application to the identification of mycorrhizae and rusts. *Molecular Ecology* 2:113–18.

276. Gebauer G, Dietrich P. 1993. Nitrogen isotope ratios in different compartments of a mixed stand of spruce, larch and beech trees and of understorey vegetation including fungi. *Isotopes in Environmental and Health Studies* 29:35–44.

277. Gehring C, Sevanto S, Patterson A, Ulrich DE, Kuske C. 2020. Ectomycorrhizal and dark septate fungal associations of pinyon pine are differentially affected by experimental drought and warming. *Frontiers in Plant Science* 11:1570.

278. Gehring CA, Theimer TC, Whitham TG, Keim P. 1998. Ectomycorrhizal fungal community structure of pinyon pines growing in two environmental extremes. *Ecology* 79:1562–72.

279. Gehring CA, Whitham TG. 1991. Herbivore-driven mycorrhizal mutualism in insect-susceptible pinyon pine. *Nature* 353:556–7.

280. Gerdemann J. 1955. Relation of a large soil-borne spore to phycomycetous mycorrhizal infections. *Mycologia* 47:619–32.

281. Gerdemann J, Trappe J. 1974. *The Endogonaceae in the Pacific Northwest.* Mycologia Memoir, No. 5. New York: New York Botanical Garden.

282. Gerdemann JW. 1964. The effect of mycorrhiza on the growth of maize. *Mycologia* 56:342–9.

283. Gherbi H, Markmann K, Svistoonoff S, Estevan J, Autran D, et al. 2008. SymRK defines a common genetic basis for plant root endosymbioses with arbuscular mycorrhiza fungi, rhizobia, and Frankiabacteria. *Proceedings of the National Academy of Sciences* 105:4928–32.

284. Giesemann P, Eichenberg D, Stöckel M, Seifert LF, Gomes SI, et al. 2020. Dark septate endophytes and arbuscular mycorrhizal fungi (Paris-morphotype) affect the stable isotope composition of "classically" non-mycorrhizal plants. *Functional Ecology* 34:2453–66.

285. Gildon A, Tinker PB. 1981. A heavy-metal tolerant strain of a mycorrhizal fungus *Transactions of the British Mycological Society* 77:648–9.

286. Glassman SI, Levine CR, DiRocco AM, Battles JJ, Bruns TD. 2016. Ectomycorrhizal fungal spore bank recovery after a severe forest fire: Some like it hot. *The ISME Journal* 10:1228–39.

287. Glassman SI, Lubetkin KC, Chung JA, Bruns TD. 2017. The theory of island biogeography applies to ectomycorrhizal fungi in subalpine tree "islands" at a fine scale. *Ecosphere* 8:e01677.

288. Glassman SI, Peay KG, Talbot JM, Smith DP, Chung JA, et al. 2015. A continental view of pine-associated ectomycorrhizal fungal spore banks: A quiescent functional guild with a strong biogeographic pattern. *New Phytologist* 205:1619–31.

289. Godbold DL, Hoosbeek MR, Lukac M, Cotrufo MF, Janssens IA, et al. 2006. Mycorrhizal hyphal turnover as a dominant process for carbon input into soil organic matter. *Plant and Soil* 281:15–24.

290. Gomez-Pompa, A., Allen MF, Fedick S, Jimenez-Osornio JJ. 2003. *Lowland Maya Area: Three Millennia at the Human–Wildland Interface*. New York: Haworth Press.

291. González-Chicas E, Cappello S, Cifuentes J. 2019. New records of Boletales (Basidiomycota) in a tropical oak forest from Mexican Southeast. *Botanical Sciences* 97:423–32.

292. Gooday GW. 1994. Physiology of microbial degradation of chitin and chitosan, pp. 279–312. In R Ratledge, ed. *Biochemistry of Microbial Degradation*. Dordrecht: Springer.

293. Goode LK, Allen MF. 2008. The impacts of Hurricane Wilma on the epiphytes of El Eden Ecological Reserve, Quintana Roo, Mexico. *Journal of the Torrey Botanical Society* 135:377–87.

294. Goode LK, Allen MF. 2009. Seed germination conditions and implications for establishment of an epiphyte, Aechmea bracteata (Bromeliaceae). *Plant Ecology* 204:179–88.

295. Goss MJ, Carvalho M, Brito I. 2017. *Functional Diversity of Mycorrhiza and Sustainable Agriculture: Management to Overcome Biotic and Abiotic Stresses*. London: Academic Press.

296. Goss RW. 1960. Mycorrhizae of Ponderosa Pine in Nebraska Grassland Soils. Research Bulletin. Nebraska Agricultural Experiment Station, 192.

297. Graham RC, Rossi AM, Hubbert KR. 2010. Rock to regolith conversion: Producing hospitable substrates for terrestrial ecosystems. *GSA Today* 20:2–9.

298. Graustein WC, Cromack K, Sollins P. 1977. Calcium oxalate: Occurrence in soils and effect on nutrient and geochemical cycles. *Science* 198:1252–4.

299. Greenall JM. 1963. The mycorrhizal endophytes of Griselinia littoralis (Cornaceae). *New Zealand Journal of Botany* 1:389–400.

300. Griffin DM. 1972. *Ecology of Soil Fungi.* Syracuse: Syracuse University Press.

301. Griffin DM. 1981. *Fungal Physiology* New York: John Wiley & Sons.

302. Grime J. 1979. *Plant Strategies and Vegetation Processes* New York: John Wiley and Sons.

303. Grime JP, Mackey JML, Hillier SH, Read DJ. 1987. Floristic diversity in a model system using experimental microcosms. *Nature* 328:420–2.

304. Gupta RS. 1995. Phylogenetic analysis of the 90 kD heat shock family of protein sequences and an examination of the relationship among animals, plants, and fungi species. *Molecular Biology and Evolution* 12:1063–73.

305. Hacskaylo E. 1967. Mycorrhizae: Indispensible invasions by fungi. *Agricultural Science Review* 5:13–20.

306. Hacskaylo E, Vozzo J. 1967. Inoculation of *Pinus caribaea* with pure cultures of mycorrhizal fungi in Puerto Rico. *Proceedings of the XVI IUFRO Congress,* Munich, 5:139–48.

307. Hadley G. 1970. Non-specificity of symbotic infection in orchid mycorrhiza. *New Phytologist* 69:1015–23.

308. Hadley G. 1984. Uptake of [^{14}C] glucose by asymbiotic and mycorrhizal orchid protocorms. *New Phytologist* 96:263–73.

309. Hall I. 1977. Species and mycorrhizal infections of New Zealand endogonaceae. *Transactions of the British Mycological Society* 68:341–56.

310. Hall SJ, McDowell WH, Silver WL. 2013. When wet gets wetter: Decoupling of moisture, redox biogeochemistry, and greenhouse gas fluxes in a humid tropical forest soil. *Ecosystems* 16:576–89.

311. Hallsworth S, Dore A, Bealey W, Dragosits U, Vieno M, et al. 2010. The role of indicator choice in quantifying the threat of atmospheric ammonia to the 'Natura 2000' network. *Environmental Science & Policy* 13:671–87.

312. Halvorson JJ, Smith JL. 2009. Carbon and nitrogen accumulation and microbial activity in Mount St. Helens pyroclastic substrates after 25 years. *Plant and Soil* 315:211–28.

313. Halvorson JJ, Smith JL, Kennedy AC. 2005. Lupine effects on soil development and function during early primary succession at Mount St. Helens, pp. 243–54. In VH Dale, FJ Swanson, CM Crisafulli, eds. *Ecological Responses to the 1980 Eruptions of Mount St. Helens.* New York: Springer-Verlag.

314. Hansen DR, Van Alfen NK, Gillies K, Powell WA. 1985. Naked dsRNA associated with hypovirulence of Endothia parasitica is packaged in fungal vesicles. *Journal of General Virology* 66:2605–14.

315. Harari YN. 2015. *Sapiens: A Brief History of Humankind.* First U.S. edition. New York: Harper.

316. Hardie K. 1985. The effect of removal of extraradical hyphae on water uptake by vesicular-arbuscular mycorrhizal plants. *New Phytologist* 101:677–84.

317. Hardie K, Leyton L. 1981. The influence of vesicular–arbuscular mycorrhiza on growth and water relations of red clover in phosphate deficient soil. *New Phytologist* 89:599–608.

318. Harley J. 1968. Presidential address: Fungal symbiosis. *Transactions of the British Mycological Society* 51:1–11.

319. Harley JL. 1959. *The Biology of Mycorrhiza*. London: Leonard Hill.

320. Harley JL. 1969. *The Biology of Mycorrhiza*, 2nd ed. London: Leonard Hill.

321. Harris D, Paul EA. 1987. Carbon requirements of vesicular–arbuscular mycorrhizae, pp. 93–105. In GE Safir, ed. *Ecophysiology of V A Mycorrhizal Plants*. Boca Raton, FL: CRC Press.

322. Harrison MJ, Dewbre GR, Liu JY. 2002. A phosphate transporter from Medicago truncatula involved in the acquisiton of phosphate released by arbuscular mycorrhizal fungi. *Plant Cell* 14:2413–29.

323. Hartig R. 1874. *Important Diseases of Forest Trees: Contributions to Mycology and Phytopathology for Botanists and Foresters*. Phytopathology Classics. Minneapolis: APS Publications.

324. Hartig T. 1840. *Vollständige Naturgeschichte der forstlichen Culturpflanzen Deutschlands*. Berlin: Förstner'sche Verlagsbuchhandlung.

325. Hartnett DC, Wilson GW. 1999. Mycorrhizae influence plant community structure and diversity in tallgrass prairie. *Ecology* 80:1187–95.

326. Hartnett DC, Wilson GW. 2002. The role of mycorrhizas in plant community structure and dynamics: Lessons from grasslands. *Plant and Soil* 244:319–31.

327. Haselwandter K, Häninger G, Ganzera M, Haas H, Nicholson G, Winkelmann G. 2013. Linear fusigen as the major hydroxamate siderophore of the ectomycorrhizal Basidiomycota *Laccaria laccata* and *Laccaria bicolor*. *Biometals* 26:969–79.

328. Haselwandter K, Read D. 1982. The significance of a root-fungus association in two *Carex* species of high-alpine plant communities. *Oecologia* 53:352–4.

329. Hasselquist N, Germino MJ, McGonigle T, Smith WK. 2005. Variability of *Cenococcum* colonization and its ecophysiological significance for young conifers at alpine–treeline. *New Phytologist* 165:867–73.

330. Hasselquist NJ, Douhan GW, Allen MF. 2011. First report of the ectomycorrhizal status of boletes on the Northern Yucatan Peninsula, Mexico determined using isotopic methods. *Mycorrhiza* 21:465–71.

331. Hasselquist NJ, Högberg P. 2014. Dosage and duration effects of nitrogen additions on ectomycorrhizal sporocarp production and functioning: An example from two N-limited boreal forests. *Ecology and Evolution* 4:3015–26.

332. Hasselquist NJ, Metcalfe DB, Marshall JD, Lucas RW, Högberg P. 2016. Seasonality and nitrogen supply modify carbon partitioning in understory vegetation of a boreal coniferous forest. *Ecology* 97:671–83.

333. Hasselquist NJ, Santiago LS, Allen MF. 2010. Belowground nitrogen dynamics in relation to hurricane damage along a tropical dry forest chronosequence. *Biogeochemistry* 98:89–100.

334. Hatch A. 1937. The physical basis of mycotrophy in *Pinus*. *Black Rock Forest Bulletin* 6:1–168.

335. Hattingh MJ, Gray LE, Gerdemann JW. 1973. Uptake and translocation of P-32-labelled phosphate to onion roots by endomycorrhizal fungi *Soil Science* 116:383–7.

336. Hayman D. 1974. Plant growth responses to vesicular arbuscular mycorrhiza: VI. Effect of light and temperature. *New Phytologist* 73:71–80.

337. Hayman D. 1975. The occurrence of mycorrhiza in crops as affected by soil fertility, pp. 495–509. In FE Sanders, B Mosse, PB Tinker, eds. *Endomycorrhizas: Proceedings of a Symposium Held at the University of Leeds, 22–25 July 1974.* New York: Academic Press.

338. He XH, Bledsoe CS, Zasoski RJ, Southworth D, Horwath WR. 2006. Rapid nitrogen transfer from ectomycorrhizal pines to adjacent ectomycorrhizal and arbuscular mycorrhizal plants in a California oak woodland. *New Phytologist* 170:143–51.

339. Heffernan JB, Soranno PA, Angilletta Jr MJ, Buckley LB, Gruner DS, et al. 2014. Macrosystems ecology: Understanding ecological patterns and processes at continental scales. *Frontiers in Ecology and the Environment* 12:5–14.

340. Heidel B, Jones G. 2006. Botanical and Ecological Characteristics of Fens in the Medicine Bow Mountains, Medicine Bow National Forest. Laramie: Wyoming Natural Diversity Database, University of Wyoming. Available from: www .uwyo.edu/wyndd/_files/docs/reports/wynddreports/u06hei06wyus.pdf. Last accessed November 16, 2021.

341. van der Heijden MGA, Klironomos JN, Ursic M, Moutoglis P, Streitwolf-Engel R, et al. 1998. Mycorrhizal fungal diversity determines plant biodiversity, ecosystem variability and productivity. *Nature* 396:69–72.

342. Heinemeyer A, Hartley IP, Evans SP, Carreira de La Fuente JA, Ineson P. 2007. Forest soil CO_2 flux: Uncovering the contribution and environmental responses of ectomycorrhizas. *Global Change Biology* 13:1786–97.

343. Helber N, Wippel K, Sauer N, Schaarschmidt S, Hause B, Requena N. 2011. A versatile monosaccharide transporter that operates in the arbuscular mycorrhizal fungus *Glomus* sp is crucial for the symbiotic relationship with plants. *Plant Cell* 23:3812–23.

344. Helm D, Allen E. 1995. Vegetation chronosequence near Exit Glacier, Kenai Fjords National Park, Alaska, USA. *Arctic and Alpine Research* 27:246–57.

345. Henkel TW, Aime MC, Chin MML, Miller SL, Vilgalys R, Smith ME. 2012. Ectomycorrhizal fungal sporocarp diversity and discovery of new taxa in Dicymbe monodominant forests of the Guiana Shield. *Biodiversity and Conservation* 21:2195–220.

346. Hernandez RR, Allen MF. 2013. Diurnal patterns of productivity of arbuscular mycorrhizal fungi revealed with the Soil Ecosystem Observatory. *New Phytologist* 200:547–57.

347. Hetrick BAD, Wilson GWT, Todd TC. 1996. Mycorrhizal response in wheat cultivars: Relationship to phosphorus. *Canadian Journal of Botany* 74:19–25.

348. Hetrick BD, Kitt DG, Wilson GT. 1986. The influence of phosphorus fertilization, drought, fungal species, and nonsterile soil on mycorrhizal growth response in tall grass prairie plants. *Canadian Journal of Botany* 64:1199–203.

349. Hetrick BD, Wilson GT, Hartnett D. 1989. Relationship between mycorrhizal dependence and competitive ability of two tallgrass prairie grasses. *Canadian Journal of Botany* 67:2608–15.

350. Hetrick BD, Wilson GT, Schwab A. 1988. Effects of soil microorganisms on mycorrhizal contribution to growth of big bluestem grass in non-sterile soil. *Soil Biology and Biochemistry* 20:501–7.

351. Heylighen F. 2008. Complexity and self-organization. In MJ Bates, MN Maack, eds. *Encyclopedia of Library and Information Sciences*. London: Taylor & Francis. Available from: http://pcp.vub.ac.be/Papers/ELIS-complexity.pdf. Last accessed November 16, 2021.

352. Hickson LD. 1993. The effects of vesicular-arbuscular mycorrhizal fungi on the light harvesting, gas exchange, and architecture of sagebrush (Artemisia tridentata ssp. tridentata). MS thesis. San Diego State University.

353. Hiiesalu I, Partel M, Davison J, Gerhold P, Metsis M, et al. 2014. Species richness of arbuscular mycorrhizal fungi: Associations with grassland plant richness and biomass. *New Phytologist* 203:233–44.

354. Hill PW, Broughton R, Bougoure J, Havelange W, Newsham KK, et al. 2019. Angiosperm symbioses with non-mycorrhizal fungal partners enhance N acquisition from ancient organic matter in a warming maritime Antarctic. *Ecology Letters* 22:2111–9.

355. Hobbie EA, Agerer R. 2010. Nitrogen isotopes in ectomycorrhizal sporocarps correspond to belowground exploration types. *Plant and Soil* 327:71–83.

356. Hobbie EA, Högberg P. 2012. Nitrogen isotopes link mycorrhizal fungi and plants to nitrogen dynamics. *New Phytologist* 196:367–82.

357. Hobbie EA, Macko SA, Shugart HH. 1999. Insights into nitrogen and carbon dynamics of ectomycorrhizal and saprotrophic fungi from isotopic evidence. *Oecologia* 118:353–60.

358. Hobbie JE, Hobbie EA. 2006. 15N in symbiotic fungi and plants estimates nitrogen and carbon flux rates in Arctic tundra. *Ecology* 87:816–22.

359. Hodge A. 2017. Accessibility of inorganic and organic nutrients for mycorrhizas, pp. 129–48. In NC Johnson, C Gehring, J Jansa, eds. *Mycorrhizal Mediation of Soil: Fertility, Structure, and Carbon Storage*. Amsterdam: Elsevier Press.

360. Hodge A, Campbell CD, Fitter AH. 2001. An arbuscular mycorrhizal fungus accelerates decomposition and acquires nitrogen directly from organic material. *Nature* 413:297–9.

361. Hoeksema JD, Chaudhary VB, Gehring CA, Johnson NC, Karst J, et al. 2010. A meta-analysis of context-dependency in plant response to inoculation with mycorrhizal fungi. *Ecology Letters* 13:394–407.

362. Högberg MN, Briones MJ, Keel SG, Metcalfe DB, Campbell C, et al. 2010. Quantification of effects of season and nitrogen supply on tree below-ground carbon transfer to ectomycorrhizal fungi and other soil organisms in a boreal pine forest. *New Phytologist* 187:485–93.

363. Högberg P, Högberg MN, Quist ME, Ekblad A, Näsholm T. 1999. Nitrogen isotope fractionation during nitrogen uptake by ectomycorrhizal and non-mycorrhizal *Pinus sylvestris*. *New Phytologist* 142:569–76.

364. Högberg P, Högbom L, Schinkel H, Högberg M, Johannisson C, Wallmark H. 1996. 15 N abundance of surface soils, roots and mycorrhizas in profiles of European forest soils. *Oecologia* 108:207–14.

365. Högberg P, Nasholm T, Franklin O, Hogberg MN. 2017. Tamm Review: On the nature of the nitrogen limitation to plant growth in Fennoscandian boreal forests. *Forest Ecology and Management* 403:161–85.

366. Högberg P, Nordgren A, Buchmann N, Taylor AFS, Ekblad A, et al. 2001. Large-scale forest girdling shows that current photosynthesis drives soil respiration. *Nature* 411:789–92.

367. Huang CH, Sun RR, Hu Y, Zeng LP, Zhang N, et al. 2016. Resolution of brassicaceae phylogeny using nuclear genes uncovers nested radiations and supports convergent morphological evolution. *Molecular Biology and Evolution* 33:394–412.

368. Huante P, Rincon E, Chapin FS. 1998. Effect of changing light availability on nutrient foraging in tropical deciduous tree-seedlings. *Oikos* 82:449–58.

369. Hubau W, Lewis SL, Phillips OL, Affum-Baffoe K, Beeckman H, et al. 2020. Asynchronous carbon sink saturation in African and Amazonian tropical forests. *Nature* 579:80–7.

370. Hughes NM, Carpenter KL, Cook DK, Keidel TS, Miller CN, et al. 2015. Effects of cumulus clouds on microclimate and shoot-level photosynthetic gas exchange in Picea engelmannii and Abies lasiocarpa at treeline, Medicine Bow Mountains, Wyoming, USA. *Agricultural and Forest Meteorology* 201: 26–37.

371. von Humboldt A, Bonpland A. 1807. *Essai sur la géographie des plantes.* Paris: Fr. Schoell.

372. Humphreys CP, Franks PJ, Rees M, Bidartondo MI, Leake JR, Beerling DJ. 2010. Mutualistic mycorrhiza-like symbiosis in the most ancient group of land plants. *Nature Communications* 1:1–7.

373. Iversen CM, Murphy MT, Allen MF, Childs J, Eissenstat DM, et al. 2012. Advancing the use of minirhizotrons in wetlands. *Plant and Soil* 352:23–39.

374. Iyer JG, Corey R, Wilde S. 1980. Mycorrhizae: Facts and fallacies. *Journal of Arboriculture* 6:213–20.

375. Jackson R, Moore L, Hoffmann W, Pockman W, Linder C. 1999. Ecosystem rooting depth determined with caves and DNA. *Proceedings of the National Academy of Sciences* 96:11387–92.

376. Jacobs R. 2019. *The Truffle Underground.* New York: Clarkson Potter.

377. Janos DP. 1980. Mycorrhizae influence tropical succession. *Biotropica* 12:56–64.

378. Janos DP. 1980. Vesicular–arbuscular mycorrhizae affect lowland tropical rain forest plant growth. *Ecology* 61:151–62.

379. Janos DP, Sahley CT, Emmons LH. 1995. Rodent dispersal of vesicular–arbuscular mycorrhizal fungi in Amazonian Peru. *Ecology* 76:1852–8.

380. Jansa J, Treseder K. 2017. Introduction: Mycorrhizas and the carbon cycle, pp. 343–55. In NC Johnson, C Gehring, J Jansa, eds. *Mycorrhizal Mediation of Soil: Fertility, Structure, and Carbon Storage.* Amsterdam: Elsevier Press.

381. Janse J. 1897. Les endophytes radicaux de quelques plantes javanaises. *Annales du Jardin Botanique de Buitenzorg* 14:53–201.

382. Jaroszewicz B, Cholewińska O, Gutowski JM, Samojlik T, Zimny M, Latałowa M. 2019. Białowieża forest: A relic of the high naturalness of European forests. *Forests* 10:849.

383. Jiménez-Osornio JJ, Rorive VM, Gomez-Pompa A, Tiessen H, Allen MF. 2008. Thinking outside the box: Tropical conservation in both protected areas and the surrounding matrix, pp. 134–40. In H Tiessen, JWB Stewart, eds.

Applying Ecological Knowledge to Landuse Decisions. Proceedings of the International Workshop held in Costa Rica by the IAI, SCOPE and IICA. Available from: www.iai.int/index.php?option=com_content&view=article& id=24:scientific-publications&catid=24:scientific-publications&Itemid=38. Last accessed November 16, 2021.

384. Jimenez-Osornio JJ, Ruenes-Morales MdR, Aké-Gómez A. Mayan. 2003. Home gardens: Sites for in situ conservation of agricultural diversity, pp. 9–15. In DI Jarvis, R Seilla-Panizo, JL Chávez-Servia, T Hodgkin, eds. *Proceedings of the Seed Systems and Crop Genetic Diversity On-Farm Proceedings of a Workshop, Pucallpa, Peru, 16–20 September, 2003, 2005*. Rome, Italy: International Plant Genetic Resources Institute (IPGRI).

385. Johnson D, Gilbert L. 2015. Interplant signalling through hyphal networks. *New Phytologist* 205:1448–53.

386. Johnson NC. 2010. Resource stoichiometry elucidates the structure and function of arbuscular mycorrhizas across scales. *New Phytologist* 185:631–47.

387. Johnson NC, Graham JH, Smith F. 1997. Functioning of mycorrhizal associations along the mutualism–parasitism continuum. *New Phytologist* 135: 575–85.

388. Johnson NC, Hoeksema JD, Bever JD, Chaudhary VB, Gehring C, et al. 2006. From Lilliput to Brobdingnag: Extending models of mycorrhizal function across scales. *BioScience* 56:889–900.

389. Johnson NC, Jansa J. 2017. Mycorrhizas: At the interface of biological, soil, and earth sciences, pp. 1–6. In NC Johnson, C Gehring, J Jansa, eds. *Mycorrhizal Mediation of Soil: Fertility, Structure, and Carbon Storage*. Amsterdam: Elsevier Press.

390. Jones EI, Afkhami ME, Akçay E, Bronstein JL, Bshary R, et al. 2015. Cheaters must prosper: Reconciling theoretical and empirical perspectives on cheating in mutualism. *Ecology Letters* 18:1270–84.

391. Jongmans A, Van Breemen N, Lundström U, Van Hees P, Finlay R, et al. 1997. Rock-eating fungi. *Nature* 389:682–3.

392. Jumpponen A, Mattson KG, Trappe JM. 1998. Mycorrhizal functioning of Phialocephala fortinii with Pinus contorta on glacier forefront soil: Interactions with soil nitrogen and organic matter. *Mycorrhiza* 7:261–5.

393. Jun DJ, Allen EB. 1991. Physiological responses of six wheatgrass cultivars to mycorrhizae. *Journal of Range Management* 44:336–41.

394. Jurgensen MF, Richter DL, Davis MM, McKevlin MR, Craft MH. 1996. Mycorrhizal relationships in bottomland hardwood forests of the southern United States. *Wetlands Ecology and Management* 4:223–33.

395. Jurinak JJ, Dudley LM, Allen MF, Knight WG. 1986. The role of calcium-oxalate in the availability of phosphorus in soils of semi-arid regions: A thermodynamic study. *Soil Science* 142:255–61.

396. Kabir Z. 2005. Tillage or no-tillage: Impact on mycorrhizae. *Canadian Journal of Plant Science* 85:23–9.

397. Kaeppler SM, Parke JL, Mueller SM, Senior L, Stuber C, Tracy WF. 2000. Variation among maize inbred lines and detection of quantitative trait loci for growth at low phosphorus and responsiveness to arbuscular mycorrhizal fungi. *Crop Science* 40:358–64.

398. Kamienski F. 1882. (trans. SM Berch, 1985). The vegetative organs of Monotropa hypopitys L. *Proceedings of the* 6th North American Conference on Mycorrhizae, Bend, Oregon, Jun 25–29, 1984. Oregon State University. Forest Research Laboratory.

399. Kanouse BB. 1936. Studies of two species of Endogone in culture. *Mycologia* 28:47–62.

400. Kårén O, Högberg N, Dahlberg A, Jonsson L, Nylund JE. 1997. Inter- and intraspecific variation in the ITS region of rDNA of ectomycorrhizal fungi in Fennoscandia as detected by endonuclease analysis. *New Phytologist* 136:313–25.

401. Karl R, Koch MA. 2013. A world-wide perspective on crucifer speciation and evolution: Phylogenetics, biogeography and trait evolution in tribe Arabideae. *Annals of Botany* 112:983–1001.

402. Kartzinel TR, Trapnell DW, Shefferson RP. 2013. Highly diverse and spatially heterogeneous mycorrhizal symbiosis in a rare epiphyte is unrelated to broad biogeographic or environmental features. *Molecular Ecology* 22:5949–61.

403. Kaschuk G, Kuyper TW, Leffelaar PA, Hungria M, Giller KE. 2009. Are the rates of photosynthesis stimulated by the carbon sink strength of rhizobial and arbuscular mycorrhizal symbioses? *Soil Biology and Biochemistry* 41:1233–44.

404. Kidston R, Lang WH. 1921. On Old Red Sandstone plants showing structure, from the Rhynie Chert Bed, Aberdeenshire. Part V. The Thallophyta occurring in the peat-bed; the succession of the plants throughout a vertical section of the bed, and the conditions of accumulation and preservation of the deposit. *Transactions of the Royal Society of Edinburgh* 52:855–902.

405. Kiers ET, Duhamel M, Beesetty Y, Mensah JA, Franken O, et al. 2011. Reciprocal rewards stabilize cooperation in the mycorrhizal symbiosis. *Science* 333:880–2.

406. Kiers ET, van der Heijden MG. 2006. Mutualistic stability in the arbuscular mycorrhizal symbiosis: Exploring hypotheses of evolutionary cooperation. *Ecology* 87:1627–36.

407. King M, Altdorff D, Li P, Galagedara L, Holden J, Unc A. 2018. Northward shift of the agricultural climate zone under 21st-century global climate change. *Scientific Reports* 8:7904.

408. Kitajima K, Allen MF, Goulden ML. 2013. Contribution of hydraulically lifted deep moisture to the water budget in a Southern California mixed forest. *Journal of Geophysical Research-Biogeosciences* 118:1561–72.

409. Kitajima K, Anderson KE, Allen MF. 2010. Effect of soil temperature and soil water content on fine root turnover rate in a California mixed conifer ecosystem. *Journal of Geophysical Research: Biogeosciences* 115:G04032.

410. Klink S, Giesemann P, Hubmann T, Pausch J. 2020. Stable C and N isotope natural abundances of intraradical hyphae of arbuscular mycorrhizal fungi. *Mycorrhiza* 30:773–80.

411. Klironomos JN. 2002. Feedback with soil biota contributes to plant rarity and invasiveness in communities. *Nature* 417:67–70.

412. Klironomos JN. 2003. Variation in plant response to native and exotic arbuscular mycorrhizal fungi. *Ecology* 84:2292–301.

413. Klironomos JN, Allen MF, Rillig MC, Piotrowski J, Makvandi-Nejad S, et al. 2005. Abrupt rise in atmospheric CO2 overestimates community response in a model plant-soil system. *Nature* 433:621–4.

414. Klironomos JN, Kendrick WB. 1996. Palatability of microfungi to soil arthropods in relation to the functioning of arbuscular mycorrhizae. *Biology and Fertility of Soils* 21:43–52.

415. Klironomos JN, Rillig MC, Allen MF. 1996. Below-ground microbial and microfaunal responses to Artemisia tridentata grown under elevated atmospheric CO_2. *Functional Ecology* 10:527–34.

416. Klironomos JN, Rillig MC, Allen MF. 1999. Designing belowground field experiments with the help of semi-variance and power analyses. *Applied Soil Ecology* 12:227–38.

417. Klopatek CC, Friese C, Allen MF, DeBano LF, Klopatek JM. 1991. The effect of a high intensity fire on the patch dynamics of VA mycorrhizae in a pinyon–juniper woodlands, pp. 123–8. In SC Nodvin, A Waldrop, eds. *Fire and the Environment: Ecological and Cultural Perspectives*: Proceedings of an International Symposium; March 20–24, 1990, Knoxville, TN. Gen Tech Rep. SE-69. Ashville, NC: USDA Forest Service, Southeastern Forest Experiment Station.

418. Knight WG, Allen MF, Jurinak JJ, Dudley LM. 1989. Elevated carbon-dioxide and solution phosphorus in soil with Vesicular–Arbuscular Mycorrhizal Western Wheatgrass. *Soil Science Society of America Journal* 53:1075–82.

419. Knutson DM, Hutchins AS, Cromack K. 1980. The association of calcium oxalate-utilizing *Streptomyces* with conifer ectomycorrhizae. *Antonie van Leeuwenhoek* 46:611–19.

420. Kobae Y, Hata S. 2010. Dynamics of periarbuscular membranes visualized with a fluorescent phosphate transporter in arbuscular mycorrhizal roots of rice. *Plant and Cell Physiology* 51:341–53.

421. Koch K. 1996. Carbohydrate-modulated gene expression in plants. *Annual Review of Plant Biology* 47:509–40.

422. Koide RT, Huenneke LF, Mooney HA. 1987. Gopher mound soil reduces growth and affects ion uptake of two annual grassland species *Oecologia* 72:284–90.

423. Koide RT, Mooney HA. 1987. Spatial variation in inoculum potential of vesicular arbuscular mycorrhizal fungi caused by formation of gopher mounds *New Phytologist* 107:173–82.

424. Kokkoris V, Chagnon P-L, Yildirir G, Clarke K, Goh D, et al. 2021. Host identity influences nuclear dynamics in arbuscular mycorrhizal fungi. *Current Biology* 31:1531–8. e6.

425. Kokkoris V, Hart M. 2019. In vitro propagation of arbuscular mycorrhizal fungi may drive fungal evolution. *Frontiers in Microbiology* October 22, 2019. https://doi.org/10.3389/fmicb.2019.02420

426. Kokkoris V, Stefani F, Dalpé Y, Dettman J, Corradi N. 2020. Nuclear dynamics in the arbuscular mycorrhizal fungi. *Trends in Plant Science* 25:765–78.

427. Kormanik PP, Bryan WC, Schultz RC. 1980. Increasing endomycorrhizal fungus inoculum in forest nursery soil with cover crops. *Southern Journal of Applied Forestry* 4:151–3.

428. Koske R, Gemma J. 1992. Fungal reactions to plants prior to mycorrhizal formation, pp. 3–36. In MF Allen, ed.*Mycorrhizal Functioning: An Integrative Plant–Fungal Process.* New York: Chapman and Hall.

429. Kough J, Malajczuk N, Linderman R. 1983. Use of the indirect immunofluorescent technique to study the vesicular-arbuscular fungus Glomus epigaeum and other Glomus species. *New Phytologist* 94:57–62.

430. Kramer PJ, Wilbur KM. 1949. Absorption of radioactive phosphorus by mycorrhizal roots of pine. *Science* 110:8–9.

431. Krings M, Taylor TN, Dotzler N. 2013. Fossil evidence of the zygomycetous fungi. *Persoonia* 30:1–10.

432. Kropp BR, McAfee BJ, Fortin JA. 1987. Variable loss of ectomycorrhizal ability in monokaryotic and dikaryotic cultures of Laccaria bicolor. *Canadian Journal of Botany* 65:500–4.

433. Krüger C, Kohout P, Janoušková M, Püschel D, Frouz J, Rydlová J. 2017. Plant communities rather than soil properties structure arbuscular mycorrhizal fungal communities along primary succession on a mine spoil. *Frontiers in Microbiology* 8:719.

434. Krüger M, Krüger C, Walker C, Stockinger H, Schüßler A. 2012. Phylogenetic reference data for systematics and phylotaxonomy of arbuscular mycorrhizal fungi from phylum to species level. *New Phytologist* 193:970–84.

435. Kucey R, Paul E. 1982. Carbon flow, photosynthesis, and N_2 fixation in mycorrhizal and nodulated faba beans (Vicia faba L.). *Soil Biology and Biochemistry* 14:407–12.

436. Kuylenstierna J, Hicks W, Cinderby S, Cambridge H. 1998. Critical loads for nitrogen deposition and their exceedance at European scale. *Environmental Pollution* 102:591–8.

437. Lambers H, Teste FP. 2013. Interactions between arbuscular mycorrhizal and non-mycorrhizal plants: Do non-mycorrhizal species at both extremes of nutrient availability play the same game. *Plant Cell and Environment* 36:1911–15.

438. Lamont B. 1984. Specialised modes of nutrition, pp. 126–45. In JS Pate, JS Beard, eds. *Kwongan, Plant Life of the Sandplain: Biology of a South-West Australian Shrubland Ecosystem.* Perth: University of Western Australia Press.

439. Lamont BB. 2003. Structure, ecology and physiology of root clusters: A review. *Plant and Soil* 248:1–19.

440. de Landa D. 1566. *Yucatan Before and After the Conquest.* Transl and notes of William Gates 1978. New York: Dover Publications.

441. Landeweert R, Leeflang P, Kuyper TW, Hoffland E, Rosling A, et al. 2003. Molecular identification of ectomycorrhizal mycelium in soil horizons. *Applied and Environmental Microbiology* 69:327–33.

442. Lanfranco L, Young JPW. 2012. Genetic and genomic glimpses of the elusive arbuscular mycorrhizal fungi. *Current Opinion in Plant Biology* 15:454–61.

443. Lansing JL. 2003. Comparing arbuscular and ectomycorrhizal fungal communities in seven North American forests and their response to nitrogen fertilization. PhD dissertation. University of California, Davis.

444. Leake J, Read D. 2017. Mycorrhizal symbioses and pedogenesis throughout Earth's history, pp. 9–33: In NC Johnson, C Gehring, J Jansa, eds. *Mycorrhizal*

Mediation of Soil: Fertility, Structure, and Carbon Storage. Amsterdam: Elsevier Press.

445. Lekberg Y, Hammer EC, Olsson PA. 2010. Plants as resource islands and storage units: Adopting the mycocentric view of arbuscular mycorrhizal networks. *FEMS Microbiology Ecology* 74:336–45.

446. León-Sánchez L, Nicolás E, Goberna M, Prieto I, Maestre FT, Querejeta JI. 2018. Poor plant performance under simulated climate change is linked to mycorrhizal responses in a semi-arid shrubland. *Journal of Ecology* 106:960–76.

447. Leslie AB, Beaulieu J, Holman G, Campbell CS, Mei W, et al. 2018. An overview of extant conifer evolution from the perspective of the fossil record. *American Journal of Botany* 105:1531–44.

448. Leung H-M, Zhen-Wen W, Zhi-Hong Y, Kin-Lam Y, Xiao-Ling P, Cheung K-C. 2013. Interactions between arbuscular mycorrhizae and plants in phytoremediation of metal-contaminated soils: A review. *Pedosphere* 23:549–63.

449. Levy Y, Krikun J. 1980. Effect of vesicular-arbuscular mycorrhiza on *Citrus jambhiri* water relations. *New Phytologist* 85:25–31.

450. Lewis D. 1973. Concepts in fungal nutrition and the origin of biotrophy. *Biological Reviews* 48:261–77.

451. Lewis D. 1975. Comparative aspects of the carbon nutrition of mycorrhizas, pp. 119–48. In FE Sanders, B Mosse, PB Tinker, eds. *Endomycorrhizas: Proceedings of a Symposium held at the University of Leeds, July 22–25, 1974.* New York: Academic Press.

452. Lewis SL, Maslin MA. 2015. Defining the anthropocene. *Nature* 519:171–80.

453. Li M, Zhao J, Tang N, Sun H, Huang J. 2018. Horizontal gene transfer from bacteria and plants to the arbuscular mycorrhizal fungus *Rhizophagus irregularis.* *Frontiers in Plant Science* 9:701.

454. Li T, Hu YJ, Hao ZP, Li H, Wang YS, Chen BD. 2013. First cloning and characterization of two functional aquaporin genes from an arbuscular mycorrhizal fungus *Glomus intraradices.* *New Phytologist* 197:617–30.

455. Li Y, Wang YG, Houghton R, Tang LS. 2015. Hidden carbon sink beneath desert. *Geophysical Research Letters* 42:5880–7.

456. Liebig J. 1840. *Die organische Chemie in ihrer Anwendung auf Agricultur und Physiologie.* Braunschweig: F. Vieweg und Sohn Braunschweig.

457. Lilieholm B, Dudley L, Jurinak J. 1992. Oxalate determination in soils using ion chromatography. *Soil Science Society of America Journal* 56:324–6.

458. Lilleskov E, Fahey T, Lovett G. 2001. Ectomycorrhizal fungal aboveground community change over an atmospheric nitrogen deposition gradient. *Ecological Applications* 11:397–410.

459. Lilleskov EA, Fahey TJ, Horton TR, Lovett GM. 2002. Belowground ectomycorrhizal fungal community change over a nitrogen deposition gradient in Alaska. *Ecology* 83:104–15.

460. Lilleskov EA, Kuyper TW, Bidartondo MI, Hobbie EA. 2019. Atmospheric nitrogen deposition impacts on the structure and function of forest mycorrhizal communities: A review. *Environmental Pollution* 246:148–62.

461. van der Linde S, Suz LM, Orme CDL, Cox F, Andreae H, et al. 2018. Environment and host as large-scale controls of ectomycorrhizal fungi. *Nature* 558:243–8.

462. Link HF. 1809. Observationes in ordines plantarum naturales. Dissertatio I. *Magazin der Gesellschaft Naturforschenden Freunde Berlin* 3:3–42.

463. Linnaeus C. 1758. *Systema naturae*. Stockholm: Laurentii Salvii.

464. Loescher HW, Oberbauer SF, Gholz H, Clark DB. 2003. Environmental controls on net ecosystem-level carbon exchange and productivity in a Central American tropical wet forest. *Global Change Biology* 9:396–412.

465. Lohman ML. 1926. Occurrences of mycorrhiza in Iowa forest plants. *University of Iowa Studies in Natural History* 11:33–8.

466. Loron CC, François C, Rainbird RH, Turner EC, Borensztajn S, Javaux EJ. 2019. Early fungi from the Proterozoic era in Arctic Canada. *Nature* 570:232–5.

467. Lovelock CE, Wright SF, Clark DA, Ruess RW. 2004. Soil stocks of glomalin produced by arbuscular mycorrhizal fungi across a tropical rain forest landscape. *Journal of Ecology* 92:278–87.

468. MacArthur RH, Wilson EO. 2016. *The Theory of Island Biogeography*. Princeton: Princeton University Press.

469. MacDonald R, Chandler MR, Mosse B. 1982. The occurrence of bacterium-like organelles in vesicular-arbuscular mycorrhizal fungi. *New Phytologist* 90:659–63.

470. MacMahon JA. 1998. Empirical and theoretical ecology as a basis for restoration: An ecological success story, pp. 220–46. In ML Pace, ed. *Successes, Limitations, and Frontiers in Ecosystem Science*. Berlin: Springer.

471. MacMahon JA, Phillips DL, Robinson JV, Schimpf DJ. 1978. Levels of biological organization: An organism-centered approach. *BioScience* 28:700–4.

472. MacMahon JA, Schimpf DJ, Andersen DC, Smith KG, Bayn RL. 1981. An organism-centered approach to some community and ecosystem concepts. *Journal of Theoretical Biology* 88:287–307.

473. MacMahon JA, Warner N. 1984. Dispersal of mycorrhizal fungi: Processes and agents, pp. 28–41. In SE Williams, MF Allen, eds. *VA Mycorrhizae and Reclamation of Arid and Semiarid Lands*. Laramie: University of Wyoming Press.

474. Maddison JA, Krzic M, Simard S, Adderly C, Khan S. 2018. Shroomroot: An action-based digital game to enhance postsecondary teaching and learning about mycorrhizae. *American Biology Teacher* 80:11–20.

475. Magurno F, Malicka M, Posta K, Wozniak G, Lumini E, Piotrowska-Seget Z. 2019. Glomalin gene as molecular marker for functional diversity of arbuscular mycorrhizal fungi in soil. *Biology and Fertility of Soils* 55:411–17.

476. Malajczuk N, Cromack Jr K. 1982. Accumulation of calcium oxalate in the mantle of ectomycorrhizal roots of Pinus radiata and Eucalyptus marginata. *New Phytologist* 92:527–31.

477. Maltz MR, Treseder KK. 2015. Sources of inocula influence mycorrhizal colonization of plants in restoration projects: A meta-analysis. *Restoration Ecology* 23:625–34.

478. Mandyam KG, Jumpponen A. 2015. Mutualism-parasitism paradigm synthesized from results of root-endophyte models. *Frontiers in Microbiology* 5:776.

479. Manrique Caamal SA. 2016. Los hongos micorrizógenos arbusculares como un método biológico para evaluar el estado del carbono orgánico del suelo en Yucatán. BS thesis. Universidad Autónoma de Yucatán, Merida, Yucatán, Mexico.

480. Margulis L, Bermudes D. 1985. Symbiosis as a mechanism of evolution: Status of cell symbiosis theory. *Symbiosis* 1:101–24.

481. Martin F. 2017. *Molecular Mycorrhizal Symbiosis*. Wiley Online Library.

482. Martin F, Boiffin V, Pfeffer PE. 1998. Carbohydrate and amino acid metabolism in the Eucalyptus globulus Pisolithus tinctorius ectomycorrhiza during glucose utilization. *Plant Physiology* 118:627–35.

483. Martin F, Canet D, Marchal JP. 1985. C-13 nuclear magnetic resonance study of Mannitol cycle and Trehalose synthesis during glucose utilization by the ectomycorrhizal ascomycete Cenococcum graniforme. *Plant Physiology* 77:499–502.

484. Marx DH, Bryan WC. 1975. Growth and ectomycorrhizal development of loblolly pine seedlings in fumigated soil infested with the fungal symbiont *Pisolithus tinctorius*. *Forest Science* 21:245–54.

485. Marx DH, Bryan WC, Cordell CE. 1977. Survival and growth of pine seedlings with *Pisolithus* ectomycorrhizae after two years on reforestation sites in North Carolina and Florida. *Forest Science* 23:363–73.

486. Masclaux FG, Wyss T, Pagni M, Rosikiewicz P, Sanders IR. 2019. Investigating unexplained genetic variation and its expression in the arbuscular mycorrhizal fungus *Rhizophagus irregularis*: A comparison of whole genome and RAD sequencing data. *PLoS One* 14:e0226497.

487. Maser C, Nussbaum RA, Trappe JM. 1978. Fungal small mammal interrelationships with emphasis on oregon coniferous forests *Ecology* 59:799–809.

488. Maurer BA. 1999. *Untangling Ecological Complexity: The Macroscopic Perspective*. Chicago, IL: University of Chicago Press.

489. May MR, Provance MC, Sanders AC, Ellstrand NC, Ross-Ibarra J. 2009. A pleistocene clone of Palmer's oak persisting in Southern California. *PLoS One* 4:e8346.

490. May RM. 1972. Limit cycles in predator–prey communities. *Science* 177:900–2.

491. May RM. 1974. *Stability and Complexity in Model Ecosystems*. Princeton: Princeton University Press.

492. May RM. 1981. Models for two interacting populations, pp. 78–104. In RM May, ed. *Theoretical Ecology: Principles and Applications*. Sunderland, MA: Sinauer Associates.

493. McDougall W, Liebtag C. 1928. Symbiosis in a deciduous forest. III. Mycorhizal relations. *Botanical Gazette* 86:226–34.

494. McFarland JW, Ruess RW, Kielland K, Pregitzer K, Hendrick R, Allen M. 2010. Cross-ecosystem comparisons of in situ plant uptake of amino acid-N and NH_4^+. *Ecosystems* 13:177–93.

495. McIntosh RP. 1986. *The Background of Ecology: Concept and Theory*. Cambridge: Cambridge University Press.

496. McNaughton SJ. 1990. Mineral nutrition and seasonal movements of African migratory ungulates. *Nature* 345:613–15.

497. Melin E, Nilsson H. 1953. Transfer of labelled nitrogen from glutamic acid to pine seedlings through the mycelium of Boletus variegatus (Sw.) Fr. *Nature* 171:134.

498. Merckx V, Freudenstein JV. 2010. Evolution of mycoheterotrophy in plants: A phylogenetic perspective. *New Phytologist* 185:605–9.

499. Michener CH. 1979. Biogeography of bees. *Annals of the Missouri Botanical Garden* 66:277–347.

500. Mikola P. 1965. Studies on the ectendotrophic mycorrhiza of pine. *Acta Forestalia Fennica* 79:1–56.

501. Millar CI, Woolfenden WB. 2016. Ecosystems past: Vegetation prehistory, pp. 131–86. In H Mooney, E Zavaleta, eds. *Ecosystems of California*. Berkeley: University of California Press.

502. Miller R, Jastrow J. 1992. The application of VA mycorrhizae to ecosystem restoration and reclamation, pp. 438–67. In MF Allen, ed. *Mycorrhizal Functioning*. New York: Chapman and Hall Press.

503. Miller R, Jastrow J, Reinhardt D. 1995. External hyphal production of vesicular–arbuscular mycorrhizal fungi in pasture and tallgrass prairie communities. *Oecologia* 103:17–23.

504. Miller RM. 1985. Mycorrhizae. *Restoration and Management Notes* 3:14–20.

505. Miller S, Allen E. 1992. Mycorrhizae, nutrient translocation and interactions between plants, pp. 301–32. In MF Allen, ed. *Mycorrhizal Functioning*. New York: Chapman and Hall Press.

506. Miller SL, McClean TM, Stanton NL, Williams SE. 1998. Mycorrhization, physiognomy, and first-year survivability of conifer seedlings following natural fire in Grand Teton National Park. *Canadian Journal of Forest Research* 28:115–22.

507. Mishler BD. 2010. Species are not uniquely real biological entities, pp. 110–22. In F Ayala, R Arp, eds. *Contemporary Debates in Philosophy of Biology*. Wiley-Blackwell Online.

508. Molina R. 1980. Ectomycorrhizal inoculation of containerized western conifer seedlings. Research Note PNW-RN-357. Portland Department of Agriculture.

509. Molina R. 1984. Proceedings of the 6th North American Conference on Mycorrhizae. June 25–29, Bend, OR. Oregon State University. College of Forestry.

510. Molina R, Massicotte H, Trappe JM. 1992. Specificity phenomena in mycorrhizal symbioses: Community-ecological consequences and practical implications, pp. 357–423. In MF Allen, ed. *Mycorrhizal Functioning*. New York: Chapman and Hall Press.

511. Montesinos-Navarro A, Segarra-Moragues JG, Valiente-Banuet A, Verdú M. 2012. The network structure of plant–arbuscular mycorrhizal fungi. *New Phytologist* 194:536–47.

512. Montesinos-Navarro A, Verdú M, Querejeta JI, Sortibrán L, Valiente-Banuet A. 2016. Soil fungi promote nitrogen transfer among plants involved in long-lasting facilitative interactions. *Perspectives in Plant Ecology, Evolution and Systematics* 18:45–51.

513. Montesinos-Navarro A, Verdú M, Querejeta JI, Valiente-Banuet A. 2017. Nurse plants transfer more nitrogen to distantly related species. *Ecology* 98:1300–10.

514. Morgan BS, Egerton-Warburton LM. 2017. Barcoded NS31/AML2 primers for sequencing of arbuscular mycorrhizal communities in environmental samples. *Applications in Plant Sciences* 5:1700017.

515. Morris SJ, Allen MF. 1994. Oxalate-metabolizing microorganisms in sagebrush steppe soil. *Biology and Fertility of Soils* 18:255–9.

516. Morris SJ, Zink T, Conners K, Allen MF. 1997. Comparison between fluorescein diacetate and differential fluorescent staining procedures for determining fungal biomass in soils. *Applied Soil Ecology* 6:161–7.

517. Morton JB, Bentivenga SP, Bever JD. 1995. Discovery, measurement, and interpretation of diversity in arbuscular endomycorrhizal fungi (Glomales, Zygomycetes). *Canadian Journal of Botany* 73:25–32.

518. Moser M. 1967. Die ektotrophe Ernährungsweise an der Waldgrenze. *Mitt Forstl Bundesversuchsanst. Wien* 75:357–80.

519. Mosse B. 1953. Fructifications associated with mycorrhizal strawberry roots. *Nature* 171:974.

520. Mosse B. 1959. Observations on the extra-matrical mycelium of a vesicular–arbuscular endophyte. *Transactions of the British Mycological Society* 42:439-IN5.

521. Mosse B. 1973. Advances in the study of vesicular-arbuscular mycorrhiza. *Annual Review of Phytopathology* 11:171–96.

522. Mosse B, Stribley D, LeTacon F. 1981. Ecology of mycorrhizae and mycorrhizal fungi. *Advances in Microbial Ecology* 5:137–210.

523. Murat C, Vizzini A, Bonfante P, Mello A. 2005. Morphological and molecular typing of the below-ground fungal community in a natural *Tuber magnatum* truffle-ground. *FEMS Microbiology Letters* 245:307–13.

524. Nadkarni NM. 1981. Canopy roots: Convergent evolution in rainforest nutrient cycles. *Science* 214:1023–4.

525. Nakano A, Takahashi K, Kimura M. 1999. The carbon origin of arbuscular mycorrhizal fungi estimated from δ^{13}C values of individual spores. *Mycorrhiza* 9:41–7.

526. National Atmospheric Deposition Program. 2019. Annual Summary. Available from https://nadp.slh.wisc.edu. Last accessed November 16, 2021.

527. Naveh Z, Lieberman AS. 1984. *Landscape Ecology: Theory and Application.* New York: Springer-Verlag.

528. Nepstad DC, de Carvalho CR, Davidson EA, Jipp PH, Lefebvre PA, et al. 1994. The role of deep roots in the hydrological and carbon cycles of Amazonian forests and pastures. *Nature* 372:666–9.

529. Neuenkamp L, Zobel M, Lind E, Gerz M, Moora M. 2019. Arbuscular mycorrhizal fungal community composition determines the competitive response of two grassland forbs. *PLoS One* 14:e0219527.

530. Newsham K, Fitter A, Watkinson A. 1995. Arbuscular mycorrhiza protect an annual grass from root pathogenic fungi in the field. *Journal of Ecology* 83:991–1000.

531. Nicolson T. 1959. Mycorrhiza in the Gramineae: I. Vesicular-arbuscular endophytes, with special reference to the external phase. *Transactions of the British Mycological Society* 42:421–38.

532. Nicolson T. 1960. Mycorrhiza in the Gramineae: II. Development in different habitats, particularly sand dunes. *Transactions of the British Mycological Society* 43:132–45.

533. Nicolson T, Johnston C. 1979. Mycorrhiza in the Gramineae: III. Glomus fasciculatus as the endophyte of pioneer grasses in a maritime sand dune. *Transactions of the British Mycological Society* 72:261–8.

534. Nielsen KB, Kjøller R, Bruun HH, Schnoor TK, Rosendahl S. 2016. Colonization of new land by arbuscular mycorrhizal fungi. *Fungal Ecology* 20:22–9.

535. NOAA. 2021. State of the Climate: Global Climate Report April 2021. Available from www.ncdc.noaa.gov/sotc/global/202104. Last accessed November 16, 2021.

536. Nottingham AT, Fierer N, Turner BL, Whitaker J, Ostle NJ, et al. 2018. Microbes follow Humboldt: Temperature drives plant and soil microbial diversity patterns from the Amazon to the Andes. *Ecology* 99:2455–66.

537. Nouhra E, Urcelay C, Longo S, Tedersoo L. 2013. Ectomycorrhizal fungal communities associated to Nothofagus species in Northern Patagonia. *Mycorrhiza* 23:487–96.

538. Nouhra ER, Dominguez LS, Daniele GG, Longo S, Trappe JM, Claridge AW. 2008. Ocurrence of ectomycorrhizal, hypogeous fungi in plantations of exotic tree species in central Argentina. *Mycologia* 100:752–9.

539. Odum EP. 1969. The strategy of ecosystem development, an understanding of ecological succession provides a basis for resolving man's conflict with nature. *Science*:262–70.

540. Odum EP, Barrett GW. 1971. *Fundamentals of Ecology*. Philadelphia, PA: Saunders.

541. Ogawa M. 1985. Ecological characters of ectomycorrhizal fungi and their mycorrhizae: An introduction to the ecology of higher fungi. *Japan Agricultural Research Quarterly* 18:305–14.

542. Öpik M, Vanatoa A, Vanatoa E, Moora M, Davison J, et al. 2010. The online database MaarjAM reveals global and ecosystemic distribution patterns in arbuscular mycorrhizal fungi (Glomeromycota). *New Phytologist* 188:223–41.

543. Opik M, Zobel M, Cantero JJ, Davison J, Facelli JM, et al. 2013. Global sampling of plant roots expands the described molecular diversity of arbuscular mycorrhizal fungi. *Mycorrhiza* 23:411–30.

544. Orchard S, Hilton S, Bending GD, Dickie IA, Standish RJ, et al. 2017. Fine endophytes (*Glomus tenue*) are related to Mucoromycotina, not Glomeromycota. *New Phytologist* 213:481–6.

545. Orlov A. 1960. Growth and growth dependent changes in absorbing roots of Picea excelsa Link (Translated title). *Botaniceskij Zurnal, SSSR* 45:888–96.

546. Pancholy SK, Rice EL. 1973. Soil enzymes in relation to old field succession: Amylase, cellulase, invertase, dehydrogenase, and urease. *Soil Science Society of America Journal* 37:47–50.

547. Pardo LH, Fenn ME, Goodale CL, Geiser LH, Driscoll CT, et al. 2011. Effects of nitrogen deposition and empirical nitrogen critical loads for ecoregions of the United States. *Ecological Applications* 21:3049–82.

548. Parker GG. 1995. Structure and microclimate of forest canopies, pp. 73–106. In MD Lowman, NM Nadkarni, eds. *Forest Canopies*. San Diego, CA: Academic Press.

549. Parniske M. 2008. Arbuscular mycorrhiza: The mother of plant root endosymbioses. *Nature Reviews Microbiology* 6:763–75.

550. Parrent JL, Vilgalys R. 2007. Biomass and compositional responses of ectomycorrhizal fungal hyphae to elevated CO_2 and nitrogen fertilization. *New Phytologist* 176:164–74.

551. Parsons WF, Ehrenfeld JG, Handel SN. 1998. Vertical growth and mycorrhizal infection of woody plant roots as potential limits to the restoration of woodlands on landfills. *Restoration Ecology* 6:280–9.

552. Pate JS, Beard JS. 1982. *Kwongan, Plant Life of the Sandplain*. Perth, WA: University of Western Australia Press.

553. Paul E, Kucey R. 1981. Carbon flow in plant microbial associations. *Science* 213:473–4.

554. Pawlowska TE, Taylor JW. 2004. Organization of genetic variation in individuals of arbuscular mycorrhizal fungi. *Nature* 427:733–7.

555. Paz C, Öpik M, Bulascoschi L, Bueno CG, Galetti M. 2020. Dispersal of arbuscular mycorrhizal fungi: Evidence and insights for ecological studies. *Microbial Ecology* 81:283–92.

556. Pearson V, Tinker P. 1975. Measurement of phosphorus fluxes in the external hyphae of endomycorrhizas, pp. 277–87. In FE Sanders, B Mosse, PB Tinker, eds. *Endomycorrhizas. Proceedings of a Symposium held at the University of Leeds, July 22–25, 1974*. New York: Academic Press.

557. Peay KG, Bidartondo MI, Arnold AE. 2010. Not every fungus is everywhere: Scaling to the biogeography of fungal–plant interactions across roots, shoots and ecosystems. *New Phytologist* 185:878–82.

558. Peay KG, Bruns TD, Kennedy PG, Bergemann SE, Garbelotto M. 2007. A strong species–area relationship for eukaryotic soil microbes: Island size matters for ectomycorrhizal fungi. *Ecology Letters* 10:470–80.

559. Pepe A, Sbrana C, Ferrol N, Giovannetti M. 2017. An in vivo whole-plant experimental system for the analysis of gene expression in extraradical mycorrhizal mycelium. *Mycorrhiza* 27:659–68.

560. Perfecto I, Vandermeer J. 1993. Distribution and turnover rate of a population of Atta cephalotes in a tropical rainforest in Costa Rica *Biotropica* 25:316–21.

561. Perry DA, Margolis H, Choquette C, Molina R, Trappe JM. 1989. Ectomycorrhizal mediation of competition between coniferous tree species *New Phytologist* 112:501–11.

562. Peterson RL, Massicotte HB. 2004. Exploring structural definitions of mycorrhizas, with emphasis on nutrient-exchange interfaces. *Canadian Journal of Botany* 82:1074–88.

563. Peterson RL, Massicotte HB, Melville LH. 2004. *Mycorrhizas: Anatomy and Cell Biology*. Ottawa: NRC Research Press.

564. Pfeffer PE, Douds DD, Becard G, Shachar-Hill Y. 1999. Carbon uptake and the metabolism and transport of lipids in an arbuscular mycorrhiza. *Plant Physiology* 120:587–98.

565. Phillips ML, McNellis BE, Allen MF, Allen EB. 2019. Differences in root phenology and water depletion by an invasive grass explains persistence in a Mediterranean ecosystem. *American Journal of Botany* 106:1210–8.

566. Pianka ER. 1970. R-selection and k-selection *American Naturalist* 104:592.

567. Pickles BJ, Simard SW. 2017. Mycorrhizal networks and forest resilience to drought, pp. 319–39. In NC Johnson, C Gehring, J Jansa, eds. *Mycorrhizal Mediation of Soil: Fertility, Structure, and Carbon Storage*. Amsterdam: Elsevier Press.

568. Pirozynski K. 1976. Fossil fungi. *Annual Review of Phytopathology* 14:237–46.

569. Pirozynski K, Malloch D. 1975. The origin of land plants: A matter of mycotrophism. *Biosystems* 6:153–64.

570. Plassard C, Becquer A, Garcia K. 2019. Phosphorus transport in mycorrhiza: How far are we? *Trends in Plant Science* 24:794–801.

571. Plattner I, Hall I. 1995. Parasitism of non-host plants by the mycorrhizal fungus *Tuber melanosporum*. *Mycological Research* 11:1367–70.

572. Plenchette C, Clermont-Dauphin C, Meynard JM, Fortin JA. 2005. Managing arbuscular mycorrhizal fungi in cropping systems. *Canadian Journal of Plant Science* 85:31–40.

573. Poca M, Coomans O, Urcelay C, Zeballos SR, Bodé S, Boeckx P. 2019. Isotope fractionation during root water uptake by Acacia caven is enhanced by arbuscular mycorrhizas. *Plant and Soil* 441:485–97.

574. Policelli N, Bruns TD, Vilgalys R, Nunez MA. 2019. Suilloid fungi as global drivers of pine invasions. *New Phytologist* 222:714–25.

575. Porter JR, Allen MF, Lane LC, Boosalis MG. 1982. Platte valley yellows, a chlorotic condition of soybeans: Symptoms and preliminary chemical analyses. *Plant and Soil* 68:283–7.

576. Pregitzer KS, DeForest JL, Burton AJ, Allen MF, Ruess RW, Hendrick RL. 2002. Fine root architecture of nine North American trees. *Ecological Monographs* 72:293–309.

577. Pringle A, Bever JD, Gardes M, Parrent JL, Rillig MC, Klironomos JN. 2009. Mycorrhizal symbioses and plant invasions. *Annual Review of Ecology, Evolution, and Systematics* 40:699–715.

578. Pringle A, Taylor JW. 2002. The fitness of filamentous fungi. *Trends in Microbiology* 10:474–81.

579. Pritchard SG, Strand AE, McCormack ML, Davis MA, Oren R. 2008. Mycorrhizal and rhizomorph dynamics in a loblolly pine forest during 5 years of free-air-CO_2-enrichment. *Global Change Biology* 14:1252–64.

580. Pritchard SG, Taylor BN, Cooper ER, Beidler KV, Strand AE, et al. 2014. Long-term dynamics of mycorrhizal root tips in a loblolly pine forest grown with free-air CO_2 enrichment and soil N fertilization for 6 years. *Global Change Biology* 20:1313–26.

581. Propster JR, Johnson NC. 2015. Uncoupling the effects of phosphorus and precipitation on arbuscular mycorrhizas in the Serengeti. *Plant and Soil* 388:21–34.

582. Purin S, Rillig MC. 2007. The arbuscular mycorrhizal fungal protein glomalin: Limitations, progress, and a new hypothesis for its function. *Pedobiologia* 51:123–30.

583. Pyare S, Longland WS. 2002. Interrelationships among northern flying squirrels, truffles, and microhabitat structure in Sierra Nevada old-growth habitat. *Canadian Journal of Forest Research* 32:1016–24.

584. Querejeta JI. 2017. Soil water retention and availability as influenced by mycorrhizal symbiosis: Consequences for individual plants, communities, and ecosystems, pp. 299–317. In NC Johnson, C Gehring, J Jansa, eds. *Mycorrhizal Mediation of Soil: Fertility, Structure, and Carbon Storage*. Amsterdam: Elsevier Press.

585. Querejeta JI, Allen MF, Alguacil MM, Roldan A. 2007. Plant isotopic composition provides insight into mechanisms underlying growth stimulation by AM fungi in a semiarid environment. *Functional Plant Biology* 34:683–91.

586. Querejeta JI, Allen MF, Caravaca F, Roldan A. 2006. Differential modulation of host plant delta^{13}C and delta^{18}O by native and nonnative arbuscular mycorrhizal fungi in a semiarid environment. *New Phytologist* 169:379–87.

587. Querejeta JI, Barea JM, Allen MF, Caravaca F, Roldan A. 2003. Differential response of delta^{13}C and water use efficiency to arbuscular mycorrhizal infection in two aridland woody plant species. *Oecologia* 135:510–5.

588. Querejeta JI, Egerton-Warburton LM, Allen MF. 2003. Direct nocturnal water transfer from oaks to their mycorrhizal symbionts during severe soil drying. *Oecologia* 134:55–64.

589. Querejeta JI, Egerton-Warburton LM, Allen MF. 2007. Hydraulic lift may buffer rhizosphere hyphae against the negative effects of severe soil drying in a California Oak savanna. *Soil Biology & Biochemistry* 39:409–17.

590. Querejeta JI, Egerton-Warburton LM, Allen MF. 2009. Topographic position modulates the mycorrhizal response of oak trees to interannual rainfall variability. *Ecology* 90:649–62.

591. Querejeta JI, Egerton-Warburton LM, Prieto I, Vargas R, Allen MF. 2012. Changes in soil hyphal abundance and viability can alter the patterns of hydraulic redistribution by plant roots. *Plant and Soil* 355:63–73.

592. Quilliam RS, Jones DL. 2010. Fungal root endophytes of the carnivorous plant Drosera rotundifolia. *Mycorrhiza* 20:341–8.

593. Quilliam RS, Jones DL. 2012. Evidence for host-specificity of culturable fungal root endophytes from the carnivorous plant Pinguicula vulgaris (Common Butterwort). *Mycological Progress* 11:583–5.

594. Rabatin SC. 1980. The occurrence of the vesicular-arbuscular-mycorrhizal fungus *Glomus tenuis* with moss. *Mycologia* 72:191–5.

595. Radhakrishnan GV, Keller J, Rich MK, Vernié T, Mbaginda DLM, et al. 2020. An ancestral signalling pathway is conserved in plant lineages forming intracellular symbioses-forming plant lineages. *Nature Plants* 6:280–9.

596. Rayner A, Coates D, Ainsworth A, Adams T, Williams E, Todd N. 1984. Biological consequences of the individualistic mycelium, pp. 509–40. In DH Jennings, ADM Rayner, eds. *The Ecology and Physiology of the Fungal Mycelium*. Cambridge: Cambridge University Press.

597. Read D. 1983. The biology of mycorrhiza in the Ericales. *Canadian Journal of Botany* 61:985–1004.

598. Read D. 1992. The mycorrhizal mycelium, pp. 102–33. In MF Allen, ed. *Mycorrhizal Functioning*. New York: Chapman and Hall Press.

599. Read D, Perez-Moreno J. 2003. Mycorrhizas and nutrient cycling in ecosystems: A journey towards relevance? *New Phytologist* 157:475–92.

600. Read DJ. 1991. Mycorrhizas in ecosystems. *Experientia* 47:376–91.

601. Redecker D, Morton JB, Bruns TD. 2000. Ancestral lineages of arbuscular mycorrhizal fungi (Glomales). *Molecular Phylogenetics and Evolution* 14:276–84.

602. Redecker D, Raab P. 2006. Phylogeny of the glomeromycota (arbuscular mycorrhizal fungi): Recent developments and new gene markers. *Mycologia* 98:885–95.

603. Reichgelt T, D'Andrea WJ, Valdivia-McCarthy AdC, Fox BR, Bannister JM, et al. 2020. Elevated CO_2, increased leaf-level productivity, and water-use efficiency during the early Miocene. *Climate of the Past* 16:1509–21.

604. Reid C, Kidd F, Ekwebelam S. 1983. Nitrogen nutrition, photosynthesis and carbon allocation in ectomycorrhizal pine, pp. 415–31. In D Atkinson, KKS Bhat, MP Coutts, PA Mason, DJ Read, eds. *Tree Root Systems and Their Mycorrhizas*. New York: Springer.

605. Reid CPP, Woods FW. 1969. Translocation of C-14-labelled compounds in mycorrhizae and its implications in interplant nutrient cycling *Ecology* 50:179.

606. Renker C, Zobel M, Öpik M, Allen MF, Allen EB, et al. 2004. Structure, dynamics and restoration of plant communities: Does arbuscular mycorrhiza matter?, pp. 189–229. In V Temperton, T Nuttle, R Hobbs, S Halle, eds. *Assembly Rules and Restoration Ecology: Bridging the Gap between Theory and Practice*. Washington, DC: Island Press.

607. Renny M, Acosta MC, Cofré N, Domínguez LS, Bidartondo MI, Sérsic AN. 2017. Genetic diversity patterns of arbuscular mycorrhizal fungi associated with the mycoheterotroph Arachnitis uniflora Phil.(Corsiaceae). *Annals of Botany* 119:1279–94.

608. Rillig MC, Allen MF. 1998. Arbuscular mycorrhizae of Gutierrezia sarothrae and elevated carbon dioxide: Evidence for shifts in C allocation to and within the mycobiont. *Soil Biology & Biochemistry* 30:2001–8.

609. Rillig MC, Allen MF. 1999. What is the role of arbuscular mycorrhizal fungi in plant-to-ecosystem responses to Elevated atmospheric CO_2? *Mycorrhiza* 9:1–8.

610. Rillig MC, Allen MF, Klironomos JN, Field CB. 1998. Arbuscular mycorrhizal percent root infection and infection intensity of Bromus hordeaceus grown in elevated atmospheric CO2. *Mycologia* 90:199–205.

611. Rillig MC, Treseder KK, Allen MF. 2002. Global change and mycorrhizal fungi, pp. 135–60. In M van der Heijden, I Sanders, eds. *Mycorrhizal Ecology*. New York: Springer.

612. Rillig MC, Wright SF, Allen MF, Field CB. 1999. Rise in carbon dioxide changes soil structure. *Nature* 400:628.

613. Rillig MC, Wright SF, Nichols KA, Schmidt WF, Torn MS. 2001. Large contribution of arbuscular mycorrhizal fungi to soil carbon pools in tropical forest soils. *Plant and Soil* 233:167–77.

614. Rimington WR, Pressel S, Duckett JG, Field KJ, Bidartondo MI. 2019. Evolution and networks in ancient and widespread symbioses between Mucoromycotina and liverworts. *Mycorrhiza* 29:551–65.

615. Ritz K, Newman E. 1984. Movement of 32 P between intact grassland plants of the same age. *Oikos* 43:138–42.

616. Ritz K, Newman E. 1985. Evidence for rapid cycling of phosphorus from dying roots to living plants. *Oikos* 45:174–80.

617. Robin A, Pradier C, Sanguin H, Mahé F, Lambais GR, et al. 2019. How deep can ectomycorrhizas go? A case study on Pisolithus down to 4 meters in a Brazilian eucalypt plantation. *Mycorrhiza* 29:637–48.

618. Robson AD, Abbott LK, Malajczuk N, eds. 1994. Management of mycorrhizas in agriculture, horticulture, and forestry. In *Proceedings of an International Symposium on the Management of Mycorrhizas in Agriculture, Horticulture, and Forestry (1992: University of Western Australia)*. Boston: Kluwer Academic.

619. Romell LG. 1939. The ecological problem of mycotrophy. *Ecology* 20:163–7.

620. Ropars J, Toro KS, Noel J, Pelin A, Charron P, et al. 2016. Evidence for the sexual origin of heterokaryosis in arbuscular mycorrhizal fungi. *Nature Microbiology* 1:1–9.

621. Rosendahl S. 2008. Communities, populations and individuals of arbuscular mycorrhizal fungi. *New Phytologist* 178:253–66.

622. Rosenzweig ML. 1995. *Species Diversity in Space and Time.* Cambridge: Cambridge University Press.

623. Rosinger C, Sandén H, Matthews B, Mayer M, Godbold DL. 2018. Patterns in ectomycorrhizal diversity, community composition, and exploration types in european beech, pine, and spruce forests. *Forests* 9:445.

624. Rossow LJ, Bryant JP, Kielland K. 1997. Effects of above-ground browsing by mammals on mycorrhizal infection in an early successional taiga ecosystem. *Oecologia* 110:94–8.

625. Ruehle JL, Marx DH. 1979. Fiber, food, fuel, and fungal symbionts. *Science* 206:419–22.

626. Ruess R, Seagle S. 1994. Landscape patterns in soil microbial processes in the Serengeti National Park, Tanzania. *Ecology* 75:892–904.

627. Ruess RW, Hendrick RL, Burton AJ, Pregitzer KS, Sveinbjornssön B, et al. 2003. Coupling fine root dynamics with ecosystem carbon cycling in black spruce forests of interior Alaska. *Ecological Monographs* 73:643–62.

628. Rundel PW, Graham EA, Allen MF, Fisher JC, Harmon TC. 2009. Environmental sensor networks in ecological research. *New Phytologist* 182:589–607.

629. Safir GR, Boyer JS, Gerdemann JW. 1972. Nutrient status and mycorrhizal enhancement of water transport in soybean *Plant Physiology* 49:700.

630. Saks U, Davison J, Opik M, Vasar M, Moora M, Zobel M. 2014. Root-colonizing and soil-borne communities of arbuscular mycorrhizal fungi in a temperate forest understorey. *Botany* 92:277–85.

631. Salisbury FB, Ross CM. 1978. *Plant Physiology.* Belmont, CA: Wadsworth.

632. Sanders FE, Mosse B, Tinker PB, eds. 1975. *Endomycorrhizas. Proceedings of a Symposium, July 22–25, 1974, University of Leeds.* New York: Academic Press.

633. Sanders FE, Tinker PB. 1973. Phosphate flow into mycorrhizal roots. *Pesticide Science* 4:385–95.

634. Sanders IR. 1999. No sex please, we're fungi. *Nature* 399:737–8.

635. Sawada K, Wan J, Oda K, Nakano S, Aimi T, Shimomura N. 2014. Variability in nucleus number in basidiospore isolates of *Rhizopogon roseolus* and their ability to form ectomycorrhizas with host pine roots. *Mycological Progress* 13:745–51.

636. Schenck NC, Perez Y. 1990. *Manual for the Identification of VA Mycorrhizal Fungi.* Gainesville: Synergistic Publications.

637. Schlesinger W, Marion G, Fonteyn P. 1989. Stable isotope ratios and the dynamics of caliche in desert soils, pp. 309–17. In PW Rundel, JR Ehleringer, KA Nagy, eds. *Stable Isotopes in Ecological Research.* Berlin: Springer.

638. Schlesinger WH. 1985. The formation of caliche in soils of the Mojave Desert, California. *Geochimica et Cosmochimica Acta* 49:57–66.

639. Schlicht A. 1889. Arbeiten aus dem pflanzenphysiologischen Institute der Königlicher landwirtschaftlichen Hochschule in Berlin, XIII. Beitrag zur

Kenntniss der Verbreitung und der Bedeutung der Mycorhizen. *Bayerisches landwirtschaftliches Jahrbuch* 18:478–506.

640. Schmidt KT, Maltz M, Ta P, Khalili B, Weihe C, et al. 2020. Identifying mechanisms for successful ecological restoration with salvaged topsoil in coastal sage scrub communities. *Diversity* 12:150.

641. Schmidt SK, Scow KM. 1986. Mycorrhizal fungi on the Galapagos Islands. *Biotropica* 18:236–40.

642. Schmitt D, Gischler E, Anselmetti FS, Vogel H. 2020. Caribbean cyclone activity: An annually-resolved Common Era record. *Scientific Reports* 10:1–17.

643. Schoknecht J, Hattingh M. 1976. X-ray microanalysis of elements in cells of VA mycorrhizal and nonmycorrhizal onions. *Mycologia* 68:296–303.

644. van Schöll L, Kuyper TW, Smits MM, Landeweert R, Hoffland E, Van Breemen N. 2008. Rock-eating mycorrhizas: Their role in plant nutrition and biogeochemical cycles. *Plant and Soil* 303:35–47.

645. Schramm JR. 1966. Plant colonization studies on black wastes from anthracite mining in Pennsylvania. *Transactions of the American Philosophical Society* 47:1–331.

646. Schüßler A, Kluge M. 2001. Geosiphon pyriforme, an endocytosymbiosis between fungus and cyanobacteria, and its meaning as a model system for arbuscular mycorrhizal research, pp. 151–61. In B Hock, ed. *The Mycota IX*. Berlin: Springer.

647. Schüßler A, Walker C. 2010. *The Glomeromycota: A Species List with New Families and Genera*. Munich: Botanische Staatssammlung Munich.

648. Schwartz MW, Hoeksema JD. 1998. Specialization and resource trade: Biological markets as a model of mutualisms. *Ecology* 79:1029–38.

649. Schweiger PF. 2016. Nitrogen isotope fractionation during N uptake via arbuscular mycorrhizal and ectomycorrhizal fungi into grey alder. *Journal of Plant Physiology* 205:84–92.

650. Seastedt T, Knapp A. 1993. Consequences of nonequilibrium resource availability across multiple time scales: The transient maxima hypothesis. *The American Naturalist* 141:621–33.

651. Selosse MA, Setaro S, Glatard F, Richard F, Urcelay C, Weiß M. 2007. Sebacinales are common mycorrhizal associates of Ericaceae. *New Phytologist* 174:864–78.

652. Seymour V, Hinckley T, Morikawa Y, Franklin J. 1983. Foliage damage in coniferous trees following volcanic ashfall from Mt. St. Helens. *Oecologia* 59:339–43.

653. Sheail J. 1987. *Seventy-Five Years in Ecology: The British Ecological Society*. Oxford: Blackwell.

654. Shemakhanova MN. 1962. *Mycotrophy of Woody Species*. Washington, DC: US Department of Commerce. Translation TT66-51073 (1967).

655. Shevliakova E, Stouffer RJ, Malyshev S, Krasting JP, Hurtt GC, Pacala SW. 2013. Historical warming reduced due to enhanced land carbon uptake. *Proceedings of the National Academy of Sciences* 110:16730–5.

656. Shi M, Fisher JB, Brzostek ER, Phillips RP. 2016. Carbon cost of plant nitrogen acquisition: Global carbon cycle impact from an improved plant nitrogen cycle in the Community Land Model. *Global Change Biology* 22:1299–314.

657. Siddiqui ZA, Futai K. 2008. *Mycorrhizae: Sustainable Agriculture and Forestry.* Springer.com.

658. Simard SW, Beiler KJ, Bingham MA, Deslippe JR, Philip LJ, Teste FP. 2012. Mycorrhizal networks: Mechanisms, ecology and modelling. *Fungal Biology Reviews* 26:39–60.

659. Simard SW, Durall DM. 2004. Mycorrhizal networks: A review of their extent, function, and importance. *Canadian Journal of Botany* 82:1140–65.

660. Singer R. 1963. Oak mycorrhiza fungi in Colombia. *Mycopathologia et mycologia applicata* 20:239–52.

661. Sirajuddin AT. 2009. Impact of atmospheric nitrogen pollution on below-ground mycorrhizal fungal community structure and composition in the San Bernardino Mountains. PhD thesis. University of California, Riverside.

662. Sirulnik AG, Allen EB, Meixner T, Allen MF. 2007. Impacts of anthropogenic N additions on nitrogen mineralization from plant litter in exotic annual grasslands. *Soil Biology & Biochemistry* 39:24–32.

663. Sizonenko TA, Dubrovskiy YA, Novakovskiy AB. 2020. Changes in mycorrhizal status and type in plant communities along altitudinal and ecological gradients: A case study from the Northern Urals (Russia). *Mycorrhiza* 30:445–54.

664. Slankis V. 1973. Hormonal relationships in mycorrhizal development, pp. 213–98. In GC Marks, TT Kozlowski, eds. *Ectomycorrhizae: Their Ecology and Physiology.* New York: Academic Press.

665. Smith GR, Wan J. 2019. Resource-ratio theory predicts mycorrhizal control of litter decomposition. *New Phytologist* 223:1595–606.

666. Smith, FA, Smith SE. 1996. Mutualism and parasitism: Diversity in function and structure in the "arbuscular" (VA) mycorrhizal symbiosis. *Advances in Botanical Research* 22:1–43.

667. Smith SE, Read DJ. 1996. *Mycorrhizal Symbiosis*, 2nd ed. Cambridge, MA: Academic Press.

668. Smith SE, Read DJ. 2010. *Mycorrhizal Symbiosis*, 3rd ed. Cambridge, MA: Academic Press.

669. Smith SE, Smith FA. 2012. Fresh perspectives on the roles of arbuscular mycorrhizal fungi in plant nutrition and growth. *Mycologia* 104:1–13.

670. Smits WTM. 1983. Dipterocarps and mycorrhiza. An ecological adaptation and a factor in forest regeneration. *Flora Malesiana Bulletin* 36:3926–37.

671. Solaiman ZM, Mickan B. 2014. Use of mycorrhiza in sustainable agriculture and land restoration, pp. 1–15. In ZM Solaiman, LK Abbott, A Varma, eds. *Mycorrhizal Fungi: Use in Sustainable Agriculture and Land Restoration.* Berlin: Springer.

672. Song YY, Simard SW, Carroll A, Mohn WW, Zeng RS. 2015. Defoliation of interior Douglas-fir elicits carbon transfer and stress signalling to ponderosa pine neighbors through ectomycorrhizal networks. *Scientific Reports* 5:1–9.

673. Soudzilovskaia NA, van Bodegom PM, Terrer C, van't Zelfde M, McCallum I, et al. 2019. Global mycorrhizal plant distribution linked to terrestrial carbon stocks. *Nature Communications* 10:5077.

674. Southworth D, He XH, Swenson W, Bledsoe CS, Horwath WR. 2005. Application of network theory to potential mycorrhizal networks. *Mycorrhiza* 15:589–95.

675. Spatafora JW, Chang Y, Benny GL, Lazarus K, Smith ME, et al. 2016. A phylum-level phylogenetic classification of zygomycete fungi based on genome-scale data. *Mycologia* 108:1028–46.

676. St. John T, Coleman D, Reid C. 1983. Association of vesicular–arbuscular mycorrhizal hyphae with soil organic particles. *Ecology* 64:957–9.

677. Staddon PL, Ramsey CB, Ostle N, Ineson P, Fitter AH. 2003. Rapid turnover of hyphae of mycorrhizal fungi determined by AMS microanalysis of ^{14}C. *Science* 300:1138–40.

678. Stahl E. 1900. Der Sinn der Mycorhizenbildung. *Jahrbucher für Wissenschartliche Botanik* 34:539–668.

679. Stanton N, Allen M, Campion M. 1981. The effect of the pesticide carbofuran on soil organisms and root and shoot production in shortgrass prairie. *Journal of Applied Ecology* 18:417–31.

680. Stebbins Jr CL. 1950. *Variation and Evolution in Plants*. New York: Columbia University Press.

681. Steidinger BS, Crowther TW, Liang J, Van Nuland ME, Werner GD, et al. 2019. Climatic controls of decomposition drive the global biogeography of forest-tree symbioses. *Nature* 569:404–8.

682. Stevens BM, Propster J, Wilson GW, Abraham A, Ridenour C, et al. 2018. Mycorrhizal symbioses influence the trophic structure of the Serengeti. *Journal of Ecology* 106:536–46.

683. Stockinger H, Krüger M, Schüßler A. 2010. DNA barcoding of arbuscular mycorrhizal fungi. *New Phytologist* 187:461–74.

684. Strullu-Derrien C, Selosse MA, Kenrick P, Martin FM. 2018. The origin and evolution of mycorrhizal symbioses: From palaeomycology to phylogenomics. *New Phytologist* 220:1012–30.

685. Stubblefield SP, Banks HP. 1983. Fungal remains in the Devonian trimerophyte *Psilophyton dawsonii*. *American Journal of Botany* 70:1258–61.

686. Stubblefield SP, Taylor TN, Seymour RL. 1987. A possible endogonaceous fungus from the triassic of Antarctica. *Mycologia* 79:905–6.

687. Stubblefield SP, Taylor TN, Trappe JM. 1987. Vesicular-arbuscular mycorrhizae from the Triassic of Antarctica. *American Journal of Botany* 74:1904–11.

688. Sudduth EB, Perakis SS, Bernhardt ES. 2013. Nitrate in watersheds: Straight from soils to streams? *Journal of Geophysical Research: Biogeosciences* 118:291–302.

689. Sulman BN, Brzostek ER, Medici C, Shevliakova E, Menge DNL, Phillips RP. 2017. Feedbacks between plant N demand and rhizosphere priming depend on type of mycorrhizal association. *Ecology Letters* 20:1043–53.

690. Sutton J, Sheppard BR. 1976. Aggregation of sand-dune soil by endomycorrhizal fungi. *Canadian Journal of Botany* 54:326–33.

691. Svane SF, Dam EB, Carstensen JM, Thorup-Kristensen K. 2019. A multispectral camera system for automated minirhizotron image analysis. *Plant and Soil* 441:657–72.

692. Swanson AC, Schwendenmann L, Allen MF, Aronson EL, Artavia-Leon A, et al. 2019. Welcome to the Atta world: A framework for understanding the effects of leaf-cutter ants on ecosystem functions. *Functional Ecology* 33:1386–99.

693. Swanson FJ, Major JJ. 2005. Physical events, environments, and geological–ecological interactions at Mount St. Helens: March 1980–2004, pp. 27–44. In

VH Dale, FJ Swanson, CM Crisafulli, eds. *Ecological Responses to the 1980 Eruption of Mount St. Helens*. New York: Springer.

694. Szaniszlo P, Powell P, Reid C, Cline G. 1981. Production of hydroxamate siderophore iron chelators by ectomycorrhizal fungi. *Mycologia* 73:1158–74.

695. Taniguchi T, Kitajima K, Douhan GW, Yamanaka N, Allen MF. 2018. A pulse of summer precipitation after the dry season triggers changes in ectomycorrhizal formation, diversity, and community composition in a Mediterranean forest in California, USA. *Mycorrhiza* 28:665–77.

696. Tansley AG. 1935. The use and abuse of vegetational concepts and terms. *Ecology* 16:284–307.

697. Taylor AF, Fransson PM, Högberg P, Högberg MN, Plamboeck AH. 2003. Species level patterns in ^{13}C and ^{15}N abundance of ectomycorrhizal and saprotrophic fungal sporocarps. *New Phytologist* 159:757–74.

698. Taylor JW, Berbee ML. 2006. Dating divergences in the Fungal Tree of Life: Review and new analyses. *Mycologia* 98:838–49.

699. Taylor JW, Hann-Soden C, Branco S, Sylvain I, Ellison CE. 2015. Clonal reproduction in fungi. *Proceedings of the National Academy of Sciences* 112:8901–8.

700. Tedersoo L, Bahram M. 2019. Mycorrhizal types differ in ecophysiology and alter plant nutrition and soil processes. *Biological Reviews* 94:1857–80.

701. Tedersoo L, Pellet P, Koljalg U, Selosse M-A. 2007. Parallel evolutionary paths to mycoheterotrophy in understorey Ericaceae and Orchidaceae: Ecological evidence for mixotrophy in Pyroleae. *Oecologia* 151:206–17.

702. Tedersoo L, Smith ME. 2013. Lineages of ectomycorrhizal fungi revisited: Foraging strategies and novel lineages revealed by sequences from belowground. *Fungal Biology Reviews* 27:83–99.

703. Terrer C, Vicca, S., Hungate BA, Phillips RP, Prentice IC. 2016. Mycorrhizal association as a primary control of the CO_2 fertilization effect. *Science* 353:72–4.

704. Terrer C, Vicca S, Stocker BD, Hungate BA, Phillips RP, et al. 2018. Ecosystem responses to elevated CO_2 governed by plant–soil interactions and the cost of nitrogen acquisition. *New Phytologist* 217:507–22.

705. Teste FP, Jones MD, Dickie IA. 2020. Dual-mycorrhizal plants: Their ecology and relevance. *New Phytologist* 225:1835–51.

706. Teste FP, Simard SW, Durall DM, Guy RD, Jones MD, Schoonmaker AL. 2009. Access to mycorrhizal networks and roots of trees: Importance for seedling survival and resource transfer. *Ecology* 90:2808–22.

707. Thaxter R. 1922. A revision of the Endogoneae. *Proceedings of the American Academy of Arts and Sciences* 57:291–351.

708. Thoreau H. 1860. *The Succession of Forest Trees*. Boston: Houghton Mifflin.

709. Tillet M. 1755. *Dissertation on the Cause of the Corruption and Smutting of the Kernels of Wheat in the Head and on the Means of Preventing These Untoward Circumstances*, reprinted 1937. Ithaca: American Phytopathological Society.

710. Tilman D. 1982. *Resource Competition and Community Structure*. Princeton: Princeton University Press.

711. Tisserant E, Malbreil M, Kuo A, Kohler A, Symeonidi A, et al. 2013. Genome of an arbuscular mycorrhizal fungus provides insight into the oldest plant symbiosis. *Proceedings of the National Academy of Sciences* 110:20117–22.

712. Titus JH, del Moral R. 1998. The role of mycorrhizal fungi and microsites in primary succession on Mount St. Helens. *American Journal of Botany* 85:370–5.

713. Titus JH, Whitcomb S, Pitoniak HJ. 2007. Distribution of arbuscular mycorrhizae in relation to microsites on primary successional substrates on Mount St. Helens. *Canadian Journal of Botany* 85:941–8.

714. Tranquillini W. 1964. Photosynthesis and dry matter production of trees at high altitudes, pp. 505–18. In MH Zimmermann, ed. *The Formation of Wood in Forest Trees*. New York: Elsevier.

715. Trappe JM. 1977. Selection of fungi for ectomycorrhizal inoculation in nurseries. *Annual Review of Phytopathology* 15:203–22.

716. Treseder KK, Allen EB, Egerton-Warburton LM, Hart MM, Klironomos JN, et al. 2018. Arbuscular mycorrhizal fungi as mediators of ecosystem responses to nitrogen deposition: A trait-based predictive framework. *Journal of Ecology* 106:480–9.

717. Treseder KK, Allen MF. 2000. Black boxes and missing sinks: Fungi in global change research. *Mycological Research* 104:1282–3.

718. Treseder KK, Allen MF. 2000. Mycorrhizal fungi have a potential role in soil carbon storage under elevated CO_2 and nitrogen deposition. *New Phytologist* 147:189–200.

719. Treseder KK, Allen MF. 2002. Direct nitrogen and phosphorus limitation of arbuscular mycorrhizal fungi: A model and field test. *New Phytologist* 155:507–15.

720. Treseder KK, Allen MF, Ruess RW, Pregitzer KS, Hendrick RL. 2005. Lifespans of fungal rhizomorphs under nitrogen fertilization in a pinyon–juniper woodland. *Plant and Soil* 270:249–55.

721. Treseder KK, Egerton-Warburton LM, Allen MF, Cheng YF, Oechel WC. 2003. Alteration of soil carbon pools and communities of mycorrhizal fungi in chaparral exposed to elevated carbon dioxide. *Ecosystems* 6:786–96.

722. Treseder KK, Maltz MR, Hawkins BA, Fierer N, Stajich JE, McGuire KL. 2014. Evolutionary histories of soil fungi are reflected in their large-scale biogeography. *Ecology Letters* 17:1086–93.

723. Treseder KK, Masiello CA, Lansing JL, Allen MF. 2004. Species-specific measurements of ectomycorrhizal turnover under N-fertilization: Combining isotopic and genetic approaches. *Oecologia* 138:419–25.

724. Treseder KK, Morris SJ, Allen MF. 2005. The contribution of root exudates, symbionts, and detritus to carbon sequestration in the soil, pp. 145–62. In F Wright, R Zobel, eds. *Roots and Soil Management: Interactions between Roots and the Soil*. Agronomy Monograph 48. Madison: American Agronomy Society.

725. Treseder KK, Schimel JP, Garcia MO, Whiteside MD. 2010. Slow turnover and production of fungal hyphae during a Californian dry season. *Soil Biology & Biochemistry* 42:1657–60.

726. Tsilikis JD. 1959. Simplicity and elegance in theoretical physics. *American Scientist* 47:87–96.

727. Tulasne L-R, Tulasne C. 1841. Observations sur le genre Elaphomyces, et description de quelques espèces nouvelles. *Annales Science Naturale Series 2* 16:5–29.

728. Turina M, Ghignone S, Astolfi N, Silvestri A, Bonfante P, Lanfranco L. 2018. The virome of the arbuscular mycorrhizal fungus Gigaspora margarita reveals

the first report of DNA fragments corresponding to replicating non-retroviral RNA viruses in fungi. *Environmental Microbiology* 20:2012–25.

729. Turner BL II, Klepeis P, Schneider LC. 2003. Three millennia in the southern Yucatan Peninsula: Implications for occupancy, use, and carrying capacity, pp. 361–87. In A Gomez-Pompa, MF Allen, S Fedick, JJ Jimenez-Osornio, eds. *The Lowland Maya Area: Three Millennia at the Human–Wildland Interface.* New York: Haworth Press.

730. Urbano P, Urbano F. 2007. Nanobacteria: Facts or fancies? *PLoS Pathogen* 3: e55.

731. Urcelay C, Diaz S, Gurvich DE, Chapin FS, Cuevas E, Dominguez LS. 2009. Mycorrhizal community resilience in response to experimental plant functional type removals in a woody ecosystem. *Journal of Ecology* 97:1291–301.

732. Vainio EJ, Pennanen T, Rajala T, Hantula J. 2017. Occurrence of similar mycoviruses in pathogenic, saprotrophic and mycorrhizal fungi inhabiting the same forest stand. *FEMS Microbiology Ecology* 93:fix003.

733. Van Diepen LT, Lilleskov EA, Pregitzer KS. 2011. Simulated nitrogen deposition affects community structure of arbuscular mycorrhizal fungi in northern hardwood forests. *Molecular Ecology* 20:799–811.

734. Van Wijk MT, Williams M, Gough L, Hobbie SE, Shaver G. 2003. Luxury consumption of soil nutrients: A possible competitive strategy in above-ground and below-ground biomass allocation and root morphology for slow-growing arctic vegetation? *Journal of Ecology* 91:664–76.

735. Varela-Cervero S, Vasar M, Davison J, Barea JM, Öpik M, Azcón-Aguilar C. 2015. The composition of arbuscular mycorrhizal fungal communities differs among the roots, spores and extraradical mycelia associated with five Mediterranean plant species. *Environmental microbiology* 17:2882–95.

736. Varenius K, Kårén O, Lindahl B, Dahlberg A. 2016. Long-term effects of tree harvesting on ectomycorrhizal fungal communities in boreal Scots pine forests. *Forest Ecology and Management* 380:41–9.

737. Vargas R, Allen EB, Allen MF. 2009. Effects of vegetation thinning on above- and belowground carbon in a seasonally dry tropical forest in Mexico. *Biotropica* 41:302–11.

738. Vargas R, Allen MF. 2008. Diel patterns of soil respiration in a tropical forest after Hurricane Wilma. *Journal of Geophysical Research–Biogeosciences* 113: G03021.

739. Vargas R, Allen MF. 2008. Dynamics of fine root, fungal rhizomorphs, and soil respiration in a mixed temperate forest: Integrating sensors and observations. *Vadose Zone Journal* 7:1055–64.

740. Vargas R, Allen MF. 2008. Environmental controls and the influence of vegetation type, fine roots and rhizomorphs on diel and seasonal variation in soil respiration. *New Phytologist* 179:460–71.

741. Vargas R, Allen MF, Allen EB. 2008. Biomass and carbon accumulation in a fire chronosequence of a seasonally dry tropical forest. *Global Change Biology* 14:109–24.

742. Vargas R, Baldocchi DD, Allen MF, Bahn M, Black TA, et al. 2010. Looking deeper into the soil: Biophysical controls and seasonal lags of soil CO_2 production and efflux. *Ecological Applications* 20:1569–82.

743. Vargas R, Baldocchi DD, Querejeta JI, Curtis PS, Hasselquist NJ, et al. 2010. Ecosystem CO_2 fluxes of arbuscular and ectomycorrhizal dominated vegetation types are differentially influenced by precipitation and temperature. *New Phytologist* 185:226–36.

744. Vargas R, Hasselquist N, Allen EB, Allen MF. 2010. Effects of a hurricane disturbance on aboveground forest structure, arbuscular mycorrhizae and belowground carbon in a restored tropical forest. *Ecosystems* 13:118–28.

745. Vargas R, Trumbore SE, Allen MF. 2009. Evidence of old carbon used to grow new fine roots in a tropical forest. *New Phytologist* 182:710–8.

746. Vasar M, Andreson R, Davison J, Jairus T, Moora M, et al. 2017. Increased sequencing depth does not increase captured diversity of arbuscular mycorrhizal fungi. *Mycorrhiza* 27:761–73.

747. Vasek FC. 1980. Creosote bush: Long-lived clones in the Mojave desert. *American Journal of Botany* 67:246–55.

748. Vašutová M, Mleczko P, López-García A, Maček I, Boros G, et al. 2019. Taxi drivers: The role of animals in transporting mycorrhizal fungi. *Mycorrhiza* 29:413–34.

749. Verbruggen E, Kiers ET, Bakelaar PN, Röling WF, van der Heijden MG. 2012. Provision of contrasting ecosystem services by soil communities from different agricultural fields. *Plant and Soil* 350:43–55.

750. Virginia R, Jenkins M, Jarrell W. 1986. Depth of root symbiont occurrence in soil. *Biology and Fertility of Soils* 2:127–30.

751. Vitousek PM, Aber JD, Howarth RW, Likens GE, Matson PA, et al. 1997. Human alteration of the global nitrogen cycle: Sources and consequences. *Ecological Applications* 7:737–50.

752. Vlček V, Pohanka M. 2020. Glomalin: An interesting protein part of the soil organic matter. *Soil and Water Research* 15:67–74.

753. Vogt KA, Grier CC, Meier CE, Edmonds RL. 1982. Mycorrhizal role in net primary priduction and nutrient cytcling in *Abies amabilis* ecosystems in Western Washington. *Ecology* 63:370–80.

754. Vohník M. 2020. Ericoid mycorrhizal symbiosis: Theoretical background and methods for its comprehensive investigation. *Mycorrhiza* 30:671–95.

755. Voříšková A, Jansa J, Püschel D, Krüger M, Cajthaml T, et al. 2017. Real-time PCR quantification of arbuscular mycorrhizal fungi: Does the use of nuclear or mitochondrial markers make a difference? *Mycorrhiza* 27:577–85.

756. Wagner CA, Taylor TN. 1981. Evidence for endomycorrhizae in Pennsylvanian age plant fossils. *Science* 212:562–3.

757. Walker JD, Geissman JW, Bowring SA, Babcock LE, compilers. 2018. *Geological Time Scale Diagram.* Available from www.geosociety.org/GSA/Education_Careers/Geologic_Time_Scale/GSA/timescale/home.aspx. Last accessed November 16, 2021.

758. Walker MD, Wahren CH, Hollister RD, Henry GH, Ahlquist LE, et al. 2006. Plant community responses to experimental warming across the tundra biome. *Proceedings of the National Academy of Sciences* 103:1342–6.

759. Wang B, Qiu Y-L. 2006. Phylogenetic distribution and evolution of mycorrhizas in land plants. *Mycorrhiza* 16:299–363.

760. Wang G, Coleman D, Freckman D, Dyer M, McNaughton S, et al. 1989. Carbon partitioning patterns of mycorrhizal versus non-mycorrhizal plants: Real-time dynamic measurements using $^{11}CO_2$. *New Phytologist* 112:489–93.

761. Wang J, Defrenne C, McCormack ML, Yang L, Tian D, et al. 2021. Fine-root functional trait responses to experimental warming: A global meta-analysis. *New Phytologist* 230:1856–67.

762. Wang JC, Pan BR, Albach DC. 2016. Evolution of morphological and climatic adaptations in Veronica L. (Plantaginaceae). *PeerJ* 4:e2333.

763. Wang Q, Wang W, He X, Zhang W, Song K, Han S. 2015. Role and variation of the amount and composition of glomalin in soil properties in farmland and adjacent plantations with reference to a primary forest in North-Eastern China. *PLoS One* 10:e0139623.

764. Warner NJ, Allen MF, MacMahon JA. 1987. Dispersal agents of vesicular-arbuscular mycorrhizal fungi in a disturbed arid ecosystem. *Mycologia* 79:721–30.

765. Weatherley P. 1975. Water relations of the root system, pp. 397–413. In JG Torrey, DT Clarkson, eds. *The Development and Function of Roots.* New York: Academic Press.

766. Weber SE, Diez JM, Andrews LV, Goulden ML, Aronson EL, Allen MF. 2019. Responses of arbuscular mycorrhizal fungi to multiple coinciding global change drivers. *Fungal Ecology* 40:62–71.

767. Weinbaum BS, Allen MF, Allen EB. 1996. Survival of arbuscular mycorrhizal fungi following reciprocal transplanting across the Great Basin, USA. *Ecological Applications* 6:1365–72.

768. Weiss FE. 1904. A mycorhiza from the lower coal-measures. *Annals of Botany* 18:255–65.

769. Weiß M, Waller F, Zuccaro A, Selosse MA. 2016. Sebacinales: One thousand and one interactions with land plants. *New Phytologist* 211:20–40.

770. Went F, Stark N. 1968. The biological and mechanical role of soil fungi. *Proceedings of the National Academy of Sciences* 60:497–504.

771. Whitbeck JL. 2001. Effects of light environment on vesicular–arbuscular mycorrhiza development in *Inga leiocalycina*, a tropical wet forest tree. *Biotropica* 33:303–11.

772. White TJ, Bruns T, Lee S, Taylor J. 1990. Amplification and direct sequencing of fungal ribosomal RNA genes for phylogenetics. *PCR Protocols: A Guide to Methods and Applications* 18:315–22.

773. Whiteside MD, Digman MA, Gratton E, Treseder KK. 2012. Organic nitrogen uptake by arbuscular mycorrhizal fungi in a boreal forest. *Soil Biology & Biochemistry* 55:7–13.

774. Whiteside MD, Garcia MO, Treseder KK. 2012. Amino acid uptake in arbuscular mycorrhizal plants. *PLoS One* 7:e47643.

775. Whiteside MD, Treseder KK, Atsatt PR. 2009. The brighter side of soils: Quantum dots track organic nitrogen through fungi and plants. *Ecology* 90:100–8.

776. Whittaker RH. 1967. Gradient analysis of vegetation. *Biological Reviews* 42:207–64.

777. Whittaker RH. 1970. *Communities and Ecosystems*. New York: Macmillan.
778. Whittaker RH, Klomp H. 1975. The design and stability of plant communities, pp. 169–83. In WH van Dobben, RH Lowe-McConnell, eds. *Unifying Concepts in Ecology*. Wageningen: Dr. W. Junk B.V. Publishers.
779. Whittingham J, Read D. 1982. Vesicular–arbuscular mycorrhiza in natural vegetation systems: III. Nutrient transfer between plants with mycorrhizal interconnections. *New Phytologist* 90:277–84.
780. Wilde SA, Corey RB, Iyer JG. 1979. Mycorrhiza-related terminology. *Forestry Research Notes* 223:1–5.
781. Williams SE, Aldon EF. 1976. Endomycorrhizal (vesicular arbuscular) associations of some arid zone shrubs. *The Southwestern Naturalist* 20:437–44.
782. Williamson M. 1972. *The Analysis of Biological Populations*. London: Academic Press.
783. Wilson DS. 1992. Complex interactions in metacommunities, with implications for biodiversity and higher levels of selection. *Ecology* 73:1984–2000.
784. Wilson J, Tommerup I. 1992. Interactions between fungal symbionts: VA mycorrhizae, pp. 199–248. In MF Allen, ed. *Mycorrhizal Functioning*. New York: Chapman and Hall Press.
785. Winkelmann G. 2017. A search for glomuferrin: A potential siderophore of arbuscular mycorrhizal fungi of the genus Glomus. *Biometals* 30:559–64.
786. Wolfe BE, Pringle A. 2012. Geographically structured host specificity is caused by the range expansions and host shifts of a symbiotic fungus. *ISME Journal* 6:745–55.
787. Won H, Renner SS. 2006. Dating dispersal and radiation in the gymnosperm Gnetum (Gnetales): Clock calibration when outgroup relationships are uncertain. *Systematic Biology* 55:610–22.
788. Wood T, Cummings B. 1992. Biotechnology and the future of VAM commercialization, pp. 468–87. In MF Allen, ed. *Mycorrhizal Functioning*. New York: Chapman and Hall Press.
789. Woods FW, Brock K. 1964. Interspecific transfer of Ca-45 + P-32 by root systems. *Ecology* 45:886–9.
790. Wooley SC, Paine TD. 2011. Infection by mycorrhizal fungi increases natural enemy abundance on tobacco (*Nicotiana rustica*). *Environmental Entomology* 40:36–41.
791. Wright DP, Read DJ, Scholes JD. 1988. Mycorrhizal sink strength influences whole plant carbon balance of Trifolium repens L. *Plant Cell and Environment* 21:881–91.
792. Wright SF, Upadhyaya A. 1996. Extraction of an abundant and unusual protein from soil and comparison with hyphal protein of arbuscular mycorrhizal fungi. *Soil Science* 161:575–86.
793. Yildirir G, Kokkoris V, Corradi N. 2020. Parasexual and sexual reproduction in arbuscular mycorrhizal fungi: Room for both. *Trends in Microbiology* 28:517–19.
794. Yoshida LC, Allen EB. 2001. Response to ammonium and nitrate by a mycorrhizal annual invasive grass and native shrub in southern California. *American Journal of Botany* 88:1430–6.
795. Yost R, Fox RL. 1979. Contribution of mycorrhizae to P nutrition of crops growing on an Oxisol 1. *Agronomy Journal* 71:903–8.

796. Zhang HS, Wu XH, Li G, Qin P. 2011. Interactions between arbuscular mycorrhizal fungi and phosphate-solubilizing fungus (Mortierella sp.) and their effects on Kostelelzkya virginica growth and enzyme activities of rhizosphere and bulk soils at different salinities. *Biology and Fertility of Soils* 47:543–54.

797. Zhang J, Tang X, He X, Liu J. 2015. Glomalin-related soil protein responses to elevated CO_2 and nitrogen addition in a subtropical forest: Potential consequences for soil carbon accumulation. *Soil Biology and Biochemistry* 83:142–9.

798. Zhang Y, Guo L-D. 2007. Arbuscular mycorrhizal structure and fungi associated with mosses. *Mycorrhiza* 17:319–25.

799. Zhao FZ, Feng XX, Guo YX, Ren CJ, Wang J, Doughty R. 2020. Elevation gradients affect the differences of arbuscular mycorrhizal fungi diversity between root and rhizosphere soil. *Agricultural and Forest Meteorology* 284:107894.

800. Zimmer K, Hynson NA, Gebauer G, Allen EB, Allen MF, Read DJ. 2007. Wide geographical and ecological distribution of nitrogen and carbon gains from fungi in pyroloids and monotropoids (Ericaceae) and in orchids. *New Phytologist* 175:166–75.

Index